W9-ANF-307

Shop Mathematics

SIXTH EDITION

C. A. FELKER

LINDA L. THOMPSON
Consultant

 **Glencoe McGraw-Hill**

New York, New York   Columbus, Ohio   Woodland Hills, California   Peoria, Illinois

Glencoe/McGraw-Hill
A Division of The McGraw·Hill Companies

Copyright © 1984 by Glencoe Publishing Company, a division of Macmillan, Inc.

All rights reserved. No part of this book may be reproduced or transmitted in any form
or by any means, electronic or mechanical, including photocopying, recording, or by any
information storage and retrieval system, without permission in writing from the Publisher.

Send all inquiries to:
Glencoe/McGraw-Hill
21600 Oxnard Street, Suite 500
Woodland Hills, CA 91367

Library of Congress Catalog Card Number: 83-81222

Printed in the United States of America.

ISBN 0-02-816310-9

9 10 11 12 13 14 15   007   04 03 02 01 00 99 98

Contents

List of Rules

Preface

This edition of *Shop Mathematics* seeks to satisfy the mathematical needs of tradespeople in the metalworking field, as well as students preparing to join their ranks. The material in this book draws upon an industrial experience of more than thirty-four years to provide practical solutions for any normal shop problem.

The units are comprehensive in scope, while stressing uncomplicated manipulation of mathematics at a level the worker can easily grasp. Time is a controlling factor in most job assignments. A detailed analysis of mathematical operations must be secondary to the requirements of production. For this reason alone, the tradesperson must first master the skills of shop mathematics.

This edition has retained the class-proven teachability of the fifth edition, while adding a number of important features. The average shopworker is not a "book mechanic." He or she depends upon elementary rules, simply stated and easily learned, to solve problems quickly. These important rules are still given preeminence in the sixth edition. Ample numbers of working examples illustrating each rule are provided for guidance. In addition, the numerous technical illustrations of the fifth edition have been retained and many new ones have been added.

A unique feature of this edition is the use of metric problems throughout the text. Since students will be expected to use metric units as they encounter them on the job, no special emphasis is placed on the metric system after it is introduced. This approach should help dispel the uncertainty usually associated with unfamiliar units of measurement.

A unit on whole number arithmetic has been added in this edition for those students who need review of these basic concepts. In addition, coverage of ratio, proportion, and percent has been expanded, and the treatment of formulas and equations has been improved. The incorporation of this additional and expanded material gives the instructor greater flexibility in designing a course to meet the unique needs of any given class.

The sixth edition offers broader coverage of basic skills without sacrificing the treatment of more advanced topics. Yet the book has been kept to a manageable length so that the entire text can be covered in one semester or quarter of classroom work.

I concede the impracticality of including material satisfactory to the requirements of all mathematics teachers. On the other hand, I have an obligation to those who intend to use this textbook long after classroom sessions are terminated.

Linda L. Thompson

Basic Operations with Whole Numbers

Numbers play an important role in the daily work of machinists. Whether operating a lathe, measuring metal tubing, or calculating the angle of a cut, the machinist must use numbers. The machinist must be able to add, subtract, multiply, and divide without making errors. In a practical situation, errors cause delay or accidents or, at best, result in sloppy work. Pieces that are unusable because of error cost a company lost time and money.

Unit 1 introduces whole numbers and the basic operations of addition, subtraction, multiplication, and division. You will learn to apply these skills to work with exponents and to find the prime factors of a number. In addition, you will begin the study of measurement, an essential idea in the shop.

The goal of Unit 1 is to teach you to do calculations and solve problems on your own. However, many problems in the shop are suitable for solving with a calculator. In this chapter and throughout the book, problems that are especially suited for solution by calculator are marked. If you wish, you may solve these without a calculator. Pocket calculators and other computers help the machinist's speed and accuracy in solving problems, but first you must understand mathematical concepts and logic. Understanding how to work with numbers in your trade takes hard work, study, and practice.

1-1   Place Value and Rounding

OBJECTIVES

After completing this section, you should be able to:

1. Tell the value of a digit.
2. Write a number in standard or expanded form.
3. Round a whole number to a specified place value.

Place Value

Whole numbers are the numbers 0, 1, 2, 3, and so on. They are written by using one or more of the ten digits, 0, 1, 2, 3, 4, 5, 6, 7, 8, and 9. Each digit has a value, depending on its place in the number.

Look at the place value chart in Figure 1-1. The **place values** from right to left are ones, tens, hundreds, thousands, ten thousands, and so on. The number shown in the chart is read

"three million, eight hundred forty-seven thousand, two hundred sixty-one"

BILLIONS	HUNDRED MILLIONS	TEN MILLIONS	MILLIONS	HUNDRED THOUSANDS	TEN THOUSANDS	THOUSANDS	HUNDREDS	TENS	ONES
			3	8	4	7	2	6	1

Figure 1-1

The value of the digit 3 in 3,847,261 is 3 millions, or 3,000,000. The value of the digit 7 is 7 thousands, or 7000.

Notice that the digits are divided, beginning from the right, into groups of three. Each group is named by the place name at the right of the group. Commas separate each group of three, making the number easier to read.

Example 1-1   Find the value of the 2 in 80,243.

Solution   The 2 is in the hundreds place. The value of the 2 is 200.

Any number with two or more digits can be written in expanded form by the use of place value. The difference between standard form and expanded form is shown in Table 1-1.

Standard Form	Expanded Form
376	300 + 70 + 6
1735	1000 + 700 + 30 + 5

Table 1-1

Example 1-2 Write the number 32,657 in expanded form.

Solution Give the value of each digit.

> 3 ten thousands, or 30,000
> 2 thousands, or 2000
> 6 hundreds, or 600
> 5 tens, or 50
> 7 ones, or 7

Write the sum of the values.

> 30,000 + 2000 + 600 + 50 + 7

Rounding

Sometimes a shopworker finds it necessary to **round** a number because a measurement is too exact or because an estimate is needed to solve a problem. Rule 1-1 tells how to round a number.

RULE 1-1   ROUNDING NUMBERS

Step 1   Find the digit in the place to which you want to round.
Step 2   Look at the next digit to the right:
- If it is less than 5, replace all the digits to the right of the place to which you are rounding with zeros.
- If it is 5 or more, add 1 to the digit in the place to which you are rounding and replace all the digits to the right with zeros.

Example 1-3 Round 1835 to the nearest hundred.

Solution

Step 1   1835
       ↑— hundreds place

Step 2   1835
       L 3 is in the place to the right. It is less than 5.

Replace all digits to the right of 8 with zeros.

> 1835 = 1800, to the nearest hundred

Example 1-4   Round 235 to the nearest ten.

    Solution

        Step 1   235
                ⌐ tens place

        Step 2   235
                ⌐ 5 or more

           Add 1 to 3 and replace all digits to the right of 3 with zeros.

           235 = 240, to the nearest ten

Example 1-5   23,679 = 24,000   to the nearest thousand
          1,983,511 = 2,000,000   to the nearest million
               995 = 1000   to the nearest ten

Exercises 1-1

Write each number in standard form.

  1. 70 + 4

  2. 200 + 50 + 9

  3. 3000 + 60 + 5

  4. 40,000 + 1000 + 800 + 60 + 4

  5. 70,000 + 3000 + 800 + 50 + 7

  6. 60,000 + 80

  7. 800,000 + 70,000 + 400 + 1

  8. 1000 + 500,000 + 8000 + 70

Write each number in expanded form.

9. 26	**10.** 73	**11.** 125	**19.** 307
13. 1820	**14.** 73,068	**15.** 980,462	**16.** 8,098,350

Give the value of each underlined digit.

17. 54	**18.** 398	**19.** 722	**20.** 8573
21. 9045	**22.** 756,291	**23.** 858,335	**24.** 353,298,016

Round each number to the nearest ten.

25. 53	**26.** 78	**27.** 3702	**28.** 996

Round each number to the nearest hundred.

29. 762	**30.** 518	**31.** 1761	**32.** 32,582

Round each number to the nearest thousand.

33. 8127   **34.** 5821   **35.** 26,359   **36.** 39,706

Round each number to the underlined place.

37. 27,652   **38.** 4005   **39.** 298   **40.** 5,832,716

41. A factory produces 19,300 machined cylinders in one day. What is the daily output of cylinders, rounded to the nearest thousand?

42. James works 2193 hours (h) in an average year. How many hours is that, rounded to the nearest thousand?

43. A toolmaker found that 32,649 parts would be needed for a job. To the nearest hundred, how many parts is that?

44. In exercise 43, would the toolmaker's estimate be useful in ordering parts? Why or why not?

1-2 Addition and Subtraction

OBJECTIVES

After completing this section, you should be able to:

1. Add whole numbers.
2. Subtract whole numbers.

Addition and subtraction are basic operations that must be mastered before more complicated calculations can be tried. Much time, effort, and material can be wasted in the shop because of a mistake in addition or subtraction. Care must be taken at all times, even when doing the simplest of calculations.

Addition

When the two numbers 36 and 15 are added, the result is 51. The numbers 36 and 15 are called the **addends,** and the number 51 is called the **sum.**

RULE 1-2 ADDING WHOLE NUMBERS

To add whole numbers, arrange the numbers in columns and add digits in corresponding places. Begin with the right-hand column.

Example 1-6   Find the sum of 475 and 321.

Solution

	Hundreds	Tens	Ones
475	4	7	5
+321 ⟶ +	3	2	1
	7	9	6

Sometimes it is necessary to rename, or "carry" digits from one column to the next.

RULE 1-3   RENAMING IN ADDITION

If the sum of the numbers in a column is more than 9, record the right-hand digit at the bottom of that column and carry the left-hand digit to the next column on the left.

Example 1-7   Add 89,857 and 29,733.

Solution   Write the numbers in columns and then add digits in corresponding places, beginning at the right.

$$
\begin{array}{r}
\overset{1\ 1\ \ \ 1}{8\,9\,,8\,5\,7} \\
+\ \ 2\,9\,,7\,3\,3 \\
\hline
1\,1\,9\,,5\,9\,0
\end{array}
$$

7 + 3 = 10; write 0 and carry 1.
1 + 5 + 3 = 9
8 + 7 = 15; write 5 and carry 1.
1 + 9 + 9 = 19; write 9 and carry 1.
1 + 8 + 2 = 11

Example 1-8   Add: 23,648 + 8,417,263 + 500,976.

Solution

$$
\begin{array}{r}
23{,}648 \\
8{,}417{,}263 \\
+\ \ \ 500{,}976 \\
\hline
8{,}941{,}887
\end{array}
$$

Line up the digits in the same place and add each column.

Subtraction

Subtraction is the **inverse,** or opposite, of addition. If you begin with 12, subtract 7, and then add 7, you again have 12.

$12 - 7 = 5$ because $5 + 7 = 12$

In $12 - 7 = 5$, 12 is the **minuend,** 7 is the **subtrahend,** and 5 is the **difference.** The difference between two whole numbers is the number that, when added to the subtrahend, gives the minuend.

RULE 1-4   SUBTRACTING WHOLE NUMBERS

To subtract whole numbers, arrange the numbers in columns and subtract digits in corresponding places. Begin with the right-hand column.

Example 1-9   Find the difference between 754 and 231.

Solution

	Hundreds	Tens	Ones
754	7	5	4
−231 ⟶ −	2	3	1
	5	2	3

When subtracting whole numbers it is sometimes necessary to rename, or "borrow."

RULE 1-5   RENAMING IN SUBTRACTION

If a number in the subtrahend is greater than the corresponding number in the minuend, rename 1 unit in the next column to the left as 10 units in the specified column.

Example 1-10   Subtract 638 from 923.

Solution

	Hundreds	Tens	Ones
923 ⟶	9	2	3
−638 −	6	3	8

Rename 2 tens, 3 ones as 1 ten, 13 ones and subtract.

	Hundreds	Tens	Ones
	9	$\overset{1}{\cancel{2}}$	13
−	6	3	8
			5

Rename 9 hundreds, 1 ten as 8 hundreds, 11 tens and subtract.

	Hundreds	Tens	Ones
	$\overset{8}{\cancel{9}}$	11	13
−	6	3	8
	2	8	5

Short form

$$
\begin{array}{r}
11 \\
8\ \cancel{1}\ 13 \\
\cancel{9}\cancel{2}\cancel{3} \\
-6\,3\,8 \\
\hline
2\,8\,5
\end{array}
$$

Recall that $12 - 7 = 5$ because $5 + 7 = 12$. Therefore, you can *check* subtraction by adding, as the next example shows.

Example 1-11   Subtract 73,256 from 384,100.

Solution   ┌──────── Rename 4 thousands as 3 thousands,
                      10   10 hundreds.

3 ∅ 9 10 ◀──└─Rename 1 hundred as 9 tens, 10 ones.
$$
\begin{array}{r}
38\cancel{4},\cancel{1}\,\cancel{0}\,\cancel{0} \\
-\ \ 7\,3\,,2\,5\,6 \\
\hline
3\,1\,0\,,8\,4\,4
\end{array}
$$

Check $$
\begin{array}{r}
3\,1\,0\,,8\,4\,4 \\
+\ \ 7\,3\,,2\,5\,6 \\
\hline
3\,8\,4\,,1\,0\,0
\end{array}
$$

It is possible to use a calculator to perform many calculations in the shop. The exercises in this book marked with a ⬚ are especially suitable for solution with a calculator if your instructor wants you to use one. The following example shows how to use your calculator to solve a subtraction problem.

Example 1-12   Subtract: $80,002 - 35,197$.

Solution	Enter	Display
	⑧ ⓪ ⓪ ⓪ ②	80002
	⊖	80002.
	③ ⑤ ① ⑨ ⑦	35197
	＝	44805.

Most shop calculations involve the use of **measurements.** In the measurement 20 centimeters, 20 is the **measure** and centimeters is the **unit of measure.** Some commonly used units of measure are inches, millimeters (mm), pounds, and grams. Rule 1-6 tells how to add and subtract measurements.

RULE 1-6   ADDING OR SUBTRACTING MEASUREMENTS

To add or subtract measurements with like units, add or sub-tract the numbers and use the given unit of measure.

Example 1-13   Find the sum of 5 inches (in.), 19 in., and 12 in.

Solution   5 in.
19 in.
$+12$ in.
36 in.

The sum is 36 in.

Exercises 1-2

Add.

1. 27
    $+52$

2. 835
    $+240$

3. 1724
    $+8255$

4. 53,792
    $+44,207$

5. 37
    $+46$

6. 529
    $+364$

7. 3871
    $+2156$

8. 48,263
    $+35,625$

9. 368
    $+975$

10. 1856
     $+3775$

11. 27,846
     $+10,397$

12. 626,543
     $+598,072$

13. 78
     562
     39
     $+1746$

14. 287
     1,953
     32,564
     $+ 2,988$

15. 10,796
     48,255
     797,802
     $+685,741$

16. 600,423
     791,833
     662,534
     875,901
     $+127,430$

17. 86 + 597 + 3745

18. 26,827 + 8 + 3540 + 27

Subtract and check.

19. 568
     $- 37$

20. 754
     -203

21. 76
     -39

22. 243
     -139

23. 8061
     -5312

24. 8642
     -1937

25. 126,117
     $- 39,758$

26. 709,326
     $-541,855$

27. 297,400
     $-189,886$

28. $841,765
     $- 837,858$

29.   720,094
        −510,638

30.   60,000
        −43,576

31.   296 − 199

32.   700 − 38

33.   127,468 − 68,779

34.   30,000 − 5879

35.   38,761,925 − 29,847,388

36.   2,000,000 − 1,762,851

Add or subtract as indicated. Note the units of measurement: millimeters (mm), grams (g), inches (in.), and pounds (lb).

37.   46 mm
        +38 mm

38.   1200 g
        − 932 g

39.   76 in.
        +148 in.

40.   832 lb
        −109 lb

41.   A shop machine was operated 78 minutes (min) in the morning and 93 min in the afternoon. For how many minutes was it operated?

42.   A size 0 NF threaded fastener has 80 threads per inch. A size 12 fastener has 28 threads per inch. How many fewer threads per inch does the size 12 fastener have?

43.   Martha, a machine inspector, must do a great deal of traveling for her job. She drove 817 kilometers (km) one week, 1025 km the second week, 956 km the third week, and 590 km the fourth week. How many kilometers did she travel that month?

44.   A shop machine cost $1428. After two years, the machine had depreciated $639. What was the value of the machine after two years?

45.   A machinist wished to cut an 18-in. bar from a bar 40 in. long. How much metal will be left?

46.   One month, the machines in a shop used 1784 kilowatt-hours (kw-hr), and the next month, they used 1900 kw-hr. How many more kilowatt-hours of electricity were used the second month?

47.   An employee of Tucker Tool worked the following hours during one 12-week period: 38, 43, 42, 36, 35, 39, 40, 47, 39, 41, 38, 42. Find the total number of hours worked.

48.   Kevin is a carpenter. He needs pieces of wood of length 176 centimeters (cm), 48 cm, and 58 cm. Will a 300-cm piece be long enough to yield the three needed pieces?

49.   One year the Kroeger Manufacturing Company recorded 189,736 accident-free employee hours. The next year, they recorded 200,103 accident-free hours. How many more accident-free hours were recorded the second year?

50.   The length of a drill bit is 167 cm. During drilling, the bit breaks in two and a piece of it 56 cm long remains in the chuck. How long is the other piece?

51. A quality control technician inspects 1800 parts in an average week. On Wednesday, 977 parts have been inspected. How many more parts must be inspected by Friday to equal an average week?

1-3 Multiplication of Whole Numbers

OBJECTIVE

After completing this section, you should be able to:

1. Find the product of two whole numbers.

When you multiply numbers, the numbers multiplied together are called **factors,** and the answer is called the **product.** In $5 \times 2 = 10$, the factors are 5 and 2, and the product is 10. The product of 5×2 can also be shown as $5 \cdot 2$, $(5)(2)$, or $5(2)$.

RULE 1-7   MULTIPLYING A WHOLE NUMBER BY A ONE-DIGIT NUMBER

Multiply each digit of the whole number by the one-digit number, from right to left.

Example 1-14   Multiply: 254×6.

Solution

	Hundreds	Tens	Ones
254 →	2	5	4
× 6	× 6	6	6
	12	30	24

→ 1200 + 300 + 24 = 1524

Short form

```
 32
254
×  6
1524
```

6 × 4 ones is 24 ones. Write 4 ones, carry 2 tens.

6 × 5 tens + 2 tens is 32 tens. Write 2 tens, carry 3 hundreds.

6 × 2 hundreds + 3 hundreds is 15 hundreds. Write 5 hundreds and 1 thousand.

To multiply any two whole numbers, use Rule 1-7 as many times as necessary. That is, multiply the first number by each digit of the second number, from right to left, and then add the products.

Example 1-15 Find the product of 436 × 846.

Solution
$$
\begin{array}{r}
436 \\
\times\ \ 846 \\
\hline
2\,616 \\
17\,440 \\
348\,800 \\
\hline
368{,}856
\end{array}
$$
2 616 ⟵ 436 × 6 ones
17 440 ⟵ 436 × 4 tens
348 800 ⟵ 436 × 8 hundreds

RULE 1-8 MULTIPLYING BY 0 OR 1

- Any number multiplied by 0 equals 0.
- Any number multiplied by 1 equals itself.

Example 1-16 Multiply 2937 × 107.

Solution
$$
\begin{array}{r}
2937 \\
\times\ \ 107 \\
\hline
20\,559 \\
0 \\
293\,7 \\
\hline
314{,}259
\end{array}
$$
20 559 ⟵ 2937 × 7 ones
0 ⟵ 2937 × 0 tens
293 7 ⟵ 2937 × 1 hundred

The machinist or toolmaker often finds it necessary to multiply measurements. This must be done to find the surface area or volume of a workpiece. The machinist must know the rules for writing units when multiplying measurements.

RULE 1-9 MULTIPLYING A MEASUREMENT BY A WHOLE NUMBER

Multiply the measure by the whole number and use the same unit of measure.

Example 1-17 Multiply 16 lb by 3.

Solution
$$
\begin{array}{r}
16\ \text{lb} \\
\times\ \ 3 \\
\hline
48\ \text{lb}
\end{array}
$$

RULE 1-10   MULTIPLYING TWO (OR THREE)
MEASUREMENTS WITH THE SAME UNIT

Multiply the measures and write the units squared (or cubed).

Example 1-18   Multiply: (a) 3 cm × 8 cm, (b) 6 yards (yd) × 3 yd × 2 yd.

Solution   (a)   3 cm × 8 cm = 24 square centimeters (sq cm)
(b)   6 yd × 3 yd × 2 yd = 36 cubic yards (cu yd)

Exercises 1-3

Multiply.

1. 86
    × 9

2. 275
    × 8

3. 1785
    × 3

4. 27,645
    × 7

5. 698
    × 27

6. 413
    × 56

7. 2983
    × 74

8. 392,756
    × 83

9. 147
    ×386

10. 2758
     × 742

11. 33,545
     × 981

12. 247,912
     × 346

13. 25
     ×40

14. 386
     ×900

15. 27,438
     × 500

16. 376,852
     × 700

17. 295
     ×306

18. 408
     ×701

19. 36,925
     × 4007

20. 251,392
     × 8060

21. 38,605 × 4000

22. 127,396 × 1008

23. 5,000,000 × 392

24. 26,043 × 50,803

25. 8 meters (m) × 10 m

26. 24 feet (ft) × 17 ft

27. 9 yd × 7 yd × 17 yd

28. 14 cm × 6 cm × 18 cm

29. Jeremy drives a total of 83 km to and from work each day. How many kilometers does he drive in 5 days?

30. What is the total length of wire on 17 spools if each spool contains 75 ft of wire?

31. Twenty-five pieces of steel, each 24 cm long, are needed for a job. What is the total length of steel required?

32. A job requires 23 pieces of metal tubing 17 in. long and 7 pieces 32 in. long. How much tubing is needed in all?

1-4   Division of Whole Numbers

OBJECTIVE

After completing this section, you should be able to:

1. Find the quotient and remainder when one whole number is divided by another.

Division is the inverse, or opposite, of multiplication. If you begin with 36, divide by 2, and then multiply by 2, you again have 36.

$36 \div 2 = 18$ because $18 \times 2 = 36$

In $36 \div 2$, 36 is called the **dividend,** and 2 is called the **divisor.** The result of the operation, 18, is called the **quotient.**

It is easy to divide 36 by 2 and find the quotient. Sometimes, however, the divisor does not divide the dividend exactly, and there is a **remainder.** For example, $27 \div 5$ is 5, remainder 2. For any two numbers, long division can be used to find the quotient and the remainder.

RULE 1-11   DIVIDING WHOLE NUMBERS

Step 1   Divide the divisor into the first digit. If this is not possible, divide into the first two digits (or as many as needed). Write the trial quotient above the right-most digit used.

Step 2   Multiply the trial quotient by the divisor. Write the product under the digits used.

Step 3   Subtract.
  • If the product is too great to be subtracted, try a smaller trial quotient.
  • If the difference is greater than the divisor, try a larger trial quotient.

Step 4   Bring down the remaining digits and repeat the steps. When there are no digits left to bring down, the last difference is the remainder.

Example 1-19   Divide 379 by 8.

Solution

Step 1   Divide 8 into 37. The result is 4.
Write it above the 7.

Step 2   Multiply 4 by 8. Write 32 under 37.

Step 3   Subtract 32 from 37. Write 5.

Step 4   Since 5 is less than 8, bring down 9.

$$\begin{array}{r} 4 \\ 8\overline{)379} \\ 32 \\ \hline 59 \end{array}$$

Step 1   Divide 8 into 59. The result is 7.
          Write it above the 9.
Step 2   Multiply 7 by 8. Write 56 under 59.
Step 3   Subtract 56 from 59. Write 3.
Step 4   Since 3 is less than 8 and there is nothing
          left to bring down, the remainder is 3.

$$\begin{array}{r} 47 \\ 8)\overline{379} \\ \underline{32} \\ 59 \\ \underline{56} \\ 3 \end{array}$$

Therefore 379 ÷ 8 is 47, remainder 3.

Recall that 36 ÷ 2 = 18 because 18 × 2 = 36. Therefore, division can be checked by multiplying.

RULE 1-12   CHECKING DIVISION

Quotient × divisor + remainder = dividend

Example 1-20 shows how to check division.

Example 1-20   Divide 25,907 by 43.

Solution
$$\begin{array}{r} 602 \\ 43)\overline{25,907} \\ \underline{25\ 8} \\ 10 \\ \underline{0} \\ 107 \\ \underline{86} \\ 21 \end{array}$$

⟵ Divide 259 by 43. (Think: 25 ÷ 4 is about 6; 6 × 43 = 258.)

⟵ Divide 10 by 43. The quotient is 0.

⟵ Divide 107 by 43. (Think: 10 ÷ 4 is about 2; 2 × 43 = 86.)

Check
$$\begin{array}{r} 602 \\ \times\ \ \ \ 43 \\ \hline 1\ 806 \\ 24\ 08 \\ \hline 25,886 \\ +\ \ \ \ 21 \\ \hline 25,907 \end{array}$$

Therefore, 25,907 ÷ 43 is 602, remainder 21.

In the shop, you may need to find how many pieces of a given length are in a piece whose measurements you know. Or you may need to find the length of a rectangle, given its area and width.

RULE 1-13   DIVIDING A MEASUREMENT BY A WHOLE NUMBER

Divide the measure by the whole number and use the unit.

Example 1-21   Divide 6156 m by 18.

Solution

$$18\overline{)6156} \qquad \frac{6156\text{ m}}{18} = 342\text{ m}$$

$$\begin{array}{r} 342 \\ 18\overline{)6156} \\ \underline{54} \\ 75 \\ \underline{72} \\ 36 \\ \underline{36} \\ 0 \end{array}$$

RULE 1-14   DIVIDE ONE MEASUREMENT BY ANOTHER

- To divide one measurement by another with like units, divide the measures. The answer will have no unit.
- To divide a measurement in square units by another with a single unit, divide the measures. The answer will have a single unit of measure.
- To divide one measure by another with different units, divide the measures. Write the answer using a quotient of the units.

Example 1-22   How many 8-cm lengths are in a steel bar 128 cm long?

Solution

$$\begin{array}{r} 16 \\ 8\overline{)128} \\ \underline{8} \\ 48 \\ \underline{48} \\ 0 \end{array} \qquad \frac{128\text{ cm}}{8\text{ cm}} = 16$$

There are sixteen 8-cm lengths in a 128-cm steel bar.

Example 1-23   Find 791 square inches (sq in.) divided by 7 in.

Solution

$$\begin{array}{r} 113 \\ 7\overline{)791} \\ 7 \\ \hline 9 \\ 7 \\ \hline 21 \\ 21 \\ \hline 0 \end{array}$$

$$\frac{791 \text{ sq in.}}{7 \text{ in.}} = 113 \text{ in.}$$

Example 1-24   Find the number of miles per hour traveled if Mario traveled 357 miles (mi) in 21 h.

Solution

$$\begin{array}{r} 17 \\ 21\overline{)357} \\ 21 \\ \hline 147 \\ 147 \\ \hline 0 \end{array}$$

$$\frac{357 \text{ mi}}{21 \text{ h}} = 17 \text{ mi/h}$$

Exercises 1-4

Divide and check.

1. $9\overline{)756}$ 2. $5\overline{)82{,}325}$ 3. $7\overline{)4196}$ 4. $6\overline{)32{,}847}$

5. $24\overline{)2040}$ 6. $39\overline{)4368}$ 7. $83\overline{)1245}$ 8. $57\overline{)11{,}742}$

9. $72\overline{)5843}$ 10. $15\overline{)4379}$ 11. $38\overline{)15{,}255}$ 12. $46\overline{)98{,}374}$

13. $296\overline{)5328}$ 14. $307\overline{)34{,}691}$ 15. $588\overline{)56{,}874}$ 16. $351\overline{)578{,}501}$

17. 856 km ÷ 8   18. 429 mm ÷ 7   19. 486 ft ÷ 3

20. 8756 h ÷ 24   21. 1445 ft ÷ 85   22. 3240 min ÷ 60

23. 79,400 cm ÷ 100   24. 5475 days ÷ 365   25. 38,672 ÷ 1760

26. 85 in. ÷ 5 in.   27. 1020 in. ÷ 12 in.

28. 84,730 km ÷ 10 km   29. 540 square meters (sq m) ÷ 90 m

30. 156 sq ft ÷ 12 ft   31. 74,263 sq cm ÷ 721 cm

32. 720 km ÷ 2 days   33. 406 m ÷ 58 min

34. 207 mi ÷ 23 h

35. How many pieces of wire, each 7 mm long, can be cut from a 266-mm length of wire?

36. A steel rod 210 in. long is to be cut into 14 pieces of equal length. What will be the length of each piece?

37. A piece of wire 272 ft long is to be cut into pieces measuring 12 ft each. How many 12-ft pieces will there be? If there is a scrap piece (a piece less than 12 ft long), what will be the length of the scrap piece?

1-5 Exponents

OBJECTIVES
After completing this section, you should be able to:

1. Write a number in exponential form.
2. Find the value of an exponential expression.

There is a shortcut that can be used to write an expression such as $8 \times 8 \times 8$. It can be written in **exponential form** as

$$8 \times 8 \times 8 = 8^3$$

The small number 3 is called an **exponent.** It tells the number of times 8 is used as a factor. Example 1-25 presents expressions written in exponential form.

Example 1-25 $7^2 = 7 \times 7$
$5^5 = 5 \times 5 \times 5 \times 5 \times 5$
$3^1 = 3$
$4^3 = 4 \times 4 \times 4$

The expression 5^5 is read as "five to the fifth power." The expressions "seven to the second power" and "four to the third power" are more commonly read as "seven squared" and "four cubed." Numbers written without exponents are understood to be expressed to the first power.

RULE 1-15   EVALUATING EXPONENTIAL EXPRESSIONS

To evaluate an expression raised to a power, use the expression as a factor the number of times given by the exponent.

$$a^n = \underbrace{a \times a \times \ldots \times a}_{n \text{ factors}}$$

Example 1-26   Evaluate 4^3.

   Solution $4^3 = 4 \times 4 \times 4 = 64$

A number in which one or more factors repeat can be expressed using exponents.

Example 1-27   Express using exponents: (a) $5 \times 5 \times 5$
                                               (b) $2 \times 2 \times 3 \times 3$
                                               (c) 15

   Solution   (a) $5 \times 5 \times 5 = 5^3$
                  (b) $2 \times 2 \times 3 \times 3 = 2^2 \times 3^2$
                  (c) $15 = 15^1$

Any number to the zero power is defined to be 1. Note that 0^0 is not defined.

Example 1-28 $5^0 = 1$
                   $273^0 = 1$

Exercises 1-5

Write each product using exponents.

1. 8×8

2. $2 \times 2 \times 2 \times 2$

3. $51 \times 51 \times 51$

4. $7 \times 7 \times 7 \times 7 \times 7$

5. $3 \times 4 \times 4 \times 4$

6. $17 \times 17 \times 17$

7. 200×200

8. 29

9. $2 \times 2 \times 3 \times 3 \times 3$

10. $8 \times 8 \times 8 \times 7 \times 7 \times 7 \times 7$

11. $14 \times 14 \times 14 \times 3 \times 3 \times 3$

12. $20 \times 20 \times 30 \times 30$

13. $15 \times 15 \times 15 \times 15 \times 15 \times 15 \times 15 \times 15 \times 15 \times 15 \times 15 \times 15$

14. $7 \times 7 \times 7 \times 7 \times 7 \times 7 \times 7 \times 7 \times 7 \times 7 \times 7 \times 7 \times 7 \times 7 \times 7 \times 7 \times 7 \times 7 \times 7 \times 7$

15. 25 (*Hint:* $25 = 5 \times 5$)

16. 64

Evaluate.

17. 8^2	**18.** 15^2	**19.** 3^3	**20.** 7^3
21. 4^4	**22.** 2^7	**23.** 13^1	**24.** 5^0
25. $2^0 \times 4^3$	**26.** $7^0 + 18^0$	**27.** $5^2 \times 4^3$	**28.** $3 \times 5^2 \times 2^3$
29. 296^3	**30.** $30^2 \times 15^3$	**31.** $29,743^0$	**32.** 1^{756}

1-6 Prime Factors

OBJECTIVES

After completing this section, you should be able to:

1. Identify prime numbers and composite numbers.
2. Find the prime factorization of a number.

Any number can be expressed as the product of two or more factors. For example, 4 and 7 are factors of 28 because $4 \times 7 = 28$. The numbers 2 and 14 are also factors of 28, as are 1 and 28. Some numbers, such as 7, have only two factors, themselves and 1. These numbers are called **prime numbers.** Numbers other than 1 that are not prime are called **composite numbers** and 1 is called a **unit.** Table 1-2 lists all prime numbers less than 100.

TABLE OF PRIMES LESS THAN 100				
2	3	5	7	11
13	17	19	23	29
31	37	41	43	47
53	59	61	67	71
73	79	83	89	97

Table 1-2

Example 1-29 23 is prime; its only factors are 1 and 23.

39 is composite; its factors are 1, 3, 13, and 39.

Every composite number can be written as a product of prime numbers.

Example 1-30 $56 = 2 \times 2 \times 2 \times 7 = 2^3 \times 7$

$32 = 2 \times 2 \times 2 \times 2 \times 2 = 2^5$

$27 = 3 \times 3 \times 3 = 3^3$

Writing a number as a product of primes is called finding its **prime factorization.**

RULE 1-16   FINDING THE PRIME FACTORIZATION OF A
NUMBER

Step 1   Try the first prime, 2.
  ● If 2 is a factor (that is, if 2 divides the number
    with remainder 0), divide the number by 2.
    Write the quotient and repeat the process
    with 2.
  ● If 2 is not a factor of the number, go on to
    step 2.

Step 2   Try the next prime, 3. Write the quotient and repeat
the process as many times as possible with 3. Go
on to the next prime when 3 no longer divides the
quotient.

Step 3   Continue until the quotient is a prime number. Write
the prime factorization, which is the product of all
the factors.

Example 1-31   Find the prime factorization of 84.

Solution $2)\overline{84}$ Try 2. $84 \div 2 = 42$
$2)\overline{42}$ Try 2 again. $42 \div 2 = 21$
$3)\overline{21}$ 2 does not divide 21. Try 3. $21 \div 3 = 7$
7 7 is a prime.
$84 = 2 \times 2 \times 3 \times 7 = 2^2 \times 3 \times 7$

Example 1-32   Find the prime factors of 539.

Solution   2, 3, and 5 do not divide 539.
$7)\overline{539}$ Try 7. $539 \div 7 = 77$
$7)\overline{77}$ Try 7. $77 \div 7 = 11$
11 11 is prime.
$539 = 7 \times 7 \times 11 = 7^2 \times 11$

Exercises 1-6

Find the prime factorization of each number. If the number is prime, write "prime."

1. 24   **2.** 58   **3.** 72   **4.** 100
5. 57   **6.** 89   **7.** 102   **8.** 41

9.	98	**10.**	125	**11.**	75	**12.**	210
13.	123	**14.**	119	**15.**	165	**16.**	221
17.	143	**18.**	107	**19.**	258	**20.**	462
21.	301	**22.**	399	**23.**	179	**24.**	496
25.	329	**26.**	442	**27.**	592	**28.**	504
29.	703	**30.**	899	**31.**	658	**32.**	2431

Unit 1 Review Exercises

1. A machinist trainee logged 4238 h to complete training. How many hours, rounded to the nearest hundred, did the trainee log?

2. A bottom part takes 12 min + 38 min + 15 min to manufacture, while the top part takes 8 min + 47 min + 7 min to manufacture. Which part takes longer to manufacture?

3. What is the difference between the total times for manufacturing the top and bottom parts discussed in exercise 2?

4. Sam uses 79 ft of material from a piece of stock 96 ft long. How much material will remain as stock?

5. The Ajax Company assembles three-wheeled cars. How many wheels must the company order to assemble 2318 cars?

6. The drafting supervisor must divide 117 drawing projects equally among nine drafters. How many projects will each drafter be assigned?

7. If three drill bits can be sharpened in 1 min, how many can be sharpened in 15 min?

8. Sylvia machined 42 parts Monday, 47 parts Tuesday, 39 parts Wednesday, 41 parts Thursday, and 46 parts Friday. If Sylvia had machined an equal number of parts each day and still finished the same total number of parts, how many parts would she have machined each day?

9. Write $3 \times 3 \times 3 \times 2 \times 2 \times 2 \times 2$ using exponents.

10. Find the prime factorization of 156.

Basic Operations with Fractions

Technicians, machinists, and toolmakers are all skilled workers. Another thing they have in common is a working knowledge of common fractions and the ability to use them as tools in their work. As any carpenter will tell you, there is a correct method of holding a hammer to drive home a nail. There is a correct method of using fractions as well.

Any capable shopworker should be able to:

Work with mixed numbers and improper fractions.

Reduce fractions to lowest terms.

Find the least common denominator.

Add and subtract fractions.

Multiply and divide fractions.

Halve fractions.

In this unit, you will learn how to apply these operations with fractions to shopwork.

2-1 Fractions and Mixed Numbers

OBJECTIVES

After completing this section, you should be able to:

1. Write a mixed number as an improper fraction.
2. Write an improper fraction as a mixed number.

If a collection of objects or a whole is divided into equal parts, each of these parts is smaller than the whole from which it is taken. Each smaller part of the whole is called a **fraction** of the whole. In other words, a fraction is

23

any number that can be expressed as one whole number divided by another whole number (excluding zero). Thus

$$\frac{1}{2}, \quad \frac{9}{8}, \quad \frac{3}{1}, \quad \frac{7}{7}$$

are all fractions.

The circle in Figure 2-1 is divided into three equal parts. Two of the three parts are shaded, representing two-thirds of the whole. The fraction $\frac{2}{3}$ describes the amount shaded.

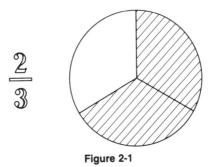

Figure 2-1

The bottom number of a fraction is called the **denominator.** It represents the total number of parts into which a whole has been divided. The top number is called the **numerator.** It represents how many of the total number of parts are under consideration. In the fraction $\frac{1}{4}$, the numerator is 1 and the denominator is 4. The fraction $\frac{1}{4}$ means that 1 part of a whole divided into 4 equal parts is being considered. The line between the numerator and the denominator signifies *division.* Additional examples are shown in Figure 2-2.

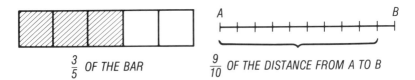

$\frac{3}{5}$ OF THE BAR $\frac{9}{10}$ OF THE DISTANCE FROM A TO B

Figure 2-2

A **proper fraction** is one in which the numerator is less than the denominator. Most fractions encountered in shopwork are proper fractions. If the numerator is equal to or greater than the denominator, the fraction is called an **improper fraction.** Fractions such as

$$\frac{3}{8}, \quad \frac{25}{32}, \quad \frac{151}{275}$$

are proper fractions, while

$$\frac{5}{3}, \quad \frac{4}{4}, \quad \frac{64}{51}$$

are improper fractions.

Improper fractions with the same numerator and denominator, such as $\frac{4}{4}$, are equal to 1.

$$\frac{1}{1} = 1 \qquad \frac{15}{15} = 1 \qquad \frac{345}{345} = 1$$

Any whole number can be written as an improper fraction by writing the number over a denominator of 1.

$$\frac{7}{1} = 7 \qquad \frac{66}{1} = 66 \qquad \frac{212}{1} = 212$$

Improper fractions can also be written as mixed numbers. A **mixed number** is a combination of a whole number and a fraction. For example, the mixed number $4\frac{5}{16}$ consists of the whole number 4 and the fraction $\frac{5}{16}$. It is read "four and five-sixteenths," since the whole number and the fraction are added together. Any mixed number can be changed to an improper fraction by using Rule 2-1.

RULE 2-1   CHANGING A MIXED NUMBER TO AN IMPROPER FRACTION

Step 1   Multiply the whole number by the denominator of the fraction.

Step 2   To the product of step 1, add the numerator of the fraction.

Step 3   Place the sum from step 2 over the denominator of the original fraction.

Example 2-1   Change $4\frac{5}{16}$ to an improper fraction.

Solution

Step 1   Multiply the whole number by the denominator of the fraction.

$$4 \times 16 = 64$$

Step 2   Add the numerator of the fraction.

$$64 + 5 = 69$$

Step 3   Place this sum over the denominator of the original fraction.

$$4\frac{5}{16} = \frac{69}{16}$$

Thus

$$4\frac{5}{16} = \frac{(4 \times 16) + 5}{16} = \frac{64 + 5}{16} = \frac{69}{16}$$

An improper fraction can be changed to a mixed number by reversing the procedure for changing a mixed number to an improper fraction, as in Rule 2-2.

RULE 2-2   CHANGING AN IMPROPER FRACTION TO A MIXED NUMBER

Step 1   Divide the numerator by the denominator.

Step 2   Write the quotient and then write the remainder (if there is one) as a fraction, using the original denominator as the denominator.

Example 2-2   Write $\frac{28}{5}$ as a mixed number.

Solution

Step 1
$$
\begin{array}{r}
5 \\
5{\overline{)28}} \\
\underline{25} \\
3
\end{array}
$$

Step 2 $\dfrac{28}{5} = 5\dfrac{3}{5}$

Exercises 2-1

Write each mixed number as an improper fraction.

1. $5\frac{1}{2}$ 2. $8\frac{2}{3}$ 3. $4\frac{3}{4}$ 4. $15\frac{2}{3}$ 5. $17\frac{3}{5}$

6. $8\frac{5}{7}$ 7. $4\frac{7}{12}$ 8. $12\frac{5}{6}$ 9. $9\frac{3}{10}$ 10. $87\frac{4}{9}$

11. $126\frac{7}{8}$ 12. $243\frac{5}{6}$

Change each improper fraction to a mixed number or whole number.

13. $\dfrac{15}{2}$ 14. $\dfrac{7}{3}$ 15. $\dfrac{11}{5}$ 16. $\dfrac{9}{3}$ 17. $\dfrac{19}{4}$

18. $\dfrac{71}{6}$ **19.** $\dfrac{127}{8}$ **20.** $\dfrac{99}{11}$ **21.** $\dfrac{41}{16}$ **22.** $\dfrac{187}{9}$

23. $\dfrac{75}{18}$ **24.** $\dfrac{175}{26}$

25. A shop helper saws a length of aluminum $9\frac{3}{4}$ ft. Write the length as an improper fraction.

26. Sam adds $\frac{13}{3}$ gallons (gal) of coolant to the coolant tank of a lathe. How much coolant did Sam add, expressed as a mixed number?

27. A certain drill is $1\frac{3}{8}$ in. in diameter. What is the diameter expressed as an improper fraction?

28. Dottie prepares $\frac{9}{4}$ as many parts manuscripts as the other N/C programmers. Express that value as a mixed number.

2-2   Equivalent Fractions

OBJECTIVES

After completing this section, you should be able to:

1. Write a fraction equivalent to a given fraction.
2. Tell if two fractions are equivalent.
3. Write a fraction in simplest form.

Fractions can be used to compare quantities. Look at Figure 2-3. Observe that $\frac{1}{3}$ of one rectangle is the same as $\frac{3}{9}$ of another rectangle of the same size. Therefore, $\frac{1}{3}$ and $\frac{3}{9}$ are equal. Fractions that are equal are called **equivalent fractions.** To show that the fractions are equivalent, write $\frac{1}{3} = \frac{3}{9}$.

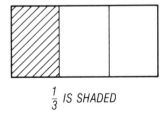
$\frac{1}{3}$ IS SHADED

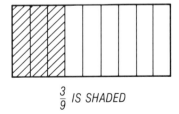
$\frac{3}{9}$ IS SHADED

Figure 2-3

The value of a fraction is not changed if the numerator and the denominator are multiplied (or divided) by the same number. For example,

$$\frac{3}{5} = \frac{3 \times 3}{5 \times 3} = \frac{9}{15}$$

Rule 2-3 tells how to find equivalent fractions.

RULE 2-3   FINDING AN EQUIVALENT FRACTION WITH A SPECIFIED DENOMINATOR

- *If the specified denominator is greater than the given denominator:*

Step 1   Divide the specified denominator by the given denominator.

Step 2   Multiply the numerator and the denominator of the given fraction by the quotient.

- *If the specified denominator is less than the given denominator:*

Step 1   Divide the given denominator by the specified denominator.

Step 2   Divide the numerator and the denominator of the given fraction by the quotient.

Example 2-3   Find a fraction equivalent to $\frac{7}{8}$ that has a denominator of 24.

Solution   The specified denominator 24 is greater than the given denominator 8.

Step 1   Divide 24 by 8.

$$24 \div 8 = 3$$

Step 2   Multiply the numerator and the denominator of the given fraction $\frac{7}{8}$ by 3.

$$\frac{7 \times 3}{8 \times 3} = \frac{21}{24}$$

Example 2-4   Find a fraction equivalent to $\frac{18}{42}$ that has a denominator of 7.

Solution   The specified denominator 7 is less than the given denominator 42.

Step 1   Divide 42 by 7.

$$42 \div 7 = 6$$

Step 2   Divide the numerator and the denominator of the given fraction $\frac{18}{42}$ by 6.

$$\frac{18 \div 6}{42 \div 6} = \frac{3}{7}$$

To tell if two fractions are equivalent, find their **cross products.** In the fractions $\frac{3}{5}$ and $\frac{9}{15}$, note that $3 \times 15 = 45$ and $9 \times 5 = 45$.

$$3 \times 15 = 45 \qquad \frac{3}{5} \times \frac{9}{15} \qquad 9 \times 5 = 45$$

The products 3×15 and 5×9 are the cross products. Cross products can be used to decide if fractions are equal.

RULE 2-4   DETERMINING EQUIVALENT FRACTIONS

Two fractions are equivalent if their cross products are equal. (Also, two fractions are not equivalent if their cross products are not equal.)

Example 2-5   Are $\frac{3}{4}$ and $\frac{9}{16}$ equivalent?

Solution   Find the cross products.

$$3 \times 16 = 48$$
$$9 \times 4 = 36$$

The cross products are not equal, so $\frac{3}{4}$ and $\frac{9}{16}$ are not equivalent.

Write $\dfrac{3}{4} \neq \dfrac{9}{16}$ to show that the fractions are not equivalent.

A fraction such as $\frac{3}{4}$ is said to be in **lowest terms** because the numerator and the denominator have no common factors other than 1. Notice that the factors of 3 are 3 and 1, and the factors of 4 are 1, 2, and 4. The only common factor is 1. In the fraction $\frac{6}{8}$, however, the factors of 6 are 1, 2, 3, and 6, and the factors of 8 are 1, 2, 4, and 8. There is a common factor of 2. Therefore, $\frac{6}{8}$ is not in lowest terms.

Example 2-6 $\frac{2}{3}$ is in lowest terms because 2 and 3 have no common factors other than 1.

$\frac{9}{12}$ is not in lowest terms because 3 is a factor of both 9 and 12.

Writing a fraction in lowest terms is called **simplifying the fraction,** or **reducing the fraction to lowest terms.**

RULE 2-5   SIMPLIFYING FRACTIONS

Step 1   Write the prime factorization of the numerator and the denominator.

Step 2   Divide the numerator and the denominator by the common prime factor, or by the product of the common prime factors if there is more than one.

Example 2-7   Simplify $\frac{21}{56}$.

Solution

Step 1   Write the prime factorization of 21 and 56.

$$21 = 3 \times 7$$
$$56 = 2 \times 2 \times 2 \times 7$$

The common factor is 7.

Step 2   Divide the numerator and the denominator by the common prime factor, 7.

$$\frac{21}{56} = \frac{21 \div 7}{56 \div 7} = \frac{3}{8}$$

Therefore, $\frac{21}{56} = \frac{3}{8}$ in lowest terms.

Example 2-8   Reduce $\frac{48}{60}$ to lowest terms.

Solution

Step 1   The common factors are 2, 2, and 3.

$$48 = 2 \times 2 \times 2 \times 2 \times 3$$
$$60 = 2 \times 2 \times 3 \times 5$$

The product of the common factors is $2 \times 2 \times 3 = 12$.

Step 2   Divide 48 and 60 by 12.

$$\frac{48}{60} = \frac{48 \div 12}{60 \div 12} = \frac{4}{5}$$

Thus, $\frac{48}{60} = \frac{4}{5}$ in lowest terms.

Short form   Divide out the pairs of common factors and then find the product of the remaining factors in the numerator and the denominator.

$$\frac{48}{60} = \frac{\overset{1}{\cancel{2}} \times \overset{1}{\cancel{2}} \times 2 \times 2 \times \overset{1}{\cancel{3}}}{\underset{1}{\cancel{2}} \times \underset{1}{\cancel{2}} \times \underset{1}{\cancel{3}} \times 5} = \frac{4}{5}$$

Exercises 2-2

Tell if the fractions in each pair are equivalent.

1. $\dfrac{6}{9}, \dfrac{14}{21}$　　**2.** $\dfrac{17}{34}, \dfrac{2}{4}$　　**3.** $\dfrac{1}{3}, \dfrac{14}{21}$　　**4.** $\dfrac{3}{4}, \dfrac{15}{20}$

5. $\dfrac{84}{7}, \dfrac{36}{3}$　　**6.** $\dfrac{1}{2}, \dfrac{4}{9}$　　**7.** $\dfrac{8}{36}, \dfrac{10}{45}$　　**8.** $\dfrac{6}{15}, \dfrac{15}{38}$

Find a fraction equivalent to the given fraction with the specified denominator.

9. $\dfrac{5}{6}(30)$　　**10.** $\dfrac{10}{15}(3)$　　**11.** $\dfrac{1}{6}(12)$　　**12.** $\dfrac{24}{27}(9)$

13. $\dfrac{14}{15}(45)$　　**14.** $\dfrac{36}{84}(7)$　　**15.** $\dfrac{12}{11}(99)$　　**16.** $\dfrac{48}{34}(17)$

17. $\dfrac{4}{1}(3)$　　**18.** $\dfrac{63}{9}(1)$　　**19.** $\dfrac{8}{45}(630)$　　**20.** $\dfrac{99}{630}(70)$

Write each fraction in lowest terms.

21. $\dfrac{7}{21}$　　**22.** $\dfrac{15}{20}$　　**23.** $\dfrac{14}{49}$　　**24.** $\dfrac{26}{39}$

25. $\dfrac{15}{9}$　　**26.** $\dfrac{21}{42}$　　**27.** $\dfrac{26}{13}$　　**28.** $\dfrac{56}{82}$

29. $\dfrac{21}{24}$　　**30.** $\dfrac{33}{55}$　　**31.** $\dfrac{36}{81}$　　**32.** $\dfrac{48}{36}$

33. $\dfrac{4000}{6000}$　　**34.** $\dfrac{27}{108}$　　**35.** $\dfrac{85}{135}$　　**36.** $\dfrac{377}{319}$

37. A machinist has two pieces of metal stock. One is $\frac{5}{8}$ in. long and the other is $\frac{15}{24}$ in. long. Are the two pieces the same length?

38. A drafter measures a model and finds it to be $\frac{14}{32}$ in. long. Rewrite the fraction in lowest terms.

39. The tool crib attendant offers a $\frac{5}{32}$-in.-diameter drill to a trainee machinist who requested a $\frac{15}{96}$-in.-diameter drill. Are the two drills the same size?

40. Karen uses a 6-in. ruler calibrated in thirty-seconds of an inch. If she measures a part $\frac{1}{4}$ in. long, how many thirty-seconds is it?

2-3 Reading a Steel Rule

OBJECTIVE
After completing this section, you should be able to:

1. Read a steel rule.

A practical application of working with fractions is the ability to read a steel rule. All readings on a steel rule must be reduced to lowest terms. A typical steel rule is shown in Figure 2-4. This steel rule is graduated in thirty-seconds of an inch, indicated by the number 32 on the left.

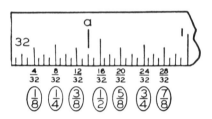

Figure 2-4

One inch is the largest graduation, followed by the half inch, quarter inches, eighth inches, sixteenth inches, and thirty-second inches. The longer graduation marks (eighths, quarters, and the half inch) are called landmarks. These landmarks are used to find smaller graduations quickly.

Example 2-9 Locate $\frac{11}{32}$ on a steel rule.

Solution Find the landmark $\frac{1}{4}$ and then change $\frac{1}{4}$ to thirty-seconds.

$$\frac{1}{4} = \frac{1 \times 8}{4 \times 8} = \frac{8}{32}$$

Count 3 thirty-seconds to the right of $\frac{8}{32}$ to find $\frac{11}{32}$.

Example 2-10 State the reading at a on the steel rule in Figure 2-4.

Solution The nearest landmark to the left of the reading is $\frac{3}{8}$. Change $\frac{3}{8}$ to thirty-seconds.

$$\frac{3}{8} = \frac{3 \times 4}{8 \times 4} = \frac{12}{32}$$

The number of additional thirty-second graduations from $\frac{12}{32}$ to the reading is 2. The reading is $\frac{14}{32}$.

Reduce $\frac{14}{32}$ to lowest terms.

$$\frac{14}{32} = \frac{7}{16}$$

The reading at a is $\frac{7}{16}$.

Exercises 2-3

Which of the following steel rule readings are expressed correctly?

1. $1\frac{1}{2}$ in.   **2.** $\frac{14}{64}$ in.   **3.** $3\frac{2}{8}$ in.   **4.** $\frac{1}{4}$ in.

5. $\frac{6}{8}$ in.   **6.** $\frac{1}{16}$ in.

How should the following steel rule readings be stated?

7. $\frac{12}{16}$ in.   **8.** $\frac{8}{32}$ in.   **9.** $2\frac{10}{64}$ in.   **10.** $\frac{2}{4}$ in.

11. $\frac{8}{8}$ in.   **12.** $\frac{6}{16}$ in.

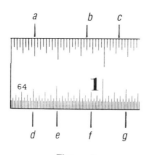

Figure A   **Figure B**

In Figure A, what is the reading:

13. at a?   **14.** at b?   **15.** at c?   **16.** at d?   **17.** at e?

18. at f?   **19.** at g?

In Figure B, what is the reading:

20. at a?   **21.** at b?   **22.** at c?   **23.** at d?   **24.** at e?

25. at f?   **26.** at g?   **27.** at h?   **28.** at i?   **29.** at j?

30. at k?   **31.** at l?   **32.** at m?   **33.** at n?   **34.** at o?

35. at p?

2-4 Finding the Least Common Denominator

OBJECTIVES
After completing this section, you should be able to:
1. Find the least common denominator of two or more fractions.
2. Change fractions with different denominators to equivalent fractions with the least common denominator.

Before fractions can be added or subtracted, they must first have the same denominator. Consider the fractions $\frac{1}{2}$, $\frac{1}{4}$, and $\frac{1}{6}$. The **lowest**, or **least, common denominator (L.C.D.)** of these fractions is the smallest number (not including zero) that the denominators 2, 4, and 6 will divide evenly. The L.C.D. is determined by finding the product of the prime factors of each of the denominators.

RULE 2-6 FINDING THE LEAST COMMON DENOMINATOR (L.C.D.)

Step 1 Write the prime factorization of each denominator in exponential form.

Step 2 Write all the different primes that occur in step 1.

Step 3 Raise each prime to the greatest power (exponent) that occurs on the prime in step 1.

Step 4 Find the product of the primes in step 3. This is the L.C.D.

Example 2-11 Find the L.C.D. for $\frac{1}{2}$, $\frac{1}{4}$, and $\frac{1}{6}$.

Solution

Step 1 Find the prime factorization of each denominator.

$$2 = 2^1$$
$$4 = 2^2$$
$$6 = 2^1 \times 3^1$$

Step 2 The different primes are

$$2, \ 3$$

Step 3 The greatest power of 2 is 2, and the greatest power of 3 is 1.

$$2^2, \ 3^1$$

Step 4   Multiply the prime from step 3.

$$2^2 \times 3^1 = 4 \times 3 = 12$$

The L.C.D. is 12.

Once the least common denominator has been found for a group of fractions, those fractions can be changed to equivalent fractions with the L.C.D.

Example 2-12   Change $\frac{5}{6}$, $\frac{7}{8}$, and $\frac{7}{10}$ to equivalent fractions using the L.C.D.

Solution   First find the L.C.D.

Step 1 $6 = 2^1 \times 3^1$
$8 = 2^3$
$10 = 2^1 \times 5^1$

Step 2   2, 3, 5

Step 3 2^3, 3^1, 5^1

Step 4 $2^3 \times 3^1 \times 5^1 = 120$

Now write equivalent fractions using the L.C.D.

$$\frac{5}{6} = \frac{5 \times 20}{6 \times 20} = \frac{100}{120} \qquad 120 \div 6 = 20$$

$$\frac{7}{8} = \frac{7 \times 15}{8 \times 15} = \frac{105}{120} \qquad 120 \div 8 = 15$$

$$\frac{7}{10} = \frac{7 \times 12}{10 \times 12} = \frac{84}{120} \qquad 120 \div 10 = 12$$

There is a shorter way to find the L.C.D. Again consider the fractions $\frac{1}{2}$, $\frac{1}{4}$, and $\frac{1}{6}$. First write the three denominators and then divide each by the lowest prime (in this case, 2) that divides at least two of the denominators. The resulting quotients are all prime numbers.

$$2\overline{)2\ 4\ 6}$$
$$\ \ \ 1\ 2\ 3$$

Therefore, no further division is necessary. All divisors (only one, in this case) are multiplied by all quotients to find the L.C.D.

L.C.D. $= 2 \times 1 \times 2 \times 3 = 12$

Example 2-13   Use a shortcut to find the L.C.D. for $\frac{13}{75}$, $\frac{2}{45}$, and $\frac{9}{125}$.

Solution   While this type of value is not typically found in machine shopwork, the principle is the same for all fractions. The first step is to select a prime number by which to divide two or more of the denominators. The number 2 will not divide any denominator evenly. Pick the next prime number, 3, which divides two of the denominators evenly.

$$\begin{array}{r} 3 \overline{)\,75\ \ 45\ \ 125} \\ 25\ \ 15\ \ 125 \end{array}$$

Observe that 125 was not divisible by 3, so it was brought down unchanged. The resulting quotients are all still divisible by another number, 5. Use 5 for the next divisor.

$$\begin{array}{r} 5 \overline{)\,25\ \ 15\ \ 125} \\ 5\ \ \ 3\ \ \ 25 \end{array}$$

Two of the resulting quotients are still divisible by 5.

$$\begin{array}{r} 5 \overline{)\,5\ \ 3\ \ 25} \\ 1\ \ 3\ \ \ 5 \end{array}$$

Note that 5 will not divide into 3, so the 3 is brought down as a repeated quotient. At this point, arrange the steps in logical order.

$$\begin{array}{r} 3 \overline{)\,75\ \ 45\ \ 125} \\ 5 \overline{)\,25\ \ 15\ \ 125} \\ 5 \overline{)\ \ 5\ \ \ 3\ \ \ 25} \\ 1\ \ \ 3\ \ \ 5 \end{array}$$

All divisors (3, 5, 5) and all quotients (1, 3, 5) are prime numbers. Multiply divisors and quotients together.

L.C.D. = $3 \times 5 \times 5 \times 1 \times 3 \times 5 = 1125$

Exercises 2-4

Find the L.C.D. for each group of fractions.

1. $\dfrac{1}{2}, \dfrac{1}{4}$ **2.** $\dfrac{3}{16}, \dfrac{1}{2}$ **3.** $\dfrac{6}{22}, \dfrac{8}{33}$

4. $\dfrac{7}{12}, \dfrac{8}{15}$ **5.** $\dfrac{3}{6}, \dfrac{5}{4}, \dfrac{3}{2}$ **6.** $\dfrac{2}{3}, \dfrac{7}{5}, \dfrac{11}{15}$

7. $\dfrac{2}{5}, \dfrac{1}{9}, \dfrac{7}{15}$ **8.** $\dfrac{3}{32}, \dfrac{2}{7}, \dfrac{1}{49}$ **9.** $\dfrac{9}{16}, \dfrac{7}{32}, \dfrac{29}{64}$

10. $\dfrac{3}{4}, \dfrac{19}{32}, \dfrac{11}{64}$ **11.** $\dfrac{1}{2}, \dfrac{3}{4}, \dfrac{7}{16}, \dfrac{5}{8}$ **12.** $\dfrac{1}{4}, \dfrac{5}{64}, \dfrac{7}{32}, \dfrac{7}{8}$

Find the L.C.D. and change to equivalent fractions.

13. $\dfrac{5}{7}, \dfrac{9}{14}$

14. $\dfrac{3}{2}, \dfrac{1}{3}, \dfrac{11}{6}$

15. $\dfrac{4}{15}, \dfrac{7}{12}$

16. $\dfrac{7}{9}, \dfrac{13}{12}, \dfrac{2}{3}$

17. $\dfrac{7}{15}, \dfrac{11}{18}$

18. $\dfrac{5}{4}, \dfrac{15}{6}$

19. $\dfrac{9}{10}, \dfrac{4}{25}$

20. $\dfrac{24}{25}, \dfrac{49}{100}$

21. $\dfrac{2}{5}, \dfrac{1}{9}, \dfrac{7}{15}$

22. $\dfrac{1}{2}, \dfrac{2}{3}, \dfrac{7}{15}$

23. $\dfrac{19}{72}, \dfrac{35}{108}, \dfrac{19}{30}$

24. $\dfrac{13}{15}, \dfrac{80}{221}, \dfrac{99}{255}$

25. Julio measured three parts to be $\frac{25}{32}$ in., $\frac{3}{8}$ in., and $\frac{7}{16}$ in. Find the L.C.D. of these fractions.

26. The shop foreman determined that $\frac{47}{51}$ of the day shift and $\frac{15}{17}$ of the night shift arrived at work on time. Find the L.C.D.

27. Change the fractions in exercise 26 to equivalent fractions.

28. Among three shop classes, $\frac{2}{3}, \frac{7}{8}$, and $\frac{5}{9}$ of the students were men. Find the L.C.D. of these fractions.

2-5 Comparing Fractions

OBJECTIVES

After completing this section, you should be able to:

1. Compare two or more fractions.
2. Compare two or more mixed numbers.

The ability to compare fractions is essential for a shopworker. For example, you will often need to compare the sizes of two items, such as drills or bolts, that are given as fractions.

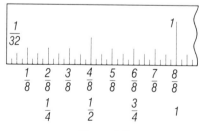

Figure 2-5

Look at the illustration of a steel rule in Figure 2-5. When two readings are compared on a steel rule, the reading to the right is greater. Notice that the larger the denominator of a fraction, the smaller its value. For example,

$\dfrac{1}{8}$ is smaller in value than $\dfrac{1}{4}$ $\left(\dfrac{2}{8} = \dfrac{1}{4}\right)$

$\dfrac{1}{64}$ is smaller in value than $\dfrac{1}{32}$ $\left(\dfrac{2}{64} = \dfrac{1}{32}\right)$

Fractions can be compared by using the symbols < (is less than) and > (is greater than).

$\dfrac{1}{8} < \dfrac{1}{4}$ $\quad$ $\dfrac{1}{8}$ is less than $\dfrac{1}{4}$

$\dfrac{1}{4} > \dfrac{1}{8}$ $\quad$ $\dfrac{1}{4}$ is greater than $\dfrac{1}{8}$

Look at Figure 2-5 again. Now notice that when fractions with the same denominator are being compared, the fraction with the greater numerator is greater. Therefore,

$\dfrac{2}{8} > \dfrac{1}{8}$ $\quad$ and $\quad$ $\dfrac{1}{4} < \dfrac{3}{4}$

RULE 2-7 COMPARING FRACTIONS

Step 1 Write equivalent fractions using the L.C.D.

Step 2 Compare numerators. The fraction with the greater numerator is greater.

Example 2-14 Which is greater, $\frac{2}{3}$ or $\frac{4}{7}$?

Solution

Step 1 The L.C.D. is 21.

$$\frac{2}{3} = \frac{2 \times 7}{3 \times 7} = \frac{14}{21}$$

$$\frac{4}{7} = \frac{4 \times 3}{7 \times 3} = \frac{12}{21}$$

Step 2 Since $14 > 12$, $\dfrac{14}{21} > \dfrac{12}{21}$. Thus

$$\frac{2}{3} > \frac{4}{7}$$

To compare mixed numbers, first compare whole numbers, then compare fractions.

RULE 2-8   COMPARING MIXED NUMBERS

Step 1   Compare the whole number parts.
- If they are not equal, the mixed number with the greater whole number part is greater.
- If they are equal, go to step 2.

Step 2   Compare the fractional parts of each mixed number, as in Rule 2-7. The mixed number with the greater fractional part is greater.

Example 2-15   Which is greater, $4\frac{3}{4}$ or $4\frac{5}{6}$?

Solution

Step 1   The whole number parts are equal.

Step 2 $\dfrac{3}{4} = \dfrac{9}{12}$ and $\dfrac{5}{6} = \dfrac{10}{12}$

Since $10 > 9$, $\dfrac{5}{6} > \dfrac{3}{4}$. Thus

$$4\frac{5}{6} > 4\frac{3}{4}$$

Exercises 2-5

Tell whether each of the following is true or false.

1. $\dfrac{1}{5} > \dfrac{1}{10}$

2. $\dfrac{2}{7} > \dfrac{2}{3}$

3. $\dfrac{1}{2} < \dfrac{1}{4}$

4. $\dfrac{5}{32} > \dfrac{5}{64}$

5. $\dfrac{3}{8} > \dfrac{2}{8}$

6. $\dfrac{9}{64} < \dfrac{15}{64}$

7. $\dfrac{1}{2} > \dfrac{2}{3}$

8. $\dfrac{7}{12} < \dfrac{9}{16}$

9. $\dfrac{11}{12} > \dfrac{9}{10}$

10. $\dfrac{3}{4} < \dfrac{5}{6}$

11. $7\dfrac{2}{3} > 7\dfrac{5}{6}$

12. $18\dfrac{17}{25} > 18\dfrac{59}{100}$

13. $15\dfrac{1}{2} > 12\dfrac{3}{5}$ **14.** $21\dfrac{9}{10} > 20\dfrac{99}{100}$ **15.** $8\dfrac{7}{8} < 8\dfrac{8}{9}$

16. $52\dfrac{2}{3} < 52\dfrac{7}{9}$

Use the appropriate symbol (> or <) to compare the following fractions.

17. $\dfrac{5}{6}, \dfrac{5}{9}$ **18.** $\dfrac{2}{11}, \dfrac{2}{13}$ **19.** $\dfrac{5}{7}, \dfrac{3}{7}$

20. $\dfrac{93}{100}, \dfrac{97}{100}$ **21.** $\dfrac{21}{64}, \dfrac{10}{32}$ **22.** $\dfrac{7}{8}, \dfrac{59}{64}$

23. $8\dfrac{5}{7}, 8\dfrac{7}{9}$ **24.** $12\dfrac{4}{11}, 12\dfrac{5}{9}$

Arrange each group of fractions in ascending order (smallest fraction first).

25. $\dfrac{3}{4}, \dfrac{2}{4}, \dfrac{7}{8}$ **26.** $\dfrac{4}{5}, \dfrac{1}{5}, \dfrac{7}{10}$ **27.** $\dfrac{2}{3}, \dfrac{3}{4}, \dfrac{5}{6}$

28. $\dfrac{9}{10}, \dfrac{2}{3}, \dfrac{7}{15}$ **29.** $6\dfrac{3}{4}, 6\dfrac{3}{5}, 6\dfrac{7}{15}$ **30.** $12\dfrac{1}{2}, 12\dfrac{2}{3}, 12\dfrac{17}{30}$

31. The fraction of the total month's work completed by the fifteenth of the month was $\frac{2}{3}$ in January, $\frac{5}{6}$ in February, and $\frac{19}{25}$ in March. Arrange the months in ascending order according to the fraction of the total month's work completed by the fifteenth.

32. A production report noted that job x is nearer completion than job y. If job x is $\frac{21}{25}$ complete and job y is $\frac{25}{32}$ complete, is the production report correct?

33. Use the appropriate symbol (> or <) to compare the sizes of a $\frac{1}{4}$-in. drill and a $\frac{19}{64}$-in. drill.

34. A machine shop has in stock three steel bars, $\frac{3}{8}$ in., $\frac{1}{2}$ in., and $\frac{3}{16}$ in. in diameter. Arrange the bars in ascending order by diameter.

2-6   Adding Fractions and Mixed Numbers

OBJECTIVES

After completing this section, you should be able to:

1. Add fractions and mixed numbers with the same denominator.

2. Add fractions and mixed numbers with unlike denominators.

You can add fractions with the same denominator simply by adding the numerators. Look at the illustration of a ruler in Figure 2-6. The illustration shows that $\frac{1}{4} + \frac{2}{4} = \frac{3}{4}$.

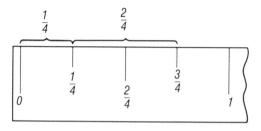

Figure 2-6

RULE 2-9   ADDING FRACTIONS WITH THE SAME
DENOMINATOR

Step 1   Add the numerators.

Step 2   Write a fraction with the sum of the numerators from step 1 as the numerator and the denominator of the given fractions as the denominator.

$$\frac{a}{c} + \frac{b}{c} = \frac{a + b}{c}$$

Step 3   Write the resulting fraction in simplest form.

Example 2-16   Find: $\frac{7}{8} + \frac{3}{8}$.

Solution $\dfrac{7}{8} + \dfrac{3}{8} = \dfrac{7 + 3}{8} = \dfrac{10}{8} = 1\dfrac{2}{8} = 1\dfrac{1}{4}$

Fractions with unlike denominators can be added after first writing the fractions with the least common denominator. Once all fractions have the same denominator, they can be added as in Rule 2-9.

RULE 2-10   ADDING FRACTIONS WITH DIFFERENT
DENOMINATORS

Step 1   Change each fraction to an equivalent fraction using the L.C.D.

Step 2   Add as in Rule 2-9.

Example 2-17   Add: $\frac{3}{4} + \frac{5}{8}$.

   Solution

      Step 1 $4 = 2^2$

                   $8 = 2^3$

               L.C.D. $= 2^3 = 8$

$$\frac{3}{4} = \frac{3 \times 2}{4 \times 2} = \frac{6}{8} \quad \text{and} \quad \frac{5}{8} = \frac{5}{8}$$

      Step 2

$$\frac{3}{4} = \frac{6}{8}$$

$$+\frac{5}{8} = \frac{5}{8}$$

$$\frac{11}{8} = 1\frac{3}{8}$$

RULE 2-11   ADDING MIXED NUMBERS

Step 1   Add the fractional parts.
Step 2   Add the whole number parts.
Step 3   Write the answer in simplest form.

Example 2-18   Find the sum of $1\frac{5}{32}$ and $3\frac{7}{16}$.

   Solution

$$1\frac{5}{32} = 1\frac{5}{32}$$

$$+3\frac{7}{16} = 3\frac{14}{32}$$

$$4\frac{19}{32}$$

Example 2-19   Add: $6\frac{3}{4} + 5\frac{2}{3}$.

   Solution

$$6\frac{3}{4} = 6\frac{9}{12}$$

$$+5\frac{2}{3} = 5\frac{8}{12}$$

$$11\frac{17}{12} = 11 + \frac{17}{12}$$

$$= 11 + 1\frac{5}{12} = 12\frac{5}{12}$$

Exercises 2-6

Find each sum. Write your answers in simplest form.

1. $\dfrac{2}{6}$
    $+\dfrac{2}{6}$

2. $\dfrac{3}{8}$
    $+\dfrac{1}{8}$

3. $\dfrac{2}{5}$
    $+\dfrac{3}{10}$

4. $\dfrac{8}{9}$
    $+\dfrac{2}{3}$

5. $\dfrac{2}{3}$
    $+\dfrac{3}{4}$

6. $\dfrac{5}{9}$
    $+\dfrac{5}{6}$

7. $\dfrac{3}{10}$
    $+\dfrac{7}{15}$

8. $\dfrac{4}{21}$
    $+\dfrac{9}{14}$

9. $5\dfrac{1}{2}$
    $+6$

10. $2\dfrac{3}{4}$
    $+7$

11. $6\dfrac{1}{3}$
    $+12\dfrac{1}{3}$

12. $12\dfrac{9}{16}$
    $+ 3\dfrac{3}{16}$

13. $3\dfrac{1}{2}$
    $+5\dfrac{1}{3}$

14. $18\dfrac{3}{5}$
    $+12\dfrac{7}{10}$

15. $9\dfrac{5}{12}$
    $+7\dfrac{7}{8}$

16. $9\dfrac{3}{10}$
    $+8\dfrac{11}{15}$

17. $46\dfrac{4}{11}$
    $+23\dfrac{8}{9}$

18. $46\dfrac{5}{12}$
    $+82\dfrac{11}{16}$

19. $127\dfrac{2}{9}$
    $+856\dfrac{3}{5}$

20. $4675\dfrac{7}{18}$
    $+3995\dfrac{11}{12}$

21. $\dfrac{1}{2} + \dfrac{1}{2}$

22. $\dfrac{3}{4} + \dfrac{5}{8}$

23. $\dfrac{9}{64} + \dfrac{1}{8}$

24. $\dfrac{3}{8} + \dfrac{1}{4} + \dfrac{2}{3}$

25. $\dfrac{7}{15} + \dfrac{9}{20} + \dfrac{4}{5}$

26. $\dfrac{5}{64} + \dfrac{5}{8}$

27. $\dfrac{5}{16} + \dfrac{7}{32} + \dfrac{11}{12}$

28. $\dfrac{2}{9} + \dfrac{1}{18} + \dfrac{2}{3}$

29. $\dfrac{3}{64} + \dfrac{9}{16} + \dfrac{1}{2}$

30. $\dfrac{3}{5} + \dfrac{1}{10} + \dfrac{2}{45}$

31. Three pieces of sheet metal are to be riveted together. If their thicknesses in inches are $\frac{1}{16}$, $\frac{1}{64}$, and $\frac{1}{32}$, what length rivet must be used, allowing $\frac{1}{16}$ extra for riveting the three pieces?

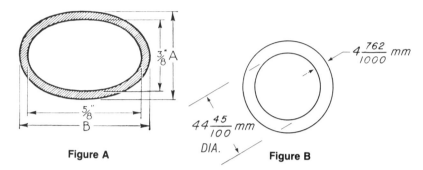

Figure A   **Figure B**

32. Find the dimensions A and B in Figure A, if the thickness of the tubing is $\frac{3}{32}$ in.

33. In Figure B, what is the outside diameter ($O.D.$) if the wall thickness is $4\frac{762}{1000}$ mm?

34. In Figure C, what is the overall length of the shaft?

35. Three pieces of steel bar are sawed off with a power saw. Each length is $4\frac{9}{64}$ in. long. If $1\frac{5}{16}$ in. remain as waste, what was the original length of the bar?

36. Study Figure D carefully. Will the four gears fit between the bearings on the shaft as an assembly?

37. One dimension on a blueprint can be found by adding $\frac{5}{8}$ in., $3\frac{1}{2}$ in., and $\frac{13}{32}$ in. What is the dimension?

38. A machinist completed one part $1\frac{5}{8}$ in. long, one part $5\frac{7}{16}$ in. long, and one part $1\frac{1}{32}$ in. long. If these parts were laid end to end, what would be the total length?

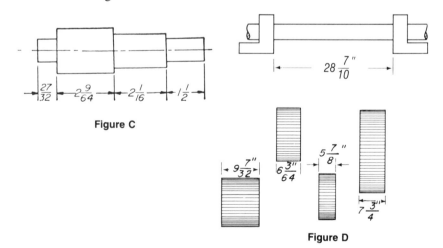

Figure C

Figure D

2-7   Subtracting Fractions and Mixed Numbers

OBJECTIVES

After completing this section, you should be able to:

1. Subtract fractions and mixed numbers with the same denominator.
2. Subtract fractions and mixed numbers with unlike denominators.

Fractions are subtracted in much the same way as they are added. Before fractions can be subtracted, they must have like denominators.

RULE 2-12   SUBTRACTING FRACTIONS

Step 1   If denominators are unlike, write equivalent fractions using the L.C.D.

Step 2   Subtract the numerators.

Step 3   Write a fraction with the difference from step 2 as the numerator and the L.C.D., or the given denominator if denominators are the same, as the denominator.

$$\frac{a}{c} - \frac{b}{c} = \frac{a-b}{c}$$

Step 4   Write the resulting fraction in simplest form.

Example 2-20   Subtract: $\frac{3}{8} - \frac{1}{4}$.

Solution

Step 1   The L.C.D. is 8.

$$\frac{1}{4} = \frac{2}{8}, \quad \frac{3}{8} = \frac{3}{8}$$

Steps 2–4

$$\frac{3}{8} - \frac{1}{4} = \frac{3}{8} - \frac{2}{8} = \frac{3-2}{8} = \frac{1}{8}$$

RULE 2-13 SUBTRACTING MIXED NUMBERS

Step 1 If denominators of the fractional parts are unlike, write equivalent fractions using the L.C.D.

Step 2 Subtract the fractional parts. If necessary, borrow 1 from the whole number part of the first mixed number, change it to an equivalent fraction using the L.C.D., and add it to the fractional part of the first mixed number.

Step 3 Subtract the remaining whole numbers.

Step 4 Write the answer in simplest form.

In Example 2-21, no renaming is needed.

Example 2-21 Find the difference between $5\frac{7}{8}$ and $3\frac{1}{8}$.

Solution

$$5\frac{7}{8}$$
$$-3\frac{1}{8}$$
$$2\frac{6}{8} = 2\frac{3}{4}$$

In subtracting $1\frac{7}{8}$ from $3\frac{1}{2}$, notice that $\frac{1}{2}$, or $\frac{4}{8}$, is less than $\frac{7}{8}$. Observe in Example 2-22 how the whole number 1 is borrowed, renamed as a fraction, and added to an existing fraction to make it large enough to subtract from.

Example 2-22 Subtract: $3\frac{1}{2} - 1\frac{7}{8}$.

Solution

$$3\frac{1}{2} = 3\frac{4}{8} = 2\frac{8}{8} + \frac{4}{8} = 2\frac{12}{8}$$
$$-1\frac{7}{8} = 1\frac{7}{8} = 1\frac{7}{8} \qquad = 1\frac{7}{8}$$
$$\rule{4cm}{0.4pt}$$
$$1\frac{5}{8}$$

When the first mixed number has no fractional part, borrow 1 and rename it as a fraction before subtracting.

Example 2-23   Find the difference between 15 and $9\frac{3}{5}$.

Solution

$$15 \; = 14\frac{5}{5}$$

$$-\; 9\frac{3}{5} = \; 9\frac{3}{5}$$

$$5\frac{2}{5}$$

Exercises 2-7

Find each difference. Write the answers in simplest form.

1.
$$\frac{7}{8}$$
$$-\frac{3}{8}$$

2.
$$\frac{9}{10}$$
$$-\frac{7}{10}$$

3.
$$\frac{2}{3}$$
$$-\frac{1}{6}$$

4.
$$\frac{15}{16}$$
$$-\frac{5}{8}$$

5.
$$\frac{7}{9}$$
$$-\frac{1}{12}$$

6.
$$\frac{4}{5}$$
$$-\frac{2}{3}$$

7.
$$\frac{16}{21}$$
$$-\frac{3}{14}$$

8.
$$\frac{13}{20}$$
$$-\frac{7}{15}$$

9.
$$5$$
$$-\frac{3}{4}$$

10.
$$12$$
$$-\frac{7}{9}$$

11.
$$23$$
$$-18\frac{5}{8}$$

12.
$$147$$
$$-\; 59\frac{5}{32}$$

13.
$$15\frac{3}{7}$$
$$-12\frac{2}{7}$$

14.
$$28\frac{27}{64}$$
$$-15\frac{13}{64}$$

15.
$$7\frac{3}{8}$$
$$-5\frac{1}{6}$$

16.
$$12\frac{7}{9}$$
$$-\; 8\frac{1}{2}$$

17.
$$23\frac{11}{12}$$
$$-17\frac{3}{4}$$

18.
$$65\frac{3}{4}$$
$$-37\frac{3}{10}$$

19.
$$26\frac{4}{9}$$
$$-16\frac{2}{3}$$

20.
$$35\frac{7}{10}$$
$$-28\frac{4}{5}$$

21.
$$88\frac{1}{5}$$
$$-59\frac{3}{4}$$

22.
$$46\frac{1}{8}$$
$$-17\frac{2}{9}$$

23.
$$708\frac{56}{247}$$
$$-539\frac{255}{323}$$

24.
$$5077\frac{865}{1024}$$
$$-3829\frac{241}{243}$$

25. $\dfrac{3}{4} - \dfrac{1}{4}$

26. $\dfrac{11}{12} - \dfrac{5}{6}$

27. $\dfrac{3}{7} - \dfrac{3}{21}$

28. $\dfrac{5}{32} - \dfrac{3}{64}$

29. $4\dfrac{2}{7} - 2\dfrac{6}{35}$

30. $5\dfrac{9}{16} - 2\dfrac{3}{4}$

31. The space between shelves in a bookcase is $14\frac{3}{4}$ in. A set of books placed on a shelf leaves a gap of $\frac{7}{32}$ in. to the bottom of the shelf above. How tall are the books in the set?

32. A machinist cuts $14\frac{5}{8}$ in. and $6\frac{2}{3}$ in. from a 28-in. piece of brass bar stock. How much material is left?

33. Half-inch end mills are ground $\frac{1}{32}$ in. undersize when dull. What will be the diameter of the $\frac{1}{2}$-in. end mill after sharpening?

34. Determine dimension M of the part shown in Figure A.

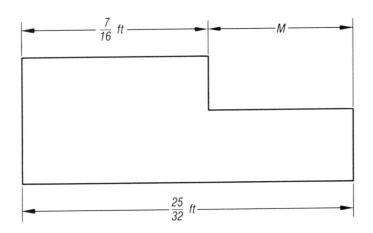

Figure A

2-8 Multiplying Fractions and Mixed Numbers

OBJECTIVES

After completing this section, you should be able to:

1. Find the product of two or more fractions or mixed numbers in simplest form.
2. Halve fractional dimensions.

Rule 2-14 gives a simple explanation of how to multiply fractions.

RULE 2-14   MULTIPLYING FRACTIONS

Place the product of all numerators over the product of all denominators and reduce to lowest terms.

Example 2-24   Multiply: $\frac{5}{6} \times \frac{7}{9}$.

Solution $\dfrac{5}{6} \times \dfrac{7}{9} = \dfrac{5 \times 7}{6 \times 9} = \dfrac{35}{54}$

Sometimes common factors can be divided out before multiplying, as shown in Example 2-25.

Example 2-25   Find the product of $\frac{2}{5}$ and $\frac{3}{8}$.

Solution $\dfrac{2}{5} \times \dfrac{3}{8} = \dfrac{\overset{1}{\cancel{2}} \times 3}{5 \times \underset{4}{\cancel{8}}} = \dfrac{3}{20}$

To multiply mixed numbers, change each mixed number to an improper fraction and multiply as fractions. Whole numbers can be changed to improper fractions by writing the whole number over a denominator of 1.

RULE 2-15   MULTIPLYING MIXED NUMBERS

Change all mixed or whole numbers to improper fractions and multiply the resulting fractions as in Rule 2-14.

Example 2-26   Multiply $2\frac{1}{2}$ by $3\frac{3}{4}$.

Solution $2\dfrac{1}{2} \times 3\dfrac{3}{4} = \dfrac{5}{2} \times \dfrac{15}{4} = \dfrac{75}{8} = 9\dfrac{3}{8}$

Example 2-27   Multiply: $5\frac{1}{2} \times 2\frac{3}{4} \times 6\frac{2}{3}$.

Solution $5\dfrac{1}{2} \times 2\dfrac{3}{4} \times 6\dfrac{2}{3} = \dfrac{11}{2} \times \dfrac{11}{\underset{1}{\cancel{4}}} \times \dfrac{\overset{5}{\cancel{20}}}{3} = \dfrac{605}{6} = 100\dfrac{5}{6}$

Example 2-28   Find the product of $\frac{1}{8}$ and 8.

Solution $\dfrac{1}{8} \times 8 = \dfrac{1}{\underset{1}{\cancel{8}}} \times \dfrac{\overset{1}{\cancel{8}}}{1} = 1$

Halving a number is the same as multiplying it by $\frac{1}{2}$. The machinist, however, uses a shortcut to halve a fractional dimension. If, for example, he or she wishes to take $\frac{1}{2}$ of $\frac{7}{16}$, the machinist simply *doubles* the denominator and gets $\frac{7}{32}$. Notice that doubling the denominator gets the same result as multiplying the fraction by $\frac{1}{2}$:

$$\frac{1}{2} \times \frac{7}{16} = \frac{7}{32}$$

Therefore, to halve a fraction, simply double the denominator.

Sometimes doubling the denominator is not the most practical method of halving a fractional measurement. For example, the smallest graduation on an ordinary steel rule is $\frac{1}{64}$ inch. So, when halving sixty-fourths, it would be of no practical value to change them to one-hundred-twenty-eighths. Instead, halve the numerator. Half of $\frac{15}{64}$ would be $7\frac{1}{2}/64$, which is visible to the naked eye on a steel rule.

RULE 2-16   HALVING FRACTIONS OR MIXED NUMBERS

Step 1   If necessary, write the mixed number as a fraction.

Step 2   Double the denominator or halve the numerator, whichever is most practical.

One practical application of halving fractions is scribing a bolt circle. When a bolt circle is to be scribed on work in preparation for drilling holes, the machinist must obtain the radius of the circle from the diameter dimension given on the blueprint. To obtain the radius, the diameter must be halved.

Example 2-29   Find the radius of a bolt circle $1\frac{19}{32}$ in. in diameter.

Solution

Step 1   Write as an improper fraction.

$$1\frac{19}{32} = \frac{51}{32}$$

Step 2   Double the denominator.

$$\frac{51}{32 \times 2} = \frac{51}{64}$$

Example 2-30   Halve $\frac{61}{64}$ in.

   Solution   Halve the numerator.

$$\frac{61 \div 2}{64} = \frac{30\frac{1}{2}}{64}$$

Exercises 2-8

Find the product and reduce to lowest terms.

1. $\frac{1}{2} \times \frac{3}{5}$

2. $\frac{3}{4} \times \frac{5}{7}$

3. $\frac{5}{8} \times \frac{2}{3}$

4. $\frac{7}{9} \times \frac{5}{6}$

5. $\frac{2}{3} \times \frac{6}{7}$

6. $\frac{1}{2} \times \frac{14}{15}$

7. $\frac{5}{9} \times \frac{7}{10}$

8. $\frac{4}{5} \times \frac{7}{8}$

9. $\frac{9}{10} \times \frac{5}{21}$

10. $\frac{3}{22} \times \frac{4}{9}$

11. $\frac{7}{15} \times \frac{9}{21}$

12. $\frac{5}{3} \times \frac{6}{25}$

13. $12 \times \frac{3}{4}$

14. $\frac{5}{7} \times 21$

15. $\frac{1}{9} \times 27$

16. $45 \times \frac{7}{9}$

17. $7\frac{1}{2} \times 10$

18. $15 \times 3\frac{2}{5}$

19. $7\frac{1}{4} \times 24$

20. $16 \times 5\frac{3}{8}$

21. $1\frac{1}{2} \times 4\frac{2}{9}$

22. $2\frac{5}{8} \times 5\frac{1}{7}$

23. $12\frac{1}{2} \times 3\frac{1}{3}$

24. $3\frac{3}{8} \times 2$

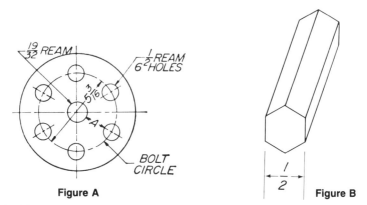

Figure A   **Figure B**

25. In Figure A, what radius is needed to scribe the bolt circle?

26. What is the total length of 14 pieces of cold-rolled steel, each $7\frac{9}{32}$ in. long?

27. Five pieces of hot-rolled steel, each $2\frac{7}{8}$ in. long, are sawed from a bar $24\frac{1}{2}$ in. long. What length of the original bar remains?

28. If five lengths of tubing, each $319\frac{87}{1000}$ cm long, are cut from a bar 1790 cm long, what length is left over?

29. The gauge on a diesel fuel tank with a capacity of 20 gal indicates that $\frac{3}{4}$ of a tank remains. How many gallons does this represent?

30. Figure B illustrates a hexagonal steel bar that is $\frac{1}{2}$ in. across the flats. From this bar, exactly 21 chisels can be made, each $6\frac{21}{64}$ in. long. How long is this bar?

31. From a hexagonal steel bar similar to the one shown in Figure B, a machinist's helper saws 35 pieces, each $5\frac{9}{32}$ in. long, for center punch parts. How long must the bar be to yield this many pieces?

32. Determine the length D in Figure C.

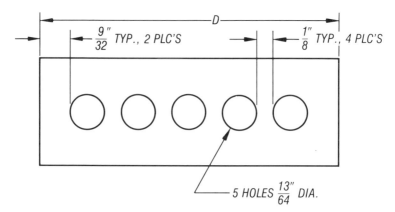

Figure C

2-9   Dividing Fractions and Mixed Numbers

OBJECTIVES

After completing this section, you should be able to:

1.   Write the reciprocal of a fraction.
2.   Find the quotient of two fractions or mixed numbers.

Before discussing how to divide fractions, it is necessary to define what is meant by reciprocals. Two numbers are **reciprocals** of each other if their product is 1.

Example 2-31

$$\frac{9}{64} \times \frac{64}{9} = \frac{576}{576} = 1 \qquad \frac{9}{64} \text{ and } \frac{64}{9} \text{ are reciprocals}$$

$$\frac{6}{1} \times \frac{1}{6} = \frac{6}{6} = 1 \qquad 6 \text{ and } \frac{1}{6} \text{ are reciprocals}$$

$$\frac{2}{3} \times \frac{3}{2} = \frac{6}{6} = 1 \qquad \frac{2}{3} \text{ and } \frac{3}{2} \text{ are reciprocals}$$

RULE 2-17   WRITING RECIPROCALS

To write the reciprocal of a fraction, exchange the numerator and denominator. This is called inverting the fraction.

Example 2-32   Write the reciprocal of $\frac{5}{8}$.

Solution   Exchange the numerator and denominator.

$$\frac{8}{5}$$

Check $\frac{8}{5} \times \frac{5}{8} = 1$, so $\frac{8}{5}$ is the reciprocal of $\frac{5}{8}$.

When two fractions are divided, the quotient can be written as a fraction.

$$\frac{7}{8} \div \frac{3}{16} \quad \text{is the same as} \quad \frac{\dfrac{7}{8}}{\dfrac{3}{16}}$$

The numerator and the denominator can then be multiplied by the same number, $\frac{16}{3}$, without changing the value of the fraction.

$$\frac{7}{8} \div \frac{3}{16} = \frac{\dfrac{7}{8}}{\dfrac{3}{16}} = \frac{\dfrac{7}{8} \times \dfrac{16}{3}}{\dfrac{3}{16} \times \dfrac{16}{3}} = \frac{\dfrac{7}{8} \times \dfrac{16}{3}}{1}$$

$$= \frac{7}{8} \times \frac{16}{3} = \frac{7 \times \overset{2}{\cancel{16}}}{\underset{1}{\cancel{8}} \times 3} = \frac{14}{3} = 4\frac{2}{3}$$

Observe that to divide, the first fraction was multiplied by the reciprocal of the second fraction.

RULE 2-18   DIVIDING FRACTIONS

To divide one fraction by another, invert the divisor (or second fraction) and multiply.

Example 2-33   Divide $\frac{7}{8}$ by $\frac{3}{16}$.

Solution $\dfrac{7}{8} \div \dfrac{3}{16} = \dfrac{7}{8} \times \dfrac{16}{3} = \dfrac{7 \times \overset{2}{\cancel{16}}}{\underset{1}{\cancel{8}} \times 3} = \dfrac{14}{3} = 4\dfrac{2}{3}$

RULE 2-19   DIVIDING MIXED NUMBERS

Change any mixed or whole numbers to improper fractions and divide as in Rule 2-18.

Example 2-34   Divide: $3\frac{3}{5} \div 2\frac{7}{10}$.

Solution $3\dfrac{3}{5} \div 2\dfrac{7}{10} = \dfrac{18}{5} \div \dfrac{27}{10}$ Write the mixed numbers as improper fractions.

$= \dfrac{18}{5} \times \dfrac{10}{27}$ Multiply by the reciprocal of the second fraction

$= \dfrac{\overset{2}{\cancel{18}} \times \overset{2}{\cancel{10}}}{\underset{1}{\cancel{5}} \times \underset{3}{\cancel{27}}}$

$= \dfrac{4}{3} = 1\dfrac{1}{3}$

Exercises 2-9

Divide and reduce to lowest terms.

1. $\dfrac{3}{4} \div \dfrac{6}{7}$ 2. $\dfrac{2}{3} \div \dfrac{8}{9}$ 3. $\dfrac{3}{4} \div \dfrac{11}{12}$

4. $\dfrac{5}{8} \div \dfrac{9}{10}$ 5. $\dfrac{2}{5} \div \dfrac{2}{3}$ 6. $\dfrac{2}{3} \div \dfrac{2}{5}$

7. $8 \div \dfrac{4}{5}$ 8. $15 \div \dfrac{3}{7}$ 9. $25 \div \dfrac{10}{11}$

10. $7 \div \dfrac{2}{3}$ 11. $\dfrac{7}{8} \div 2$ 12. $\dfrac{8}{11} \div 4$

13. $\dfrac{5}{6} \div 10$ 14. $\dfrac{15}{16} \div 3$ 15. $2\dfrac{1}{2} \div \dfrac{5}{8}$

16. $6\dfrac{2}{3} \div \dfrac{7}{9}$ 17. $\dfrac{7}{11} \div 4\dfrac{1}{5}$ 18. $\dfrac{3}{4} \div 2\dfrac{1}{10}$

19. $3\dfrac{3}{4} \div 6\dfrac{1}{2}$ 20. $1\dfrac{2}{3} \div 6\dfrac{2}{3}$ 21. $4\dfrac{1}{8} \div 4\dfrac{2}{5}$

22. $22\dfrac{1}{2} \div 4\dfrac{2}{7}$ 23. $4\dfrac{1}{2} \div 8\dfrac{2}{3}$ 24. $3\dfrac{2}{5} \div 2\dfrac{1}{2}$

25. If sheets of aluminum are $\frac{1}{16}$ in. thick, how many sheets are in a pile $2\frac{7}{8}$ in. high?

26. If one bolt weighs $\frac{1}{5}$ ounce (oz), how many such bolts are in a pile of bolts weighing 25 oz?

27. In Figure A, what is the value of dimension d?

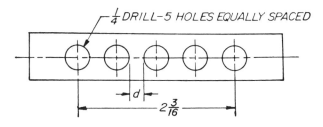

Figure A

28. How many tie rods are in $110\frac{1}{4}$ lb if the weight of one rod is $1\frac{5}{16}$ lb?

29. How many pieces of tubing $234\frac{1}{10}$ mm long can be cut from a bar $3657\frac{6}{10}$ mm long if $1\frac{6}{10}$ mm are wasted for each saw cut?

30. The following lengths of 1-by-19-strand control cable are required for an electrical job: $12\frac{2}{3}$ ft, $15\frac{1}{2}$ ft, $31\frac{3}{4}$ ft, $7\frac{5}{12}$ ft, and $18\frac{5}{8}$ ft. What is the total length of cable required for this job?

31. If a workpiece $2\frac{1}{2}$ in. in diameter is placed in a lathe and a cut $\frac{1}{32}$ in. deep is taken, what is the finished diameter?

32. If 16 aluminum main body castings stack $53\frac{3}{4}$ in. high, how thick is each casting?

Unit 2 Review Exercises

1. Write $37\frac{5}{9}$ as an improper fraction.

2. Write $\frac{21}{49}$ in simplest form.

3. Add: $\frac{6}{13} + \frac{23}{26} + \frac{43}{52} + 1\frac{11}{13}$.

4. Find the L.C.D. of $\frac{1}{2}$, $\frac{3}{4}$, $\frac{5}{7}$, and $\frac{9}{14}$.

5. Multiply: $\frac{15}{32} \times \frac{4}{5} \times \frac{16}{4}$.

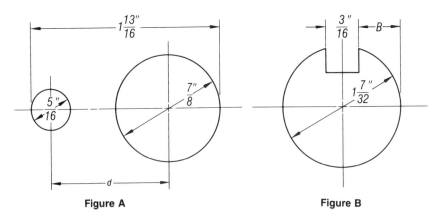

Figure A   Figure B

6. In Figure A, what is the value of dimension d?

7. To locate the key seat in Figure B, what is measurement B?

8. What is the difference in diameters between a steel bolt of diameter $\frac{11}{16}$ in. and a steel bolt of diameter $\frac{9}{32}$ in.?

9. A certain epoxy resin requires $\frac{7}{64}$ oz of resin and $\frac{1}{8}$ oz of hardener. Which is the smaller amount?

10. One piece $1\frac{3}{10}$ ft was cut from a board $4\frac{2}{3}$ ft long. How much of the board was left?

Basic Operations with Decimals

This is one of the most important units in the book, because all trades-people must learn to work with decimals. A fraction is often the best tool to use for quick figuring. However, the decimal form is easier to use in most cases involving several fractions.

3-1 Place Value and Decimal Equivalents

OBJECTIVES

After completing this section, you should be able to:

1. Read and write decimals.
2. Write the fractional equivalent of a decimal.
3. Write the decimal equivalent of a fraction.

A **decimal** is a fraction with a denominator of 10 or one of the powers of 10. The denominator is not written, but is understood to exist. A **decimal point** is used to indicate that what follows this point is a fraction. The figures that appear to the right of this decimal point form the numerator of a decimal fraction.

To read and write decimals, you must have an understanding of place value. The value of a decimal depends on how many figures are to the right of the decimal point. Each place in a decimal has a value ten times as large as the next place to the right. Figure 3-1 shows the decimal places and their relative values. The decimals decrease by a factor of $\frac{1}{10}$ in moving to the right. They increase by a factor of 10 in moving to the left of the decimal point. That is, in moving to the right, $\frac{1}{100}$ is $\frac{1}{10}$ of $\frac{1}{10}$, and in moving to the left, 100 is 10×10.

HUNDRED THOUSANDS	TEN THOUSANDS	THOUSANDS	HUNDREDS	TENS	UNITS	DECIMAL POINT	TENTHS	HUNDREDTHS	THOUSANDTHS	TEN THOUSANDTHS	HUNDRED THOUSANDTHS
9	7	5	3	1	8	.	8	6	4	2	1

Figure 3-1

To read a decimal, use the place value of the last digit to the right as the denominator. Thus 0.46 is read "forty-six hundredths" and 0.0325 is read "three hundred twenty-five ten-thousandths." The zero to the left of the decimal point is not read. It indicates that there is no whole number part. Table 3-1 tells how to read a decimal fraction.

Fraction	Decimal	Read
$\frac{5}{10}$	0.5	five-tenths
$\frac{38}{100}$	0.38	thirty-eight hundredths
$\frac{476}{1000}$	0.476	four hundred seventy-six thousandths
$\frac{8}{1000}$	0.008	eight-thousandths
$\frac{324,501}{1,000,000}$	0.324501	three hundred twenty-four thousand, five hundred one millionths

Table 3-1

A number that combines a whole number and a fraction, such as 23.85, is called a **mixed decimal,** or simply a decimal. The decimal 23.85 is read "twenth-three *and* eighty-five hundredths." To avoid errors in reading mixed decimals, the decimal can be read as digits with the word *point* telling you where the decimal point occurs. Thus 23.85 also can be read "two, three, point, eight, five."

Whole numbers can be written as decimals as shown in Rule 3-1.

RULE 3-1   WRITING WHOLE NUMBERS AS DECIMALS

To write a whole number as a decimal, add a decimal point to the right of the number and annex as many zeros as desired.

Example 3-1 $23 = 23.0 = 23.00 = 23.000 = \ldots$

Fractions also can be written as decimals. Rules 3-2 and 3-3 tell how to find the decimal equivalent of a fraction.

RULE 3-2   WRITING FRACTIONS AS DECIMALS WHEN THE DENOMINATOR IS A POWER OF 10

To find the decimal equivalent of a fraction whose denominator is a power of 10, replace the denominator with a decimal point and add the appropriate number of zeros between the number and the decimal point.

Example 3-2   Find the decimal equivalents of $\frac{53}{100}$ and $\frac{18}{1000}$.

Solution $\dfrac{53}{100} = 0.53$

$\dfrac{18}{1000} = 0.018$

RULE 3-3   WRITING ANY FRACTION AS A DECIMAL

To find the decimal equivalent of any common fraction, divide the denominator into the numerator.

Example 3-3   Find the decimal equivalent of $\frac{5}{16}$.

Solution

$$
\begin{array}{r}
0.3125 \\
16\overline{)5.0000} \\
\underline{4\ 8} \\
20 \\
\underline{16} \\
40 \\
\underline{32} \\
80 \\
\underline{80} \\
0
\end{array}
$$

Write 5 as 5.0000.
Put the decimal point in the quotient above the one in the dividend.

Example 3-4   Find the decimal equivalent of the fraction $\frac{3}{8}$.

Solution
$$
\begin{array}{r}
0.375 \\
8\overline{)3.000} \\
\underline{2\,4} \\
60 \\
\underline{56} \\
40 \\
\underline{40} \\
0
\end{array}
$$

The metalworker rarely needs to carry division further than four decimal places, since most shop measuring tools are not calibrated for less than one ten-thousandth of an inch. When working with measuring tools that do not measure closer or less than one-thousandth of an inch, the number of ten-thousandths must be estimated.

The machinist expresses all decimals in thousandths. In math classes you were taught to express a number such as 0.1472 as "one thousand four hundred seventy-two ten-thousandths." The machinist or toolmaker would shorten this to "one hundred forty-seven and two-tenths (thousandths)."

Machinists use expressions such as, "The shaft is one and three tenths oversize." This is written 0.0013 on a blueprint. Machinists also say, "The work is half a thousandth oversize." If this were written down, it would appear as 0.0005. The machinist emphasizes, in very precise language, the units of greatest interest—thousandths and ten-thousandths of an inch.

The capable metalworker learns the decimal equivalents of quarters, eighths, and sixteenths of an inch through constant use. This makes it easy for a machinist to identify common fractions that match such decimals. The decimal equivalents listed in Table 3-2 are those most used in shopwork. You should memorize them.

$\frac{1}{4} = 0.250$	$\frac{1}{8} = 0.125$	$\frac{5}{8} = 0.625$	
$\frac{3}{4} = 0.750$	$\frac{3}{8} = 0.375$	$\frac{7}{8} = 0.875$	
$\frac{1}{16} = 0.0625$	$\frac{5}{16} = 0.3125$	$\frac{9}{16} = 0.5625$	$\frac{13}{16} = 0.8125$
$\frac{3}{16} = 0.1875$	$\frac{7}{16} = 0.4375$	$\frac{11}{16} = 0.6875$	$\frac{15}{16} = 0.9375$

Table 3-2

There are 32 thirty-seconds and 64 sixty-fourths in 1 inch. Learning the decimal equivalents of all these would be quite an accomplishment, but unnecessary and a waste of time. Every shop has one or more decimal equivalent charts (Figure 3-2).

Countries on the metric system use a simplified decimal equivalent chart of common fractions, similar to the one shown in Table 3-3. Occasionally, in gearing and indexing work, the need arises for a toolmaker to change a decimal to the fraction from which it is derived.

DECIMAL EQUIVALENTS
of fraction, wire gauge, letter and metric sizes

SIZE	DECIMAL INCHES	SIZE	DECIMAL INCHES	SIZE	DECIMAL INCHES	SIZE	DECIMAL INCHES	SIZE	DECIMAL INCHES	SIZE	DECIMAL INCHES
97	.0059	59	.0410	2.75mm	.1083	5mm	.1969	N	.3020	13mm	.5118
96	.0063	1.05mm	.0413	7/64	.1094	8	.1990	7.7mm	.3031	33/64	.5156
95	.0067	58	.0420	35	.1100	5.1mm	.2008	7.75mm	.3051	17/32	.5312
94	.0071	57	.0430	2.8mm	.1102	7	.2010	7.8mm	.3071	13.5mm	.5315
93	.0075	1.1mm	.0433	34	.1110	13/64	.2031	7.9mm	.3110	35/64	.5469
92	.0079	1.15mm	.0453	33	.1130	6	.2040	5/16	.3125	14mm	.5512
.2mm	.0079	56	.0465	2.9mm	.1142	5.2mm	.2047	8mm	.3150	9/16	.5625
91	.0083	3/64	.0469	32	.1160	5	.2055	O	.3160	14.5mm	.5709
90	.0087	1.2mm	.0472	31	.1200	5.25mm	.2067	8.1mm	.3189	37/64	.5781
.22mm	.0087	1.25mm	.0492	3.1mm	.1220	5.3mm	.2087	8.2mm	.3228	15mm	.5906
89	.0091	1.3mm	.0512	3.2mm	.1260	4	.2090	P	.3230	19/32	.5938
88	.0095	55	.0520	3.25mm	.1280	5.4mm	.2126	8.25mm	.3248	39/64	.6094
.25mm	.0098	1.35mm	.0531	30	.1285	3	.2130	21/64	.3281	15.5mm	.6102
87	.0100	54	.0550	3.3mm	.1299	5.5mm	.2165	8.4mm	.3307	5/8	.6250
86	.0105	1.4mm	.0551	3.4mm	.1339	7/32	.2188	Q	.3320	16mm	.6299
85	.0110	1.45mm	.0571	29	.1360	5.6mm	.2205	8.5mm	.3346	41/64	.6406
.28mm	.0110	1.5mm	.0591	3.5mm	.1378	2	.2210	8.6mm	.3386	16.5mm	.6496
84	.0115	53	.0595	28	.1405	5.7mm	.2244	R	.3425	21/32	.6562
.3mm	.0118	1.55mm	.0610	9/64	.1406	5.75mm	.2264	8.7mm	.3438	17mm	.6693
83	.0120	1/16	.0625	3.6mm	.1417	1	.2280	11/32	.3445	43/64	.6719
82	.0125	1.6mm	.0630	27	.1440	5.8mm	.2283	8.75mm	.3465	11/16	.6875
.32mm	.0126	52	.0635	26	.1470	5.9mm	.2323	8.8mm	.3465	17.5mm	.6890
81	.0130	1.65mm	.0650	3.75mm	.1476	A	.2340	S	.3480	45/64	.7031
80	.0135	51	.0670	25	.1495	15/64	.2344	8.9mm	.3504	18mm	.7087
.35mm	.0138	1.75mm	.0689	3.8mm	.1496	6mm	.2362	9mm	.3543	23/32	.7188
79	.0145	50	.0700	24	.1520	B	.2380	T	.3580	18.5mm	.7283
1/64	.0156	1.8mm	.0709	3.9mm	.1535	6.1mm	.2402	9.1mm	.3583	47/64	.7344
.4mm	.0157	1.85mm	.0728	23	.1540	C	.2420	23/64	.3594	19mm	.7480
78	.0160	49	.0730	5/32	.1562	6.2mm	.2441	9.2mm	.3622	3/4	.7500
.45mm	.0177	48	.0760	22	.1570	6.25mm	.2461	9.25mm	.3642	49/64	.7656
77	.0180	5/64	.0781	21	.1590	D	.2460	9.3mm	.3661	19.5mm	.7677
.5mm	.0197	47	.0785	20	.1610	1/4	.2500	U	.3680	25/32	.7812
76	.0200	2mm	.0787	4.1mm	.1614	E	.2500	9.4mm	.3701	20mm	.7874
75	.0210	2.05mm	.0807	19	.1660	6.3mm	.2480	9.5mm	.3740	51/64	.7969
.55mm	.0217	46	.0810	4.25mm	.1673	6.4mm	.2520	V	.3770	20.5mm	.8071
74	.0225	45	.0820	4.3mm	.1693	6.5mm	.2559	9.6mm	.3780	13/16	.8125
.6mm	.0236	44	.0860	18	.1695	F	.2570	3/8	.3750	21mm	.8268
73	.0240	2.2mm	.0866	11/64	.1719	6.6mm	.2598	9.7mm	.3819	53/64	.8281
72	.0250	2.25mm	.0886	17	.1730	G	.2610	9.8mm	.3858	27/32	.8438
.65mm	.0256	43	.0890	4.4mm	.1732	6.7mm	.2638	W	.3860	21.5mm	.8465
71	.0260	2.3mm	.0906	16	.1770	17/64	.2656	9.9mm	.3898	55/64	.8594
.7mm	.0276	2.35mm	.0925	4.5mm	.1772	6.75mm	.2657	25/64	.3906	22mm	.8661
70	.0280	42	.0935	15	.1800	H	.2660	10mm	.3937	7/8	.8750
69	.0292	3/32	.0938	4.6mm	.1811	6.9mm	.2717	X	.3970	22.5mm	.8858
.75mm	.0295	2.4mm	.0945	14	.1820	I	.2720	Y	.4040	57/64	.8906
68	.0310	41	.0960	13	.1850	7mm	.2756	13/32	.4062	23mm	.9055
1/32	.0312	2.45mm	.0965	4.7mm	.1850	J	.2770	Z	.4130	29/32	.9062
.8mm	.0315	40	.0980	3/16	.1875	7.1mm	.2795	10.5mm	.4134	59/64	.9219
67	.0320	2.5mm	.0984	4.75mm	.1870	K	.2810	27/64	.4219	23.5mm	.9252
66	.0330	39	.0995	12	.1890	9/32	.2812	11mm	.4331	15/16	.9375
.85mm	.0335	38	.1015	4.8mm	.1890	7.25mm	.2854	7/16	.4375	24mm	.9449
65	.0350	2.6mm	.1024	11	.1910	7.3mm	.2874	11.5mm	.4528	61/64	.9531
.9mm	.0354	37	.1040	4.9mm	.1929	L	.2900	29/64	.4531	24.5mm	.9646
64	.0360	2.7mm	.1063	10	.1935	7.4mm	.2913	15/32	.4688	31/32	.9688
63	.0370	36	.1065	9	.1960	M	.2950	12mm	.4724	25mm	.9843
.95mm	.0374					7.5mm	.2953	31/64	.4844	63/64	.9844
62	.0380					19/64	.2969	12.5mm	.4921	1	1.0000
61	.0390					7.6mm	.2992	1/2	.5000		
1mm	.0394										
60	.0400										

CLEVELAND TWIST DRILL
CUTTING and THREADING TOOLS
PRECISION SLIDES
MACHINES and ACCESSORIES

Figure 3-2 Decimal equivalent chart. *(Courtesy of Cleveland Twist Drill)*

$\dfrac{1}{2} = 0.500$	$\dfrac{1}{3} = 0.333$	$\dfrac{2}{3} = 0.667$	$\dfrac{1}{4} = 0.250$
$\dfrac{1}{5} = 0.200$	$\dfrac{2}{5} = 0.400$	$\dfrac{3}{5} = 0.600$	$\dfrac{4}{5} = 0.800$
$\dfrac{1}{6} = 0.1667$	$\dfrac{1}{10} = 0.1000$	$\dfrac{1}{14} = 0.0714$	$\dfrac{1}{18} = 0.0556$
$\dfrac{1}{7} = 0.1429$	$\dfrac{1}{11} = 0.0909$	$\dfrac{1}{15} = 0.0667$	$\dfrac{1}{19} = 0.0526$
$\dfrac{1}{8} = 0.1250$	$\dfrac{1}{12} = 0.0833$	$\dfrac{1}{16} = 0.0625$	$\dfrac{1}{20} = 0.0500$
$\dfrac{1}{9} = 0.1111$	$\dfrac{1}{13} = 0.0769$	$\dfrac{1}{17} = 0.0588$	

Table 3-3

RULE 3-4 CHANGING DECIMAL FRACTIONS TO
EQUIVALENT (OR NEARLY EQUIVALENT)
FRACTIONS

- *If the decimal has only one place,* use the digit as the numerator of a fraction with denominator of 10.
- *If the decimal has two places,* use the digits as the numerator of a fraction with denominator 100.
- *If the decimal has three or more places,* use the first three digits as the numerator of a fraction with denominator 1000.
- Reduce the resulting fraction to its lowest terms.

Example 3-5 $0.6 = \dfrac{6}{10} = \dfrac{3}{5}$

$0.25 = \dfrac{25}{100} = \dfrac{1}{4}$

$0.750 = \dfrac{750}{1000} = \dfrac{3}{4}$

$0.025 = \dfrac{25}{1000} = \dfrac{1}{40}$

$0.2503 = \dfrac{250}{1000} = \dfrac{1}{4}$

Most fractions are not this simple to reduce to lowest terms. Consider the decimal 0.2173. By the method given above, it is about equal to $\frac{217}{1000}$. This is about equal to $\frac{5}{23}$, but you might have difficulty in finding this unless you used *continued fractions.* Continued fractions involve a special technique that requires much practice and has limited applications. (See Chapter 15 on gearing calculations.)

More than 90 percent of a machinist's work involves converting fractions to decimals, because it is much easier, in most cases, to work with decimals than with fractions.

Exercises 3-1

Express each decimal in words.

1. 0.21	**2.** 0.468	**3.** 0.01	**4.** 0.6
5. 0.0435	**6.** 5.238	**7.** 9.10032	**8.** 26.478

Write each of the following in decimal form.

9. six-tenths

10. eighty-four hundredths

11. one hundred seventeen-thousandths

12. fifty-eight thousandths

13. two and seven-tenths

14. thirty-one and sixteen ten-thousandths

15. one hundred and sixty-five thousandths

16. ninety-two and one hundred three thousandths

Write each decimal as a fraction.

17. 0.3   **18.** 0.015   **19.** 0.850   **20.** 0.3125

21. 0.375   **22.** 0.625   **23.** 0.0625   **24.** 0.9385

Write each fraction as a decimal.

25. $\dfrac{9}{10}$ **26.** $\dfrac{57}{1000}$ **27.** $\dfrac{9}{100}$ **28.** $\dfrac{95}{100}$

29. $\dfrac{9}{10,000}$ **30.** $\dfrac{375}{1000}$ **31.** $\dfrac{9}{1000}$ **32.** $\dfrac{531}{10,000}$

Find the decimal equivalent of each fraction.

33. $\dfrac{1}{16}$ **34.** $\dfrac{19}{32}$ **35.** $\dfrac{45}{64}$ **36.** $\dfrac{7}{32}$ **37.** $\dfrac{1}{8}$

38. $\dfrac{15}{16}$ **39.** $\dfrac{21}{32}$ **40.** $\dfrac{13}{16}$ **41.** $\dfrac{27}{64}$ **42.** $\dfrac{5}{8}$

43. What fractional-size drill would a machinist request from the tool crib to drill a hole 0.3438 in. in diameter?

44. A blueprint dimension specifies a hole $\frac{11}{16}$ in. in diameter. What decimal size is equivalent?

45. A design engineer asked a prototype machinist to turn a steel shaft $\frac{7}{8}$ in. in diameter. What would be the equivalent decimal size?

3-2 Significant Digits and Rounding Decimals

OBJECTIVES

After completing this section, you should be able to:

1. Round decimals.

2. Give the number of significant digits in a decimal.

Because shopwork rarely involves decimals with more than four places, it is often necessary to round decimals. Rounding decimals is similar to rounding whole numbers.

RULE 3-5 ROUNDING DECIMALS

Step 1 Find the digit in the place to which you wish to round.

Step 2 Look at the next digit to the right:
- If it is less than 5, drop all the digits to the right of the place to which you are rounding.
- If it is 5 or more, add 1 to the digit in the place to which you are rounding and drop all digits to the right of that place.

Example 3-6 Round 0.863 to the nearest hundredth.

Solution

Step 1 0.863
          ∟___ hundredths place

Step 2 0.863
          ∟___ 3 is in the place to the right. It is less than 5. Drop all digits to the right of 6.

0.863 = 0.86 to the nearest hundredth

Example 3-7 Round 0.78952 to the nearest thousandth.

Solution

Step 1 0.78952
          ∟_____ thousandths place

Step 2 0.78952
          ∟_____ 5 or more. Add 1 to 9 and drop all digits to the right of 9.

0.78952 = 0.790 to the nearest thousandth.

Example 3-8 Round each decimal to the underlined place.

Solution 0.687 = 0.7 to the nearest tenth
0.0539 = 0.054 to the nearest thousandth
1.99885 = 2.00 to the nearest hundredth.

The measurements with which a shopworker deals are not exact. Those digits in a measurement known to be reliable are called **significant digits.**

RULE 3-6 DETERMINING SIGNIFICANT DIGITS

The following digits in a measurement are significant:
- All nonzero digits.
- All zeros between nonzero digits.
- In a decimal, all zeros between the decimal point and a nonzero digit to the left of the decimal point, and all zeros that follow nonzero digits to the right of the decimal point.

Example 3-9 26,000 has two significant digits.

48,030 has four significant digits.

80,000.1 has six significant digits.

0.02 has one significant digit.

0.3500 has four significant digits.

26.0 has three significant digits.

When working with measurements or other approximate numbers, answers can be no more accurate than the least accurate number in the problem. For example, the sum of 6.25 and 8.4 is 14.65. If these numbers were measurements, the answer would be rounded to 14.7, since the least accurate measurement is accurate to one *decimal place* (places to the right of the decimal point.)

When multiplying or dividing, round to the least number of significant digits. However, if one of the numbers is exact, its significant digits are not counted. If a machinist makes 4 rivets, each 1.2486 inch long, the total thickness is 4 × 1.2486 inch, or 4.9944 inch. Because 4 is an exact number, only the number of significant digits in 1.2486 is considered when giving the answer.

To avoid giving answers that appear to be more reliable than they are, use the following rule.

RULE 3-7 ROUNDING MEASUREMENTS AND APPROXIMATE NUMBERS

- In addition or subtraction, round to the least number of decimal places occurring in any number.
- In multiplication or division, round to the least number of significant digits occurring in any number, but do not count exact numbers.
- In finding powers or square roots, round to the same number of significant digits as in the given number.

Example 3-10   The sum of 29.63 cm and 18.5 cm is 48.13 cm. Round the answer to the appropriate place.

Solution   The answer is 48.1 cm, rounded to the least number of decimal places, which is three.

Example 3-11   The product of 21.9 ft and 1.8 ft is 39.42 square feet (sq ft). Round the answer to the appropriate place.

Solution   The answer is 39 sq ft, rounded to two significant digits.

Exercises 3-2

Round each number to the nearest tenth.

1. 0.19	**2.** 0.34	**3.** 0.75	**4.** 4.291
5. 15.952	**6.** 27.008	**7.** 99.75	**8.** 431.728

Round each number to the nearest hundredth.

9. 14.786	**10.** 5.9273	**11.** 9.992	**12.** 476.285
13. 0.1726	**14.** 0.9752	**15.** 67.9711	**16.** 429.213

Round each number to the nearest thousandth.

17. 0.8435	**18.** 0.0912	**19.** 0.1008	**20.** 2.63542
21. 10.89962	**22.** 102.86354	**23.** 0.0008	**24.** 23.21953

Give the number of significant digits in each number.

25. 256	**26.** 2560	**27.** 2506	**28.** 25600
29. 38.0020	**30.** 107.0	**31.** 20,000.63	**32.** 10,000,001
33. 88,792	**34.** 43,000	**35.** 0.0074	**36.** 0.02000
37. 5,000,000,000		**38.** 5,000,000,000.0	
39. 12,308		**40.** 600.02050	

In exercises 41–50, round each answer to the appropriate number of digits.

41.
$$\begin{array}{r} 4.786 \\ 5.1 \\ +27.98 \\ \hline 37.866 \end{array}$$

42.
$$\begin{array}{r} 29.00 \\ -\ 5.436 \\ \hline 23.564 \end{array}$$

43.
$$\begin{array}{r} 168.712 \\ 15.9036 \\ +5998.39 \\ \hline 6183.0056 \end{array}$$

44.
$$\begin{array}{r} 7863 \\ -\ 15.81 \\ \hline 7847.19 \end{array}$$

45.
$$\begin{array}{r} 2.501 \\ \times\ 3.25 \\ \hline 8.12825 \end{array}$$

46.
$$\begin{array}{r} 588.3 \\ \times\ 20 \\ \hline 11,766 \end{array}$$

47.
$$\begin{array}{r} 588.3 \\ \times\ 2.0 \\ \hline 1176.6 \end{array}$$

48.
$$6.9\overline{)332.58}\quad 48.2$$

49. A thrust washer is 0.3437 in. thick. Eight such washers have a combined thickness of 2.7496 in.

50. The circumference of a circle is about 3.1416 times the diameter of the circle. For a diameter of 8.60 m, the circumference is 27.01776.

3-3 Adding and Subtracting Decimals

OBJECTIVES

After completing this section, you should be able to:

1. Add decimals.
2. Subtract decimals.

The same rules that apply to adding and subtracting whole numbers apply to decimals. There is only one precaution needed for decimals: All decimal points *must* be placed *under* one another.

RULE 3-8   ADDING OR SUBTRACTING DECIMALS

Step 1   Write the decimals in a column. Be sure decimal points are aligned.

Step 2   Add or subtract the numbers in corresponding places.

Step 3   Write the decimal point in the answer below the decimal points in the numbers above.

Step 4   If the numbers are approximate, round the answer to the appropriate place.

Example 3-12   Add: 8.5 + 2.07 + 4.2705.

    Solution
$$\begin{array}{r} 8.5 \\ 2.07 \\ +\ 4.2705 \\ \hline 14.8405 \end{array}$$

Example 3-13   Add: 15.18 + 2.763 + 0.75 + 1.02.

    Solution
$$\begin{array}{r} 15.18 \\ 2.763 \\ 0.75 \\ +\ 1.02 \\ \hline 19.713 \end{array}$$

Although it has not been done in these examples, zeros may be added to fill in empty spaces for ease in lining up numbers. If this were done, Example 3-13 would look like this:

$$
\begin{array}{r}
15.180 \\
2.763 \\
0.750 \\
+\ 1.020 \\
\hline
19.713
\end{array}
$$

Remember, however, not to count these zeros as significant digits when rounding the answers.

Example 3-14 The spacers on a milling arbor have thicknesses 0.875 cm, 1.0324 cm, 1.02 cm, and 0.923 cm. Find the total thickness.

Solution
$$
\begin{array}{r}
0.875 \\
1.0324 \\
1.02 \\
+0.923 \\
\hline
3.8504
\end{array}
$$

Because the given numbers are measurements, round to the fewest number of places: two. The sum is 3.85 cm.

Example 3-15 Subtract: 12.405 − 6.71.

Solution
$$
\begin{array}{r}
12.405 \\
-\ 6.71 \\
\hline
5.695
\end{array}
$$

Example 3-16 Subtract: 2.015625 − 0.03125.

Solution
$$
\begin{array}{r}
2.015625 \\
-0.03125 \\
\hline
1.984375
\end{array}
$$

Example 3-17 The outside diameter of a pipe is 0.695 in. and the inside diameter is 0.6 in. How thick is the pipe?

Solution
$$
\begin{array}{r}
0.695 \\
-0.6 \\
\hline
0.095
\end{array}
$$

Because the given numbers are approximate, round to the fewest number of places: one. The pipe is 0.1 in. thick.

Exercises 3-3

Find each sum or difference.

1.	1.4 +2.8	2.	7.92 +6.6	3.	8.03 +9.216	4.	178.8 + 39.6

| 5. | 4.76
+3.98 | 6. | 5.08
+7.361 | 7. | 8.325
+3.95 | 8. | 7.603
+4.591 |

| 9. | 156.735
+ 98.246 | 10. | 547.8
+302.75 | 11. | 1.4325
+9.873 | 12. | 2768.006
+4382.57 |

| 13. | 1.43
0.6
+9.862 | 14. | 4.339
6.847
11.421
+ 9.076 | 15. | 48.2601
56.97
+28 | 16. | 5.2
15.0635
248.371
+ 9.11 |

| 17. | 0.96
−0.88 | 18. | 0.60
−0.16 | 19. | 0.3
−0.21 | 20. | 0.87
−0.19 |

| 21. | 4.0
−3.6 | 22. | 5.2
−1.83 | 23. | 4.8632
−1.918 | 24. | 16.23
− 5.794 |

| 25. | 4.736
−1.8521 | 26. | 8
−7.6324 | 27. | 4.0031
−2.76 | 28. | 6.3406
−1.9271 |

| 29. | 14.36
− 9.827 | 30. | 270.6
− 83.98 | 31. | 176
− 38.9651 | 32. | 27.0803
− 4.29 |

33. Spacers are shown mounted on a milling arbor in Figure A. They are the following thicknesses: 0.25 in., 0.3125 in., 0.5 in., 0.375 in., 0.8125 in., and 0.9375 in. Find the overall thickness.

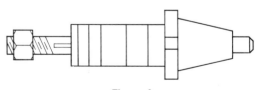

Figure A

34. In Figure B, parts of a gear tooth are shown. The addendum equals 1.3333 in. and the dedendum is 1.5428 in. Once the gear cutter touches the circumference of the gear blank, how much must the table be raised to cut the correct depth of tooth?

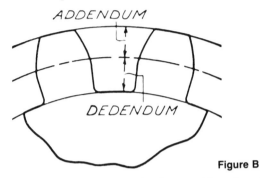

Figure B

35. In Figure C, what is dimension R in thousandths of an inch?

36. If the $1\frac{1}{8}$-in. dimension in Figure C is changed to 1.050 in., what would be dimension R?

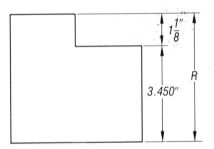

Figure C

37. What diameter hole results when $\frac{3}{16}$ in. is removed from each side of a $\frac{3}{4}$-in. hole? Express your answer in decimal form.

38. A flat gauge mounted on the magnetic chuck of a surface grinder is shown in Figure D. What will its finished thickness be?

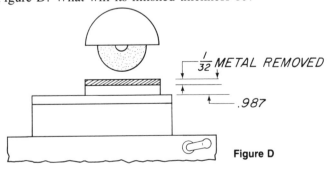

Figure D

39. A plate 2.172 in. thick will have $\frac{3}{32}$ in. milled off each side. Figure E demonstrates this situation. What will the finished plate thickness be?

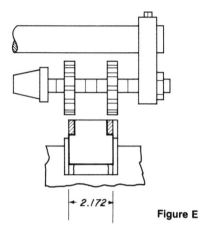

|← 2.172 →|

Figure E

40. Figure F illustrates a stamping for a French machine part. Find dimensions L, a, and b. (All dimensions are in millimeters.)

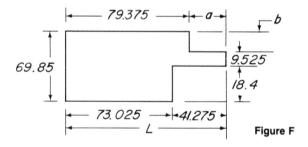

Figure F

41. In Figure G, if a chip 3.175 mm thick is cut on a shaft 85.725 mm in diameter, what will the finished diameter be?

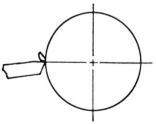

Figure G

42. In Figure H, find diameters *a* and *b* and length *L*. All dimensions are in millimeters.

43. If the outside diameter (*O.D.*) of a piece of tubing is 1.875 in. and its wall thickness is 0.0625 in., what is the inside diameter (*I.D.*)?

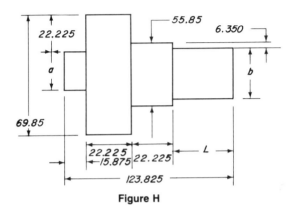

Figure H

44. The end of a German steam pipe is shown in Figure I. What is its outside diameter?

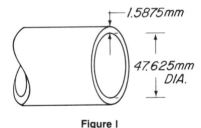

Figure I

3-4   Gauge Blocks

OBJECTIVE

After completing this section, you should be able to:

1. Select proper gauge block combinations.

Adding and subtracting decimals becomes extremely important in the proper selection of gauge blocks to give a desired dimension. Gauge blocks are precise rectangular blocks of steel that are used for measuring or setting other gauges. They are made of steel that has been alternately heated and frozen until stresses within the steel that might cause it to change size are eliminated. At this point, the steel is said to be *stabilized*.

Two opposing edges of each block are ground and lapped to a superior degree of flatness and size. Selected-size blocks are pressed or "wrung" together to exclude air between them. Two or more blocks, when properly

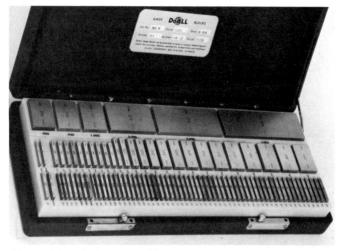

Figure 3-3   92-piece gauge block set.   *(Photo Courtesy of DoALL Company)*

0.0501 in.	0.0510 in.	0.0500 in.	0.1500 in.
0.0502 in.	0.0520 in.	0.0600 in.	0.1000 in.
0.0503 in.	0.0530 in.	0.0700 in.	0.2000 in.
0.0504 in.	0.0540 in.	0.0800 in.	0.3000 in.
0.0505 in.	0.0550 in.	0.0900 in.	0.4000 in.
0.0506 in.	0.0560 in.	0.1100 in.	0.5000 in.
0.0507 in.	0.0570 in.	0.1200 in.	1.0000 in.
0.0508 in.	0.0580 in.	0.1300 in.	2.0000 in.
0.0509 in.	0.0590 in.	0.1400 in.	

Table 3-4

fitted together, will cling to each other with a force greater than atmospheric pressure.

A standard inspection set of 92 pieces is shown in Figure 3-3. Less expensive 36-piece sets are also available. Table 3-4 lists the sizes of blocks in a 36-piece set. The standard block sizes in Table 3-4 will be used for the remainder of this unit.

The rule for selecting blocks to build into combinations is quite simple.

RULE 3-9   SELECTING GAUGE BLOCKS

- Use as few blocks as possible, to avoid error.
- Work from right to left of the decimal to eliminate the *last* figure first.

Example 3-18   Make up a combination of gauge blocks to equal 0.9213 in.

Solution   Block 0.0503 eliminates 0.0003.

$$\begin{array}{r} 0.9213 \\ -0.0503 \\ \hline 0.8710 \text{ remainder} \end{array}$$

Block 0.0510 eliminates 0.0010.

$$\begin{array}{r} 0.8710 \\ -0.0510 \\ \hline 0.8200 \text{ remainder} \end{array}$$

Block 0.1200 eliminates 0.0200.

$$\begin{array}{r} 0.8200 \\ -0.1200 \\ \hline 0.7000 \text{ remainder} \end{array}$$

Blocks 0.5000 and 0.2000 complete the subtraction.

$$\begin{array}{r} 0.7000 \\ -0.5000 \\ \hline 0.2000 \text{ remainder} \\ 0.2000 \\ -0.2000 \\ \hline \text{no remainder} \end{array}$$

The five blocks selected are wrung together to form a rectangular combination, or "stack," equal to 0.9213 in.

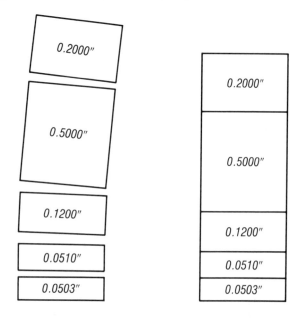

Exercises 3-4

Use the list in Table 3-4 to determine which gauge blocks should be used to make up the following lengths.

1. 1.2892 in. **2.** 0.8625 in. **3.** 2.1427 in.

4. 3.0723 in. **5.** 1.2156 in. **6.** 0.875 in.

7. 3.6363 in. **8.** 1.625 in. **9.** 0.7372 in.

10. Use the list in Table 3-4 to determine which gauge blocks should be used to inspect the width of the slot in Figure A.

11. Which gauge blocks should be used to inspect the total width of the part shown in Figure A?

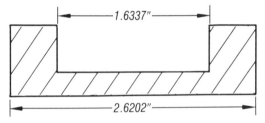

Figure A

3-5 Multiplying Decimals

OBJECTIVE

After completing this section, you should be able to:

1. Find the product of two or more decimals.

Decimals are multiplied exactly as whole numbers are, with one exception: The decimal point in the product must be placed correctly.

RULE 3-10   MULTIPLYING DECIMALS

Step 1   Multiply the two factors as if they were whole numbers.

Step 2   Count the total number of decimal places in all the factors.

Step 3   Insert a decimal point in the answer so that the number of places to the right of the decimal point equals the sum of the places in the factors. (Annex zeros to the left of the given digits if necessary.)

Step 4   If the numbers are approximate, round the answer to the appropriate place.

Example 3-19 Multiply: 0.56×0.27.

Solution
$$
\begin{array}{r}
0.56 \longleftarrow \text{factor} \\
\times \quad 0.27 \longleftarrow \text{factor} \\
\hline
392 \\
112 \\
\hline
0.1512 \longleftarrow \text{product}
\end{array}
$$

Each factor has two decimal places. This gives a sum of four places to be counted off *from the right* toward the left in the answer.

$$0.\underset{\text{4 places}}{\underline{1512}}$$

Example 3-20 Multiply: 8.25×0.063.

Solution
$$
\begin{array}{r}
8.25 \ (2 \text{ places}) \\
\times \quad 0.063 \ (3 \text{ places}) \\
\hline
2475 \\
4950 \\
\hline
0.51975
\end{array}
$$

$$0.\underset{\text{5 places}}{\underline{51975}}$$

Example 3-21 Multiply: 0.23×0.04.

Solution
$$
\begin{array}{r}
0.23 \ (2 \text{ places}) \\
\times \quad 0.04 \ (2 \text{ places}) \\
\hline
0.0092
\end{array}
$$

4 places. Annex two zeros to the left of the given digits.

Example 3-22 Use a calculator to find the product of 2.63 in. and 1.7 in.

Solution

Enter	Display
$\boxed{2}\boxed{.}\boxed{6}\boxed{3}$	2.63
$\boxed{\times}$	2.63
$\boxed{1}\boxed{.}\boxed{7}$	1.7
$\boxed{=}$	4.471

Since the given numbers are measurements, round the answer to two significant digits, the fewest number. The answer is approximately 4.5 sq in.

Exercises 3-5

Place the decimal point in each product.

1. $2.56 \times 3.829 = 980224$ 2. $26.8 \times 37.12 = 994816$
3. $9.87 \times 0.006 = 5922$ 4. $0.007 \times 0.0008 = 56$

Find each product.

5.	7	6.	16	7.	27	8.	84
	$\times 0.4$		$\times 0.8$		$\times 0.9$		$\times 0.1$

9.	0.26	10.	0.06	11.	2.1	12.	7.3
	$\times\ 0.8$		$\times\ 0.5$		$\times 0.006$		$\times 0.0001$

13.	4.93	14.	26.1	15.	38.79	16.	22.99
	$\times\ 0.8$		$\times\ 0.7$		$\times\ 0.3$		$\times\ 0.2$

17.	86.9	18.	3.71	19.	12.76	20.	43
	$\times\ 4.2$		$\times 2.91$		$\times\ 0.08$		$\times 2.76$

21.	789.3	22.	0.173	23.	27.926	24.	2.079
	$\times 0.4862$		$\times 0.841$		$\times\ 4.311$		$\times 20.79$

25. 0.6×2.84 26. 7.38×11.6 27. 4.23×18.1
28. $7.3 \times 0.1\ \times 2.6$ 29. $8.5 \times\ 9.6 \times 12.43$
30. $17.6 \times\ 0.1 \times 0.08$

31. What length of stock is required to cut four pins, each 1.583 in. long, allowing $\frac{1}{16}$ in. waste on each pin?

32. If a machinist side mills 0.067 in. per pass and takes four passes, how much material will be milled?

33. A numerical control part programmer plans to peck drill a machined part eight times (eight cycles of tool-engaging material) and advance the drill 0.045 in. per peck. How deep will the hole be?

34. If a thrust washer is 0.3437 in. thick, what is the combined thickness of six washers?

The circumference of (distance around) a circle is approximately 3.1416 multiplied by the diameter. Find the circumferences for circles with the following diameters.

35. 8.00 in.   36. 2.5 in.   37. 1.625 in.
38. 46.0375 mm   39. 130.175 mm   40. 77.7875 mm

3-6   Dividing Decimals

OBJECTIVE

After completing this section, you should be able to:

1.   Divide one decimal by another.

Dividing one decimal by a whole number or by another decimal is like doing ordinary long division, except that special attention must be paid to proper placement of the decimal point in the quotient. To do this, first eliminate the decimal point in the divisor.

RULE 3-11   MOVING DECIMAL POINTS

Moving the decimal point in a number one place to the right is the same as multiplying by 10; moving it two places is the same as multiplying by 10 × 10, or 100; and so on.

Thus to divide 59.28 by 5.2, first multiply both numbers by 10.

$$\frac{59.28}{5.2} \times \frac{10}{10} = \frac{59.28 \times 10}{5.2 \times 10} = \frac{592.8}{52}$$

This moves the decimal point the same number of places in both numbers, eliminating the decimal point in the divisor. Then divide as with whole numbers, placing the decimal point in the quotient above the decimal point in the dividend.

RULE 3-12   DIVIDING DECIMALS

Step 1   Move the decimal point the same number of places in both the divisor and the dividend to make the divisor a whole number.

Step 2   Divide as if the dividend were a whole number.

Step 3   Place the decimal point in the quotient above the new decimal point in the dividend.

Step 4   Round the answer, if needed.

Example 3-23   Divide 59.28 by 5.2.

Solution

Step 1   Move the decimal point one place in both numbers to make the divisor a whole number.

$$5.2\overline{)59.2\,8}$$

Step 2-4   Divide as with whole numbers. Place the decimal point in the quotient over the new decimal point in the dividend.

$$
5.2\overline{)59.2\,8} \quad \begin{array}{r} 11.4 \\ \hline \end{array}
$$

$$
\begin{array}{r}
11.4 \\
5.2\overline{)59.2\,8} \\
\underline{52} \\
7\,2 \\
\underline{5\,2} \\
2\,0\,8 \\
\underline{2\,0\,8}
\end{array}
$$

Sometimes it is necessary to annex zeros before dividing. Zeros often must be annexed as placeholders when the divisor has as many, or more, decimal places as the dividend.

Example 3-24   Divide 5.2 by 18.6.

Solution   First eliminate the decimal point in the divisor.

$$
18.6\overline{)5.2}
$$

Then divide. It is necessary to annex zeros.

$$
\begin{array}{r}
0.279 \\
18.6\overline{)5.2\,000}
\end{array}
$$

Example 3-25   Divide 73 by 0.0006. Round the answer to two significant digits.

Solution

$$
\begin{array}{r}
12\,1666 \\
0.0006\overline{)73.0000} \\
\underline{6} \\
13 \\
\underline{12} \\
10 \\
\underline{6} \\
40 \\
\underline{36} \\
40 \\
\underline{36} \\
40 \\
\underline{36} \\
4
\end{array}
$$

Annex zeros before moving the decimal point.

To two significant digits, 121,666 is 120,000.

Example 3-26   The area of a metal plate is 17.54 sq in. The length is 4.3 in. Find the width if the width is the area divided by the length.

Solution

$$
\begin{array}{r}
4.07 \\
4.3\overline{)17.5\,40} \\
\underline{17\ 2} \\
3\ 40 \\
\underline{3\ 01} \\
39
\end{array}
$$

Round to two significant digits. Annex a zero and carry the division to three digits.

The width is about 4.1 in.

Exercises 3-6

Place the decimal point in each answer.

1. $4.22\overline{)8.2712}$ 196

2. $0.08\overline{)26.00}$ 3 25

3. $7.9\overline{)0.00474}$ 6

4. $5\overline{)4}$ 8

Divide.

5. $64\overline{)5.12}$

6. $87\overline{)607.26}$

7. $7\overline{)16.45}$

8. $5\overline{)483.65}$

9. $0.13\overline{)9.75}$

10. $2.6\overline{)106.34}$

11. $4.5\overline{)0.09}$

12. $0.08\overline{)0.6096}$

13. $2.85\overline{)2.1375}$

14. $0.67\overline{)19.162}$

15. $9.1\overline{)0.0182}$

16. $1.3\overline{)0.000169}$

17. $0.17\overline{)33.66}$

18. $1.34\overline{)1794.26}$

19. $173.4\overline{)1768.68}$

20. $0.387\overline{)3.11535}$

Annex zeros and divide until there is a remainder of zero.

21. $20\overline{)0.8}$

22. $25\overline{)67}$

23. $80\overline{)36}$

24. $125\overline{)1}$

Round each quotient to the nearest tenth.

25. $75\overline{)96.32}$

26. $4.78\overline{)15.392}$

27. $70\overline{)30}$

28. $7.3\overline{)26.48}$

29. $15.1\overline{)7.283}$

30. $4.23\overline{)178.964}$

31. The area of a circle is obtained by dividing 3.1416 by 4 and then multiplying the answer by the square of the diameter. The formula looks like this:

$$
A = D^2\left(\frac{3.1416}{4}\right)
$$

What is the quotient of 3.1416 divided by 4?

32. Two diameters of a shaft are shown in Figure A. What is one-half the difference between the two diameters?

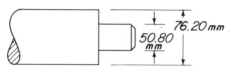

Figure A

33. If the small diameter in Figure A were 0.6875 in. and the large diameter were 1.125 in., what would be one-half the difference between the two diameters? (Round to the nearest thousandth.)

34. The shop teacher assigned Jim to saw a 10.375-ft-long bar of cold-rolled steel into 32 equal pieces for student shop projects. Allowing a $\frac{1}{16}$-in. loss per piece, how long will each piece be? (Round to the nearest thousandth.)

Unit 3 Review Exercises

1. Convert $\frac{41}{64}$ to decimal form.

2. Round the answer from exercise 1 to ten-thousandths.

3. Add 23.567, 3.48, and 12.6, and round the answer to the appropriate number of digits.

4. If each part is 0.750 cm thick and five parts are clamped together in a vise for machining operation, how wide is the stack of parts?

5. Three parallel spacers position a part above the clamps of a fixture. If the spacers are 1.000 in., 0.5000 in., and 0.1875 in., respectively, how high is the part raised by the spacers?

6. Rough machining a block whose finished thickness should be 0.4219 in., the machine operator leaves $\frac{1}{32}$ in. stock. How thick, in decimal form, is the stock block?

7. The machining department inspector needs a stack of gauge blocks 2.3142 in. high. Using the set described in Table 3-4, which blocks should be used?

8. Five boxes, each measuring 16.345 mm, are stacked one upon the other. How tall is the stack of five boxes?

9. What is the circumference of a vertical turret lathe worktable with a diameter of 6.25 ft? (Circumference = 3.1416 × diameter.)

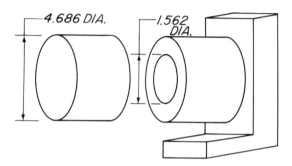

Figure A

10. The shaft in Figure A is 4.686 in. in diameter. It must fit snugly into a bearing that has a hole 1.562 in. in diameter. How much larger is the shaft than the hole?

Measurement

More than 90 percent of the world's industrialized countries have adopted the metric system of measuring because of its simplicity and ease of use. Many American manufacturers have converted to the metric system in order to compete successfully in the world market.

Americans are seeing more and more metric measures. The weights of packaged goods are often expressed in grams as well as pounds and ounces. Eyeglass lenses are calibrated in millimeters. The drug industry is fully converted to metric measures. Some road signs give distances in both miles and kilometers, and motorists often purchase so many liters of gas, although they ask for gallons. Figure 4-1 illustrates metric measurements for clothing pattern sizes.

It is probable that the United States eventually will convert to the metric system of measurement. The more you practice using the metric system and converting between English and metric measurements, the easier it will be for you to make the transition.

4-1 The English System

OBJECTIVES

After completing this section, you should be able to:

1. Use the English system of measurement.
2. Convert from one unit to another within the English system.

In any system of measurement, units are needed to measure length, capacity, and mass (weight). Any measurement is made up of a **unit of measure,** such as inches or feet, and the **measure,** which is a number. In the measurement 20 inches, the unit of measure is inches and the measure is 20.

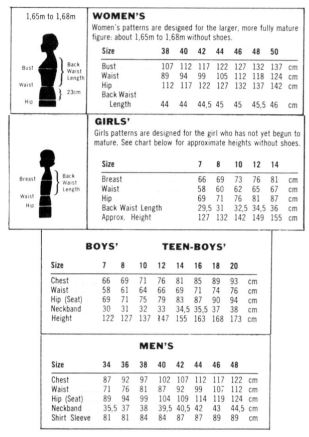

1,65m to 1,68m	**WOMEN'S**
	Women's patterns are designed for the larger, more fully mature figure: about 1,65m to 1,68m without shoes.

Size	38	40	42	44	46	48	50	
Bust	107	112	117	122	127	132	137	cm
Waist	89	94	99	105	112	118	124	cm
Hip	112	117	122	127	132	137	142	cm
Back Waist Length	44	44	44,5	45	45	45,5	46	cm

GIRLS'

Girls patterns are designed for the girl who has not yet begun to mature. See chart below for approximate heights without shoes.

Size	7	8	10	12	14	
Breast	66	69	73	76	81	cm
Waist	58	60	62	65	67	cm
Hip	69	71	76	81	87	cm
Back Waist Length	29,5	31	32,5	34,5	36	cm
Approx. Height	127	132	142	149	155	cm

BOYS'				**TEEN-BOYS'**			

Size	7	8	10	12	14	16	18	20	
Chest	66	69	71	76	81	85	89	93	cm
Waist	58	61	64	66	69	71	74	76	cm
Hip (Seat)	69	71	75	79	83	87	90	94	cm
Neckband	30	31	32	33	34,5	35,5	37	38	cm
Height	122	127	137	147	155	163	168	173	cm

MEN'S								

Size	34	36	38	40	42	44	46	48	
Chest	87	92	97	102	107	112	117	122	cm
Waist	71	76	81	87	92	99	107	112	cm
Hip (Seat)	89	94	99	104	109	114	119	124	cm
Neckband	35,5	37	38	39,5	40,5	42	43	44,5	cm
Shirt Sleeve	81	81	84	84	87	87	89	89	cm

Figure 4-1. Commercial use of metric system.
(The McCall Pattern Company)

Length

In the English system, length is measured by units such as feet or inches. English units of length, or linear measure, are given in Table 4-1.

Shopworkers often find it necessary to convert from one unit of measure to another, such as feet to inches or inches to feet. The numbers in Table 4-1 are called **conversion factors.** To change one unit to another, it is necessary to multiply or divide a given measurement by a conversion factor.

LINEAR MEASURE
12 inches (in.) = 1 foot (ft)
3 feet (36 in.) = 1 yard (yd)
$5\frac{1}{2}$ yards ($16\frac{1}{2}$ ft) = 1 rod (rd)
5280 feet (1760 yd) = 1 mile (mi)

Table 4-1

RULE 4-1   CONVERTING UNITS OF MEASURE

- To change from a smaller unit of measure to a larger one, divide the given measurement by the conversion factor.
- To change from a larger unit of measure to a smaller one, multiply the given measurement by the conversion factor.

Example 4-1   Convert 20 yd to inches.

Solution   The conversion factor is 36, since 1 yd = 36 in. Multiply, because the change is from a larger unit (yards) to a smaller unit (inches).

$$\begin{array}{r} 20 \\ \times\ 36 \\ \hline 120 \\ 600 \\ \hline 720 \end{array}$$

Therefore, 20 yd = 720 in.

Example 4-2   Maria ran the 880-yd event in 2 min 15 seconds (s). Find what part of a mile she ran.

Solution   The conversion factor is 1760, since 1 mi = 1760 yd. Divide, because the conversion is from a smaller unit to a larger unit.

$$\begin{array}{r} 0.5 \\ 1760\overline{)880.0} \\ \underline{880\ 0} \\ 0 \end{array}$$

880 yd = 0.5 mi, so Maria ran 0.5 mi.

Sometimes it is necessary to convert measurements involving two different units to a measurement with only a single unit. To do this, simply convert one measurement into the units of the other and add the two measures with the same units.

Example 4-3   Write 12 ft 8 in. as inches.

Solution   12 ft is 12 × 12 = 144 in. Thus 12 ft 8 in. = 144 + 8 in. = 152 in.

Example 4-4 Change the marathon distance of 26 mi 385 yd to miles.

Solution 1 mi = 1760 yd, so

$$385 \text{ yd} = \frac{385}{1760} \text{ mi} = \frac{7}{32} \text{ mi}$$

Thus 26 mi 385 yd = $26\frac{7}{32}$ mi.

Capacity

The English system of measurement uses two sets of units to measure capacity: one for dry measure and one for liquid measure. The two sets of units and their conversion factors are shown in Table 4-2. Apply Rule 4-1 to convert from one unit to another within each set.

CAPACITY	
Liquid Measure	**Dry Measure**
16 fluid ounces (fl oz) = 1 pint (pt)	2 pints = 1 quart
2 pints = 1 quart (qt)	8 quarts = 1 peck (pk)
4 quarts = 1 gallon (gal)	4 pecks = 1 bushel (bu)

Table 4-2

The tables of units of measure in this book do not list every possible conversion factor. However, they give you enough information to solve a problem, even if the conversion factor you need is not listed. For example, you can change gallons to pints even though the conversion factor is not given. First change gallons to quarts, for which the conversion factor is given. Then change the number of quarts to pints, for which the conversion factor also is given.

Example 4-5 Find the number of pints in 6 gal.

Solution First multiply by the conversion factor of 4 to convert to quarts.

$$6 \text{ gal} \longrightarrow 4 \times 6 = 24 \text{ qt}$$

Next multiply by the conversion factor of 2 to convert to pints.

$$24 \text{ qt} \longrightarrow 2 \times 24 = 48 \text{ pt}$$

Therefore, 6 gal = 48 pt.

Example 4-6 Find the number of pecks in 12 qt.

Solution The conversion factor is 8. Divide.

$$
\begin{array}{r}
1.5 \\
8)\overline{12.0} \\
-\,8 \\
\hline
4\,0 \\
-\,4\,0 \\
\hline
0
\end{array}
$$

Thus 12 qt = 1.5 pk.

Weight

Ounces, pounds, and tons are the three most commonly used units of weight in the English system. They are listed, along with their conversion factors, in Table 4-3. Use Rule 4-1 to convert from one unit to another.

WEIGHT
16 ounces (oz) = 1 pound (lb)
2000 pounds = 1 (short) ton (T)

Table 4-3

Example 4-7 How many pounds are there in 56 oz?

Solution The conversion factor is 16. Divide.

$$
\begin{array}{r}
3.5 \\
16)\overline{56.0} \\
48 \\
\hline
8\,0 \\
8\,0 \\
\hline
0
\end{array}
$$

Thus 56 oz = 3.5 lb.

Example 4-8 A piece of metal weighs 5 lb 3 oz. How many ounces is that?

Solution To change pounds to ounces, multiply by the conversion factor of 16.

$$16 \times 5 = 80 \text{ oz}$$

5 lb = 80 oz, so 5 lb 3 oz = (80 + 3) oz = 83 oz.

Time

The same units of time are used in both the English and metric systems. These units are listed, along with their conversion factors, in Table 4-4. Apply Rule 4-1 to convert units of time.

TIME
60 second(s) = 1 minute (min)
60 minutes = 1 hour (h)
24 hours = 1 day (da)
7 days = 1 week (wk)
52 weeks = 1 year (yr)
12 months (mo) = 1 year

Table 4-4

Example 4-9 How many hours are there in 8½ days?

Solution Since you are changing to a smaller unit, multiply by the conversion factor of 24.

$$24 \times 8\frac{1}{2} = 24 \times \frac{17}{2} = 204$$

Thus $8\frac{1}{2}$ da = 204 h.

Example 4-10 How many hours are there in 9000 s?

Solution First change seconds to minutes. Divide by the conversion factor of 60.

$$\begin{array}{r} 150 \\ 60\overline{)9000} \\ \underline{60} \\ 300 \\ \underline{300} \\ 0 \end{array}$$ 9000 s = 150 min

Change minutes to hours. Divide by the conversion factor of 60.

$$\begin{array}{r} 2.5 \\ 60\overline{)150.0} \\ \underline{120} \\ 30\ 0 \\ \underline{30\ 0} \\ 0 \end{array}$$

Thus, 9000 s = 2.5 h.

Exercises 4-1

Find the number of inches in each measurement.

1. 12 ft　　**2.** 8 rd　　**3.** $\frac{3}{5}$ yd　　**4.** 4 ft 8 in.

Find the number of feet in each measurement.

5. 480 in.　　**6.** 17 yd　　**7.** $2\frac{1}{2}$ mi　　**8.** 5 yd 2 ft

Find the number of yards in each measurement.

9. 15 ft　　**10.** 148 in.　　**11.** 3.4 mi　　**12.** 8 rd 3 yd

Find the number of pints in each measurement.

13. 4 qt　　**14.** 4 pk　　**15.** 72 fl oz　　**16.** $5\frac{1}{2}$ gal

Find the number of quarts in each measurement.

17. $11\frac{1}{4}$ gal　　**18.** 19 pt　　**19.** 160 fl oz　　**20.** 5 gal 3 qt

Find the number of pecks in each measurement.

21. 64 pt　　**22.** $4\frac{1}{2}$ bu　　**23.** 28 qt　　**24.** 2 bu 3 pk

Find the number of pounds in each measurement.

25. 5 T　　**26.** 84 oz　　**27.** 1 T 837 lb　　**28.** 0.385 T

Find the number of seconds in each measurement.

29. 25.7 min　　**30.** 8.6 h　　**31.** 1 h 53 min　　**32.** 1 da

Find the number of minutes in each measurement.

33. 720 s　　**34.** 5.2 h　　**35.** 1 da　　**36.** 5 h 35 min

Find the number of days in each measurement.

37. 8 h　　**38.** 103.2 h　　**39.** 4 wk　　**40.** 8 wk 4 da

Complete each of the following.

41. (a) 5 mi = ? yd　　　　(b) 46,464 ft = ? mi

42. (a) 88 oz = ? lb　　　　(b) 128 fl oz = ? pt

43. (a) 2000 fl oz = ? gal　　(b) 18 qt = ? gal

44. (a) 9.2 h = ? s　　　　(b) 48 wk = ? yr

45. (a) $4\frac{1}{2}$ yr = ? wk　　(b) $4\frac{1}{2}$ yr = ? mo

46. A ball of string contains 450 ft of string. How many yards of string is that? How many inches?

47. During a 60-min television program, there were 12 min of commercials. What fractional part of an hour was devoted to the program?

48. How many ounces of water must be added to a 6-oz can of frozen orange concentrate to make 1.5 pt of orange juice?

49. An end dump truck has a capacity of 22.5 T. How many pounds is that?

50. John agreed to repay a loan for his automobile in 42 mo. How many years is that?

4-2 The Metric System

OBJECTIVES

After completing this section, you should be able to:

1. Use the metric system of measurement.

2. Convert from one unit to another within the metric system.

In the metric system, units of measure are based on the number 10 and its powers. All units can be divided by 2 and 5, the factors of 10. The metric system uses a basic unit for each type of measurement, such as the meter for length or the liter for capacity. Other units of measurement within each type are powers of 10 of the basic unit. The following prefixes are used to designate the powers of 10.

milli = one-thousandth
centi = one-hundredth
deci = one-tenth
deka = ten times
hecto = one hundred times
kilo = one thousand times

Using the meter as the basic unit, Table 4-5 shows how these prefixes are used and gives their relative values.

Prefix	+	Basic Unit	=	Unit	Relative Value
milli		meter		millimeter (mm)	0.001 meter
centi		meter		centimeter (cm)	0.01 meter
deci		meter		decimeter (dm)	0.10 meter
		meter (m)			1 meter
kilo		meter		kilometer (km)	1000 meters

Table 4-5

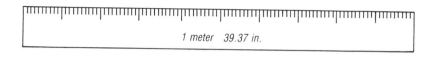

Figure 4-2

Length

The **meter** is the fundamental building block of the metric system and the basic unit of length. It was established after the French Revolution of 1789 as one ten-millionth of the distance along a meridian that extended from the North Pole, through Paris, to the equator. The English equivalent of this distance is 39.37 inches, approximately $3\frac{3}{8}$ inches longer than the yard. Figure 4-2 illustrates the relationship between the meter and the English yard.

The most common units of length formed by adding metric prefixes to the meter are the millimeter, the centimeter, the decimeter, and the kilometer. The meter can be combined with other prefixes to form additional units of length, as shown in Table 4-6. Hectometer and dekameter, however, are not often used. Table 4-6 also gives conversion factors. Follow Rule 4-1 to change from one unit to another in the metric system.

Multiples			Unit	Subdivisions		
Kilometer						
10 hm	Hectometer					
100 dam	10 dam	Dekameter				
1000 m	100 m	10 m	Meter			
			10 dm	Decimeter		
			100 cm	10 cm	Centimeter	
			1000 mm	100 mm	10 mm	Millimeter
km	hm	dam	m	dm	cm	mm

Table 4-6

Example 4-11   How many centimeters are there in 8.6 m?

Solution   The conversion factor is 100 because 1 m = 100 cm. Since you are converting to a smaller unit, multiply.

8.6 × 100 = 860

Thus 8.6 m = 860 cm.

Example 4-12   How many centimeters are there in 98 mm?

Solution   The conversion factor is 10, since 1 cm = 10 mm. You are converting to a larger unit, so divide.

98 ÷ 10 = 9.8

Thus 98 mm = 9.8 cm.

A measurement such as 8 meters 7 decimeters 4 centimeters can be written as a single decimal, 8.74 meters, since 7 decimeters = 0.7 meter and 4 cm = 0.04 meter.

Figure 4-3 shows a **metric steel rule.** Each of the larger subdivisions is a centimeter. Each centimeter is divided into smaller divisions, millimeters. Read the measurements shown by the letters.

A.   2.6 cm, or 2 cm 6 mm, or 26 mm
B.   3.5 cm, or 3 cm 5 mm, or 35 mm
C.   5.1 cm, or 5 cm 1 mm, or 51 mm
D.   6.2 cm, or 6 cm 2 mm, or 62 mm

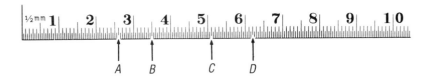

Figure 4-3

CAPACITY MEASURE
1 milliliter (mL) = $\frac{1}{10}$ cL = $\frac{1}{1000}$ liter (L) 10 mL = 1 centiliter (cL) = $\frac{1}{100}$ L 10 cL = 1 deciliter (dL) = $\frac{1}{10}$ L 10 dL = 1 liter = 100 cL = 1000 mL 10 L = 1 dekaliter (daL) 10 daL = 1 hectoliter (hL) = 100 L 10 hL = 1 kiloliter (kL) = 1000 L

Table 4-7

Capacity

Another unit of the metric system is the **liter.** It measures liquid capacity. Liquids such as milk, wine, and gasoline are sold by the liter. Large containers of liquids such as casks and drums are designated in hectoliters. A hectoliter is equivalent to 100 liters, since the prefix *hecto* means "hundred." Table 4-7 shows some of the units of measure of capacity and their relationships. The liter and the milliliter are most commonly used. Use Rule 4-1 to convert from one unit to another.

Example 4-13   How many milliliters are there in 4.3 L?

   Solution   The conversion factor is 1000, since 1 L = 1000 mL. Multiply, because you are converting to a smaller unit.

   $$4.3 \times 1000 = 4300$$

   Thus 4.3 L = 4300 mL.

Mass (Weight)

The standard units for measuring mass in the metric system are the gram and kilogram. One kilogram equals about 2.20 pounds.

The same prefixes are used with grams as with meters and liters. For instance, a kilogram is 1000 grams and a centigram is $\frac{1}{100}$ gram.

Table 4-8 shows some units of mass and their relationships.

MASS
1 milligram (mg) = $\frac{1}{10}$ cg = $\frac{1}{1000}$ gram (g)
10 mg = 1 centigram (cg) = $\frac{1}{100}$ gram
10 cg = 1 decigram (dg) = $\frac{1}{10}$ gram
10 dg = 1 gram = 100 cg = 1000 mg
10 g = 1 dekagram (dag)
10 dag = 1hectogram (hg) = 100 grams
10 hg = 1 kilogram (kg) = 1000 grams

Table 4-8

Example 4-14   How many decigrams are there in 8.37 dag?

   Solution   The conversion factor is 100, since 1 dag = 100 dg. Multiply, because you are changing to a smaller unit.

   $$8.37 \times 100 = 837$$

   Thus 8.37 dag = 837 dg.

In popular usage, the word *kilogram* is shortened to *kilo* (plural *kilos*). The metric ton (1000 kilograms) is used only for very heavy objects, such as bulk containers or machinery.

Figure 4-4 shows the relationship of the liter to the decimeter and the kilogram. If a container that measured 1 decimeter on each side—in other words, a cubic decimeter—were to be filled completely with pure water, it would be equivalent to 1 kilogram in mass and 1 liter in liquid capacity.

Table 4-9 summarizes the common metric units of length, capacity, and mass.

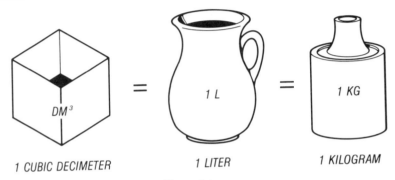

1 CUBIC DECIMETER   1 LITER   1 KILOGRAM

Figure 4-4

	0.001	**0.01**	**0.1**	**Basic Unit**	**10**	**100**	**1000**
Length	millimeter mm	centimeter cm	decimeter dm	meter m	dekameter dam	hectometer hm	kilometer km
Capacity	milliliter mL	centiliter cL	deciliter dL	liter L	dekaliter daL	hectoliter hL	kiloliter kL
Mass	milligram mg	centigram cg	decigram dg	gram g	dekagram dag	hectogram hg	kilogram kg

Table 4-9

Exercises 4-2

Find the number of meters in each measurement.

1. 10.5 km   **2.** 430 cm   **3.** 1000 mm   **4.** 10 dam 8 m

Find the number of centimeters in each measurement.

5. 48 mm   **6.** 14.7 m   **7.** 8.07 km   **8.** 93 m 28 cm

Find the number of kilometers in each measurement.

9. 6905 m   **10.** 10,000 cm   **11.** 50 hm   **12.** 40 km 1 hm

Find the number of liters in each measurement.

13. 50.2 kL   **14.** 3000 mL   **15.** 485 cL   **16.** 87 kL 4 hL

Find the number of milliliters in each measurement.

17. 2.3 L   **18.** 108 cL   **19.** 5.8 kL   **20.** 4 cL 8 mL

Find the number of grams in each measurement.

21. 70 cg   **22.** 84.5 dag   **23.** 8.92 kg   **24.** 5 dag 4 g 8 dg

Find the number of kilograms in each measurement.

25. 10,580 g   **26.** 478 hg   **27.** 910,000 cg   **28.** 8 kg 3 hg

Write each measure as a number of meters.

29. 4 m 3 dm 8 cm   **30.** 8 dam 4 m 9 cm

Write each measure as a number of liters.

31. 8 L 52 cL   **32.** 4 hL 7 daL 5 L

Write each measure as a number of grams.

33. 4 g 8 dg 7 cg   **34.** 2 kg 4 hg

Find the length of each segment in centimeters; in millimeters; in centimeters and millimeters.

35. _____

36. _____

37. _____

38. _____

39. _____

40. The distance from Paris to Berlin is 1092 km. How many meters is that?

41. A can of fruit weighs 832 g. How many kilograms is that?

42. Give the readings for positions *c*, *d*, *e*, *f*, *g*, and *h* of Figure A in both millimeters and centimeters.

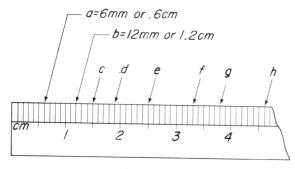

Figure A

43. A measurement on a blueprint is shown in centimeters. What is the conversion factor to change the measurement to millimeters? To meters? To kilometers?

44. The weight of a steel bar is 23.67 kg. After milling, the bar weighs 22.96 kg. What is the initial weight, and the weight of the bar after milling, in grams?

45. The capacity of a metal drum is 12 hectoliters. How many liters would that be?

4-3 Metric–English Conversion

OBJECTIVES

After completing this section, you should be able to:

1. Convert English units to metric units.

2. Convert metric units to English units.

The following comparisons of English and metric units should help the student and shopworker to "think metric."

A millimeter is slightly more than $\frac{1}{32}$ inch.

A centimeter is approximately $\frac{3}{5}$ inch.

A meter is about $1\frac{1}{10}$ yards.

A kilometer is about $\frac{5}{8}$ mile.

In conformance with metrication, industrial designers will begin placing both metric and English dimensions on blueprints. All metric dimensions, no matter how large, will be in millimeters.

$$1 \text{ millimeter} = 0.0394 \text{ inch}$$

$$\frac{1}{10} \text{ millimeter} = 0.00394 \text{ inch}$$

$$\frac{1}{100} \text{ millimeter} = 0.00039 \text{ inch}$$

$$\frac{1}{1000} \text{ millimeter} = 0.00004 \text{ inch}$$

Table 4-10 gives metric equivalents of some common units of length in the English system. These equivalents are used as conversion factors when changing from one system of measurement to the other.

Metric measures and their English equivalents are equally important for everyday use in a shop. Figure 4-5 is a reproduction of a decimal equivalent chart, in English and metric measurements, that might be carried in a worker's pocket or, in larger form, hung on a shop wall for quick reference.

METRIC EQUIVALENTS, LENGTH			
	mm	cm	m	km
in.	25.4	2.54	0.0254	
ft	304.8	30.48	0.3048	
yd	914.4	91.44	0.9144	
mi			1609.3	1.609

Table 4-10

DECIMAL EQUIVALENTS – MILLIMETERS *to* INCHES

1^m/m thru 20^m/m		21^m/m thru 40^m/m		41^m/m thru 60^m/m		61^m/m thru 80^m/m		81^m/m thru 100^m/m	
MM	INCH	MM	INCH	MM	INCH	MM	INCH	MM	INCH
1	= 0.0394	21	= 0.8268	41	= 1.6142	61	= 2.4016	81	= 3.1890
2	= 0.0787	22	= 0.8661	42	= 1.6535	62	= 2.4409	82	= 3.2283
3	= 0.1181	23	= 0.9055	43	= 1.6929	63	= 2.4803	83	= 3.2677
4	= 0.1575	24	= 0.9449	44	= 1.7323	64	= 2.5197	84	= 3.3071
5	= 0.1968	25	= 0.9842	45	= 1.7716	65	= 2.5590	85	= 3.3464
6	= 0.2362	26	= 1.0236	46	= 1.8110	66	= 2.5984	86	= 3.3858
7	= 0.2756	27	= 1.0630	47	= 1.8504	67	= 2.6378	87	= 3.4252
8	= 0.3150	28	= 1.1024	48	= 1.8898	68	= 2.6772	88	= 3.4646
9	= 0.3543	29	= 1.1417	49	= 1.9291	69	= 2.7165	89	= 3.5039
10	= 0.3937	30	= 1.1811	50	= 1.9685	70	= 2.7559	90	= 3.5433
11	= 0.4331	31	= 1.2205	51	= 2.0079	71	= 2.7953	91	= 3.5827
12	= 0.4724	32	= 1.2598	52	= 2.0472	72	= 2.8346	92	= 3.6220
13	= 0.5118	33	= 1.2992	53	= 2.0866	73	= 2.8740	93	= 3.6614
14	= 0.5512	34	= 1.3386	54	= 2.1260	74	= 2.9134	94	= 3.7008
15	= 0.5905	35	= 1.3779	55	= 2.1653	75	= 2.9527	95	= 3.7401
16	= 0.6299	36	= 1.4173	56	= 2.2047	76	= 2.9921	96	= 3.7795
17	= 0.6693	37	= 1.4567	57	= 2.2441	77	= 3.0315	97	= 3.8189
18	= 0.7087	38	= 1.4961	58	= 2.2835	78	= 3.0709	98	= 3.8583
19	= 0.7480	39	= 1.5354	59	= 2.3228	79	= 3.1102	99	= 3.8976
20	= 0.7874	40	= 1.5748	60	= 2.3622	80	= 3.1496	100	= 3.9370

MM	INCH	MM	INCH	MM	INCH
0.01	= 0.00039	200	= 7.8740	700	= 27.5590
0.10	= 0.0039	300	= 11.8110	800	= 31.4960
1.00	= 0.0394	400	= 15.7480	900	= 35.4330
10.00	= 0.3937	500	= 19.6850	1000	= 39.3700
100.00	= 3.9370	600	= 23.6220		

THE du MONT CORPORATION

GREENFIELD, MASSACHUSETTS 01301, U.S.A. · TEL. 413-773-3674

Figure 4-5 *(Courtesy of The DuMont Corporation)*

Shopworkers in many industries must work with thousandths and ten-thousandths of an inch for precise work. Table 4-11 lists metric equivalents of precision dimensions.

in.	mm	in.	mm
0.0001	0.0025	0.0035	0.0889
0.0002	0.0051	0.004	0.1016
0.0003	0.0076	0.0045	0.1143
0.0004	0.0102	0.005	0.1270
0.0005	0.0127	0.0055	0.1397
0.0006	0.0152	0.006	0.1524
0.0007	0.0178	0.0065	0.1651
0.0008	0.0203	0.007	0.1778
0.0009	0.0229	0.0075	0.1905
0.001	0.0254	0.008	0.2032
0.0015	0.0381	0.0085	0.2154
0.002	0.0508	0.009	0.2286
0.0025	0.0635	0.0095	0.2413
0.003	0.0762	0.010	0.2540

Table 4-11

Such conversion, or metrication, aids will become increasingly common in business and industrial establishments. Nevertheless, there will be times when you need to make on-the-spot conversions from English to metric measurements and a chart may not be available. There is a rule that can be used in such circumstances.

RULE 4-2 CONVERTING FROM ENGLISH TO METRIC UNITS

Multiply the English unit by the conversion factor listed in Table 4-10.

Example 4-15 Convert the following from English to metric units: (a) 5 in., (b) $\frac{3}{4}$ in., (c) $2\frac{1}{2}$ in.

Solution (a) 1 in. = 2.54 cm = 25.4 mm. Therefore,

$$5 \text{ in.} = 5 \times 2.54 = 12.70 \text{ cm}$$
$$= 5 \times 25.4 = 127.0 \text{ mm}$$

(b) $\frac{3}{4}$ in. = 0.75 × 2.54 = 1.905 cm

$$= 0.75 \times 25.4 = 19.050 \text{ mm}$$

(c) $2\frac{1}{2}$ in. = 2.5 × 2.54 = 6.350 cm

$$= 2.5 \times 25.4 = 63.50 \text{ mm}$$

Example 4-16   Convert 1.5 mi to kilometers.

Solution   The conversion factor is 1.609. Multiply.

$$
\begin{array}{r}
1.609 \\
\times \quad 1.5 \\
\hline
8045 \\
1609 \\
\hline
2.4135
\end{array}
$$

1.5 mi $=$ 2.4135 km, which is about 2.4 km to two significant digits.

Although the shopworker needs to convert primarily from English to metric units, you can also convert from metric to English units. Rule 4-3 tells how.

RULE 4-3   CONVERTING FROM METRIC TO ENGLISH UNITS

Divide the metric unit by the conversion factor listed in Table 4-10.

Example 4-17   Convert 40 km to miles.

Solution   The conversion factor is 1.609. Divide.

Enter   Display

4 0 ÷ 1 . 6 0 9   24.86016159

Therefore, 40 km is about 25 mi.

Table 4-12 shows the conversion factors for capacity and weight.

METRIC EQUIVALENTS	
Mass (Weight)	**Capacity**
1 oz $=$ 28.35 g 1 lb $=$ 453.6 g $=$ 0.4536 kg 100 lb $=$ 45.36 kg 1 T (2000 lb) $=$ 907.19 kg $=$ 0.907 metric ton	1 (liquid) pt $=$ 0.473 L 1 (liquid) qt $=$ 0.946 L 1 gal $=$ 3.785 L

Table 4-12

Example 4-18 Convert 5 lb to metric measure.

  Solution 1 lb = 453.6 g

  Therefore,

$$5 \text{ lb} = 5 \times 453.6 = 2268 \text{ g}$$
$$= 2.268 \text{ kg.}$$

Example 4-19 How many gallons are there in 70.5 L?

  Solution The conversion factor is 3.785. Divide.

Enter	Display
$\boxed{7}\boxed{0}\boxed{.}\boxed{5}\boxed{\div}\boxed{3}\boxed{.}\boxed{7}\boxed{8}\boxed{5}$	18.62615588

  Thus 70.5 L is about 18.6 gal.

Example 4-20 How many liters are there in 4 gal?

  Solution 1 gal = 3.785 L

  Therefore,

$$4 \text{ gal} = 4 \times 3.785 = 15.14 \text{ L}$$

Exercises 4-3

Convert each measurement to millimeters.

1. 0.029 in.   **2.** 0.3755 in.   **3.** $\frac{13}{32}$ in.

4. 0.581 in.   **5.** $\frac{5}{8}$ in.   **6.** $4\frac{3}{4}$ in.

Convert each measurement to centimeters.

7. $1\frac{3}{8}$ in.   **8.** $4\frac{3}{4}$ in.   **9.** 6 in.

10. 4 in.   **11.** 27 in.   **12.** 39.37 in.

13. $1\frac{1}{2}$ ft   **14.** 2 ft   **15.** $3\frac{3}{4}$ ft

16. $7\frac{1}{4}$ ft   **17.** 8.5 ft   **18.** 9.75 ft

Convert each measurement to meters.

19. $2\frac{1}{2}$ ft   **20.** 3 ft   **21.** 6 ft

22. 7 ft 9 in.   **23.** 14.5 ft   **24.** 5280 ft

Convert each measurement to kilometers.

25. $\frac{3}{4}$ mi **26.** 5 mi. **27.** 320 mi **28.** 55 mi

Convert to grams.

29.	8 oz	**30.**	24 oz	**31.**	3 oz
32.	12 oz	**33.**	5 oz	**34.**	6.5 oz
35.	8 lb	**36.**	4.5 lb	**37.**	20 lb
38.	6.75 lb	**39.**	7 lb	**40.**	11 lb

Convert to kilograms.

41.	200 lb	**42.**	150 lb	**43.**	2500 lb
44.	1500 lb	**45.**	2 lb	**46.**	5000 lb

Convert to liters.

47.	2 gal	**48.**	9 gal	**49.**	10 gal	**50.**	4 qt
51.	15 qt	**52.**	5 qt	**53.**	7 pt	**54.**	12 pt

Convert to the indicated unit.

55. 5.8 m; feet **56.** 3.7 km; miles

57. 825 cm; inches **58.** 7000 m; miles

59. 320 m; yards **60.** 2000 cm; feet

61. 141.75 g; ounces **62.** 760 kg; pounds

63. 2903 kg; pounds **64.** 50 kg; ounces

65. 3.874 L; pints **66.** 15 L; gallons

67. 10.0276 L; quarts **68.** 5 kL; gallons

69. A broken metric bolt that was originally 1.77 cm long must be replaced with a bolt of the same length. How many inches long must the replacement bolt be?

70. The gas tank of a new foreign car holds 50 L. What is the capacity of the gas tank in gallons?

4-4   Working with Measurements Expressed in Two or More Units

OBJECTIVE

After completing this section, you should be able to:

1. Add, subtract, multiply, and divide measurements expressed in two or more units.

While working with measurements expressed in two or more units, such as 11 feet 3 inches or 34 meters 8 centimeters, it is sometimes necessary to add or subtract these measurements. It is fairly easy to add such measurements—simply find the sum of the inches and the sum of the feet, or of the centimeters and of the meters, and then simplify if needed. Study Example 4-21.

Example 4-21

4 m 3 cm	9 lb 8 oz	5 gal 2 pt
8 m 26 cm	15 lb 2 oz	6 gal 1 pt
12 m 29 cm	24 lb 10 oz	11 gal 3 pt

In the following examples, observe that you need to simplify your answers.

Example 4-22   Add 10 yd 2 ft and 18 yd 2 ft.

Solution   10 yd 2 ft
18 yd 2 ft
28 yd 4 ft = 29 yd 1 ft

Remember: 3 ft = 1 yd, so 4 ft = 1 yd + 1 ft.

Example 4-23   The first leg of Kirsten's flight took 2 h 23 min. She had a layover of 18 min in Chicago and then spent 1 h 56 min traveling. How long did the trip take?

Solution   Add.

2 h 23 min
18 min
1 h 56 min
3 h 97 min = 4 h 37 min   (60 min = 1 h)

Her trip took 4 h 37 min.

To subtract one measurement from another, subtract each part. Study the subtractions in Example 4-24.

Example 4-24

10 yd 2 ft 8 in.	31 m 8 dm	14 lb 13 oz
8 yd 1 ft 4 in.	7 m	6 lb 5 oz
2 yd 1 ft 4 in.	24 m 8 dm	8 lb 8 oz

In Example 4-25, you must rename one of the larger units before you can subtract.

Example 4-25   Find the difference between 8 kg 7 hg and 3 kg 9 hg.

Solution $\begin{array}{r} 8 \text{ kg } 7 \text{ hg} \\ 3 \text{ kg } 9 \text{ hg} \end{array} = \begin{array}{r} 7 \text{ kg } 17 \text{ hg} \\ 3 \text{ kg } 9 \text{ hg} \\ \hline 4 \text{ kg } 8 \text{ hg} \end{array}$ (1 kg = 10 hg)

You also may need to multiply or divide measurements expressed in two or more units of measure. Remember to use the rules from Unit 1 to find the appropriate unit of measure for the answer.

Example 4-26   Multiply 8 yd 2 ft by 4.

Solution
$$\begin{array}{r} 8 \text{ yd } 2 \text{ ft} \\ \times \phantom{8 \text{ yd } 2} 4 \\ \hline 32 \text{ yd } 8 \text{ ft} \end{array} = 32 \text{ yd} + 2 \text{ yd } 2 \text{ ft}$$
$$(8 \text{ ft} = 6 \text{ ft} + 2 \text{ ft} = 2 \text{ yd } 2 \text{ ft})$$
$$= 34 \text{ yd } 2 \text{ ft}$$

Example 4-27   A piece of metal 3 ft 7 in. long is to be cut into two equal parts. Find the length of each part.

Solution   Rewrite 3 ft 7 in. as 2 ft 19 in., since there would be a remainder when dividing 3 by 2.

$$3 \text{ ft } 7 \text{ in.} = (2 \text{ ft} + 1 \text{ ft}) + 7 \text{ in.}$$
$$= 2 \text{ ft} + 12 \text{ in.} + 7 \text{ in.} \qquad (1 \text{ ft} = 12 \text{ in.})$$
$$= 2 \text{ ft } 19 \text{ in.}$$

Then divide.

$$\begin{array}{r} 1 \text{ ft } 9\frac{1}{2} \text{ in.} \\ 2\overline{)2 \text{ ft } 19 \text{ in.}} \\ \underline{2 \text{ ft } 18 \text{ in.}} \\ 1 \end{array}$$

Write a fraction with the remainder (1) as the numerator and the divisor (2) as the denominator.

The pieces will be 1 ft $9\frac{1}{2}$ in. long.

Example 4-28   To make 48 parts takes 14 h 18 min of machine time. Three machines are to be used. How much time will be required on each machine?

Solution   14 h 18 min = (12 h + 2 h) + 18 min
                       = 12 h + 120 min + 18 min (1 h = 60 min)
                       = 12 h 138 min

$$\begin{array}{r} 4\text{ h}\quad 46\text{ min} \\ \hline 3\overline{)12\text{ h } 138\text{ min}} \end{array}$$

It will take 4 h 46 min on each machine.

Exercises 4-4

Add.

1.   7 ft 3 in.
     +8 ft 5 in.

2.   1 mi 875 yd
     +2 mi 300 yd

3.   73 km 80 m
     +52 km 71 m

4.   5 yd 2 ft 7 in.
     +3 yd 1 ft 2 in.

5.   2 qt 3 pt
     +5 qt 2 pt

6.   7 g 8 dg
     +9 g 4 dg

7.   6 da 12 h 40 min
     +9 da 15 h 53 min

8.   5 T   905 lb
     +8 T 1763 lb

9.   20 km 5 hm 8 dam
     +13 km 9 hm 7 dam

Subtract.

10.   4 yr 271 da
      −2 yr 165 da

11.   7 yd 8 ft 4 in.
      −3 yd 2 ft 1 in.

12.   12 m 13 cm
      − 8 m   5 cm

13.   10 mi
      − 5 mi 628 yd

14.   28 da
      −17 da 17 h

15.   2 lb   5 oz
      −   15 oz

16.   8 bu
      −4 bu 2 pk

17.   11 gal 1 qt
      − 5 gal 3 qt

18.   700 kg
      −340 kg 7 hg

Add or subtract.

19.   14 cm 2 mm
      − 6 cm 8 mm

20.   15 yd 2 ft   8 in.
      +11 yd   17 in.

Multiply.

21.   3 yd 2 ft
      ×   6

22.   1 yd 1 ft 17 in.
      ×   8

23.   28 mi 385 yd
$\times$ _____ 7

24.   5 h 18 min 37 s
$\times$ _____ 2

25.   8 m $\times$ 10 m

26.   24 ft $\times$ 17 ft

27.   9 yd $\times$ 7 yd $\times$ 17 yd

28.   14 cm $\times$ 6 cm $\times$ 18 cm

Divide.

29.   (2 h 26 min) $\div$ 2

30.   (15 cm 8 mm) $\div$ 4

31.   (18 yd 1 ft) $\div$ 5

32.   (11 ft 8 in.) $\div$ 4

33.   (27 L 36 dL) $\div$ 9

34.   (176 kg 480 g) $\div$ 10

35.   Three machine parts weigh 2 lb 7 oz, 1 lb 3 oz, and 4 lb 9 oz. Find the total weight of the three parts.

36.   A shop machine was used 4 da 7 h 18 min one week and 3 da 5 h 52 min the following week. How many days, hours, and minutes was the machine used over the two-week period?

37.   A block 123 cm 3 mm long is placed on top of a second block 46 cm 7 mm long. What is the total height of the two blocks?

38.   If the milling on a part requires 1 h 12 min, how long will it take to do the milling on three of these parts?

39.   Three machine parts weigh 3 cg 12 dg, 142 dg, and 5 cg 78 dg. Find the total weight of seven of each of the three parts.

40.   Find the cutting speed in feet per minute of a boring mill that cuts 123 ft of cast iron in $2\frac{1}{4}$ min.

41.   The shop instructor asked Kiko to prepare blanks for four students from a piece of leftover stock 1 m 20 cm 9 mm long. If Kiko used up all the stock and made each blank the same size, how long would each blank be?

42.   A blueprint calls for six equally spaced holes to be drilled in a metal plate 1 ft 9 in. long (see Figure A). How far apart must the centers of the holes be if the center of the first hole and the last hole are the same distance from the ends of the plate? (*Hint:* Into how many equal spaces will the six holes divide the plate?)

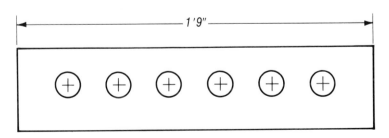

Figure A

Unit 4 Review Exercises

1. How many inches are in the following measurements: (a) 12 ft, (b) 4 ft 8 in., (c) $\frac{3}{5}$ yd?

2. How many quart bottles can be filled from a 5-gal jug?

3. A machine can produce three parts every $4\frac{1}{4}$ min. How many parts can be produced in an 8-h day?

4. A coil of telephone wire stretches $4\frac{1}{2}$ mi. How far is that in yards?

5. How many 7-cm lengths are in a steel bar 130 cm long?

6. Convert the following English units to metric units: (a) 6 in., (b) $\frac{7}{8}$ in., (c) $3\frac{5}{8}$ in.

7. The gas station attendant pumped 15.4 L into the gas tank of a customer's car. How many gallons did the attendant pump?

8. It takes Ernest 2 h 43 min to machine a single part for an aircraft turbine assembly. How long will it take him to machine 25 such parts?

9. A typical steel I-beam weighs 42 lb $14\frac{1}{2}$ oz per foot. How much would a 4-in. section weigh?

10. A machinist working on a lathe cuts 36 mm off the end of an aluminum bar 180 cm long. The remainder is to be cut into eight pieces of equal length. How long will each piece be?

Equations, Formulas, and Square Root

A formula is a rule expressed as an equation. Many machine shop problems can be solved by using formulas. They are essential in such operations as designing gear trains and calculating feeds and speeds.

Finding the square root of a number is a useful tool that enables a machinist, sheet metal worker, or toolmaker to solve layout problems in jobs such as metal forming, jig boring, and fixture and jig construction. In finding the length of the side of a square or the diameter of a circle, it may be necessary to find the square root of a number.

This unit will show you how to solve problems by using formulas and equations and how to find square roots.

5-1 Properties of Equations

OBJECTIVES

After completing this section, you should be able to:

1. Use the properties of equations.
2. Solve simple equations.

An **equation** is a statement that two quantities are equal. The following are all equations.

$$C = r^2$$

$$x + 7 = 12$$

$$x^2 = 9$$

The equations covered in this book will be limited to **linear equations** such as $x + 7 = 12$ or $5x - 9 = 16$. In linear equations, the **variable,** or quantity to be found, is to the first degree or power. Any letter can be used as a variable. When a number and a variable, or two variables, are written together, multiplication is understood. Thus $5x$ means 5 times x.

A **solution** is a number that makes an equation true. The solution to $x + 7 = 12$ is 5, because $5 + 7 = 12$ is a true statement. On the other hand, 3 is not a solution, because $3 + 7 = 12$ is not a true statement. Equations with the same solution are called **equivalent equations.** The following are all equivalent equations.

$$x + 7 = 12$$

$$3x = 15$$

$$x = 5$$

Finding the solution to an equation is called **solving the equation.** It involves finding the number, or quantity, that can replace the variable to make the equation true. There are four basic rules, or properties of equations, that can be used to solve an equation.

PROPERTY 1 The same quantity can be added to both members of an equation to produce an equivalent equation.

Example 5-1 Solve $x - 8 = 5$.

$$\begin{aligned}
\textbf{Solution} \quad x - 8 &= 5 \\
x - 8 + 8 &= 5 + 8 \qquad \text{Property 1} \\
x &= 13
\end{aligned}$$

$$\begin{aligned}
\textbf{Check} \quad 13 - 8 &\overset{?}{=} 5 \\
5 &= 5
\end{aligned}$$

PROPERTY 2 The same quantity can be subtracted from both members of an equation to produce an equivalent equation.

Example 5-2 Solve $26 = 18 + a$.

$$\begin{aligned}
\textbf{Solution} \quad 26 &= 18 + a \\
26 - 18 &= 18 + a - 18 \qquad \text{Property 2} \\
8 &= a
\end{aligned}$$

$$\begin{aligned}
\textbf{Check} \quad 26 &\overset{?}{=} 18 + 8 \\
26 &= 26
\end{aligned}$$

PROPERTY 3 Both members of an equation can be multiplied by the same nonzero quantity to produce an equivalent equation.

Example 5-3   Solve $\frac{z}{5} = 10$.

Solution $\dfrac{z}{5} = 10$

$5 \cdot \dfrac{z}{5} = 5 \cdot 10$ Property 3

$z = 50$

Check $\dfrac{50}{5} \overset{?}{=} 10$

$10 = 10$

PROPERTY 4   Both members of an equation can be divided by the same nonzero number to produce an equivalent equation.

Example 5-4   Solve $2.4x = 3.84$.

Solution $2.4x = 3.84$

$\dfrac{2.4x}{2.4} = \dfrac{3.84}{2.4}$ Property 4

$x = 1.6$

Check $(2.4)(1.6) \overset{?}{=} 3.84$

$3.84 = 3.84$

Sometimes it is necessary to use more than one property to solve an equation. Always use Property 1 or Property 2 first.

Example 5-5   Solve $3.1z + 2.5 = 4.36$.

Solution $3.1z + 2.5 = 4.36$

$3.1z + 2.5 - 2.5 = 4.36 - 2.5$ Property 2

$3.1z = 1.86$

$\dfrac{3.1z}{3.1} = \dfrac{1.86}{3.1}$ Property 4

$z = 0.6$

Check $3.1(0.6) + 2.5 \overset{?}{=} 4.36$

$1.86 + 2.5 \overset{?}{=} 4.36$

$4.36 = 4.36$

Exercises 5-1

Solve each equation and check.

1. $x + 18 = 29$

2. $\dfrac{1}{2} + x = \dfrac{5}{8}$

3. $6.9 + y = 43.76$

4. $a - 12 = 98$

5. $z - \dfrac{5}{6} = 12\dfrac{3}{8}$

6. $x - 12.7 = 20.635$

7. $\dfrac{w}{14} = 58$

8. $\dfrac{1}{3}x = 15.3$

9. $\dfrac{z}{9.1} = 7.62$

10. $5m = 73$

11. $4\dfrac{4}{7}t = 24$

12. $9.5x = 126.54$

13. $36x - 5 = 139$

14. $5 + 2x = 17$

15. $3.7a - 5 = 5.36$

16. $\dfrac{a}{15} - 7 = 21$

17. $2.56 + 7.4a = 212.78$

18. $4\dfrac{1}{2} = \dfrac{t}{7} + 2$

5-2   Solving Equations

OBJECTIVE

After completing this section, you should be able to:

1. Solve linear equations with one variable.

One very useful property used to solve equations and formulas is the distributive property of multiplication over addition.

RULE 5-1   DISTRIBUTIVE PROPERTY OF MULTIPLICATION OVER ADDITION

$a(b + c) = ab + ac$

for all numbers *a*, *b*, and *c*.

The distributive property also works for subtraction.

$a(b - c) = ab - ac$

Example 5-6 $2(3 + 5) = 6 + 10 = 16$ and $2(3 + 5) = 2(8) = 16$

$5(x - 2) = 5x - 10$

$7(4 + a) = 28 + 7a$

$5x + 3x = (5 + 3)x = 8x$

Like terms are terms such as $3x$ and $5x$, in which the variable is the same. As shown in the examples above, like terms can be combined using the distributive property.

RULE 5-2   SOLVING LINEAR EQUATIONS

Step 1   If there are any fractions, multiply both sides by the L.C.D.

Step 2   Remove all parentheses by using the distributive property.

Step 3   Combine like terms on each side of the equation, using the distributive property.

Step 4   Use properties 1 and 2 as many times as necessary to combine the terms containing the variable if they occur on both sides of the equation.

Step 5   Use properties 1–4 to complete the solution.

Example 5-7   Solve $3(x + 1) = 11 + x$.

Solution
$$3(x + 1) = 11 + x$$
$$3x + 3 = 11 + x \qquad \text{Step 2}$$
$$3x - x + 3 = 11 + x - x \qquad \text{Step 4 (Property 2)}$$
$$2x + 3 = 11$$
$$2x = 8 \qquad \text{Step 5 (Property 2)}$$
$$x = 4 \qquad \text{Step 5 (Property 4)}$$

Check
$$3(4 + 1) \stackrel{?}{=} 11 + 4$$
$$3(5) \stackrel{?}{=} 15$$
$$15 = 15$$

Example 5-8   Solve $\frac{1}{2}(2 - x) = \frac{3}{4}$.

Solution
$$\frac{1}{2}(2 - x) = \frac{3}{4}$$
$$4\left[\frac{1}{2}(2 - x)\right] = 4 \cdot \frac{3}{4} \qquad \text{Step 1 (L.C.D. = 4)}$$
$$2(2 - x) = 3$$

$$4 - 2x = 3 \qquad \text{Step 2}$$
$$4 - 2x + 2x = 3 + 2x \qquad \text{Step 5 (Property 1)}$$
$$4 = 3 + 2x$$
$$1 = 2x \qquad \text{Step 5 (Property 2)}$$
$$\frac{1}{2} = x \qquad \text{Step 5 (Property 4)}$$

Check $\quad \frac{1}{2}\left(2 - \frac{1}{2}\right) \overset{?}{=} \frac{3}{4}$

$$\frac{1}{2}\left(1\frac{1}{2}\right) \overset{?}{=} \frac{3}{4}$$

$$\frac{1}{2}\left(\frac{3}{2}\right) \overset{?}{=} \frac{3}{4}$$

$$\frac{3}{4} = \frac{3}{4}$$

Example 5-9   Solve $5(x + 3.5) - 2.6 = 21.9$.

Solution $\quad 5(x + 3.5) - 2.6 = 21.9$

$$5x + 17.5 - 2.6 = 21.9 \qquad \text{Step 2}$$
$$5x + 14.9 = 21.9 \qquad \text{Step 3}$$
$$5x = 7 \qquad \text{Step 5 (Property 2)}$$
$$x = 1.4 \qquad \text{Step 5 (Property 4)}$$

Check $\quad 5(1.4 + 3.5) - 2.6 \overset{?}{=} 21.9$

$$5(4.9) - 2.6 \overset{?}{=} 21.9$$
$$24.5 - 2.6 \overset{?}{=} 21.9$$

Exercises 5-2

Solve and check.

1. $2(x - 4) = 2$
2. $5 + 4x - 1 = 16$
3. $7y - 18 = 2y + 17$
4. $5 - 3a = a + 1$
5. $2x - 9 = 3x - 21$
6. $\dfrac{z}{25} = 10$
7. $\dfrac{a}{5} - 13 = 4$
8. $3(x - 9) + 2 = 16$
9. $2.1x - 1.73 = 15.91$
10. $1.7 = \dfrac{x}{2.4}$

11. $\frac{3}{5}a + 3 = 9$ **12.** $\frac{c}{3} - \frac{c}{7} = 16$

13. $\frac{5}{6}(c - 2) = 7$ **14.** $2.5 = 0.4c - 6.82$

15. $5.8(x + 1) - 3x = 16.216$ **16.** $\frac{1}{10} = \frac{x - 4}{2} - \frac{1}{5}$

17. $6x + 2(3x - 9) = 3(x - 3)$ **18.** $2(3x + 5) = 3(x - 1) + 14$

19. $2(z + 7.6) + 18.09 = 3(z - 5.31)$

20. $\frac{1}{3}(x - 5) + \frac{3}{5}(2x + 1) = x + 2\frac{2}{3}$

5-3 Working with Formulas

OBJECTIVE
After completing this section, you should be able to:

1. Use a formula to find a specified value.

Many machine shop problems are solved by formulas. A formula is an equation that gives a rule in a mathematical shorthand language. It substitutes letters for values and is very useful for simplifying a long, wordy rule.

Consider the following rule: *The circumference of a circle equals pi multiplied by the diameter of the circle.* As a formula, this rule can be written $C = \pi d$, where C represents the circumference, d represents the diameter, and π(pi) is a constant about equal to 3.1416.

Figure 5-1 shows a square, which is a plane figure with four equal sides and four $90°$ angles. Assume that you are interested in the perimeter of this figure. The perimeter P of any figure is equal to the sum of its sides. For a square, this relationship can be stated as

$$P = s + s + s + s$$

or

$$P = 4s$$

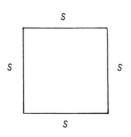

Figure 5-1

where s is the length of one side. Therefore, in Figure 5-1, if $a = 2$ inches, $P = 4(2) = 8$ inches.

Example 5-10   Shown here is an equilateral triangle. Since all its sides are equal by definition, its perimeter is equal to $3s$. For $s = 2.6$ cm, find the perimeter.

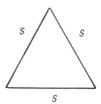

Solution $P = 3s$

$P = 3(2.6)$

$\quad = 7.8$

The perimeter is 7.8 cm.

Example 5-11   To find the area of a square, use the formula $A = s^2$. Recall that s^2 means $s \times s$. Find the area of a square if the length of a side is $2\frac{3}{4}$ ft.

Solution $A = s^2$

$$A = \left(2\frac{3}{4}\right)^2 = \left(2\frac{3}{4}\right) \times \left(2\frac{3}{4}\right)$$

$$= \frac{11}{4} \times \frac{11}{4}$$

$$= \frac{121}{16}$$

$$= 7\frac{9}{16}$$

The area is $7\frac{9}{16}$ sq ft.

Example 5-12   The formula rpm $= 4C.S./D$ is used to find revolutions per minute (rpm) of a lathe or milling machine spindle. Here $C.S.$ stands for cutting speed. Find rpm if $C.S. = 75$ and $D = 6\frac{1}{8}$.

Solution $\text{rpm} = \dfrac{4C.S.}{D}$

$$\text{rpm} = \frac{4(75)}{6\frac{1}{8}} = 300 \div 6\frac{1}{8}$$

$$= 300 \div \frac{49}{8} = 300 \times \frac{8}{49}$$

$$= \frac{2400}{49} = 48\frac{48}{49}, \text{ or about } 49$$

Exercises 5-3

Use the formula P = 4s to find the perimeter P for each value of s.

1. $s = 5$ **2.** $s = 1\dfrac{5}{8}$ **3.** $s = 2\dfrac{9}{16}$

4. $s = \dfrac{25}{32}$ **5.** $s = 0.0875$ **6.** $s = 1.75$

7. $s = \dfrac{9}{64}$ **8.** $s = \dfrac{15}{16}$

Use the formula A = s² to find the area A for each value of s.

9. $s = 10$ **10.** $s = \dfrac{1}{2}$ **11.** $s = \dfrac{5}{8}$

12. $s = 2\dfrac{1}{2}$ **13.** $s = 3\dfrac{9}{16}$ **14.** $s = 1.125$

15. $s = 2.375$ **16.** $s = 0.875$

The formula for finding the area of a circle is A = πD²/4. The Greek letter π (pi) represents a constant value of about 3.1416. The most direct way to solve this formula is to find the product of π and D² and then divide their product by 4. Find A for each value of D.

17. $D = 2$ **18.** $D = 5$ **19.** $D = 1\dfrac{1}{2}$ **20.** $D = 3.5$

The formula for the perimeter of a rectangle is $P = 2(l + w)$. *Find the perimeter P for each given length l and width w.*

21. $l = 2\frac{1}{2}; w = 4$ **22.** $l = 3\frac{1}{4}; w = 3$

23. $l = 11; w = 9$ **24.** $l = 13; w = 10$

25. $l = 5.625; w = 7.250$

The pitch of a screw thread is the distance from a point on one thread to the same point on the next thread. This distance is found by the formula $P = 1/N$, *where N is the number of threads in every inch. Find each value for P to three decimal places.*

26. $N = 8$ **27.** $N = 24$ **28.** $N = 4\frac{1}{2}$ **29.** $N = 18$

30. $N = 16$

Use the formula rpm $= 4C.S./D$ *to find each rpm.*

31. $C.S. = 150; D = 3$ **32.** $C.S. = 200; D = 0.500$

33. $C.S. = 50; D = 1$ **34.** $C.S. = 600; D = 2$

5-4　Changing the Subject in a Formula

OBJECTIVE

After completing this section, you should be able to:

1. Solve a formula for a specified variable.

If the length of a side s of a square is known, the formula $P = 4s$ can be used to find the perimeter P. In this formula, the **subject** of the formula is P, the variable standing alone. The formula is said to be solved for P.

The formula can also be used to find s, given P. Suppose the perimeter of a square is 21.6 centimeters.

$$P = 4s$$

$$21.6 = 4s$$

$$\frac{21.6}{4} = s$$

$$s = 5.4$$

The length of a side is 5.4 centimeters.

In many cases, it is easier to solve the formula for the variable you wish to find *before* substituting given values into the formula. This is called *changing the subject in a formula.*

RULE 5-3   SOLVING A FORMULA FOR A SPECIFIED VARIABLE

The properties for solving equations can be used to solve formulas for a specified variable.

Example 5-13   Solve $P = 4s$ for s.

Solution $P = 4s$

$\dfrac{P}{4} = s$ Property 4

Example 5-14   Solve $c^2 = a^2 + b^2$ for b^2.

Solution $c^2 = a^2 + b^2$

$c^2 - a^2 = a^2 - a^2 + b^2$ Property 2

$c^2 - a^2 = b^2$

Example 5-15   Solve $A = 2l + 2w$ for w.

Solution $A = 2l + 2w$

$A - 2l = 2w$ Property 2

$\dfrac{A - 2l}{2} = w$ Property 4

Example 5-16   The formula

$$O.D. = \frac{N + 2}{D.P.}$$

is used to find the outside diameter $O.D.$ of a gear when the number of teeth N and diametral pitch $D.P.$ are known. Find the number of teeth if $O.D. = 5$ in. and $D.P. = 12$.

Solution   First solve the equation for N.

$$O.D. = \frac{N + 2}{D.P.}$$

$$(O.D.)(D.P.) = N + 2$$

$$(O.D.)(D.P.) - 2 = N + 2 - 2$$

$$(O.D.)(D.P.) - 2 = N$$

Now substitute the given values in the equation.

$$N = (O.D.)(D.P.) - 2$$
$$= 5(12) - 2$$
$$= 60 - 2 = 58$$

The number of teeth is 58.

Exercises 5-4

1. Angle A plus angle B plus angle C equals $180°$. Solve for angles A and B in terms of the other angles.

2. Use the formula in exercise 1 to find angle A if angle $B = 50°$ and angle $C = 70°$.

3. $M = D - 1.5155p + 3G$ is a formula to find the measurement M over the wires in the three-wire method of measuring screw threads. Find D in terms of M, G, and p.

4. Use the formula in exercise 3 to find D, the thread diameter, if $G = 0.0321$ in., $p = \frac{1}{18}$ in., and $M = 0.3246$ in.

5. The formula $d = D - 1/N$ is a threading formula. Find D in terms of the other variables.

6. The formula in exercise 5 is used to determine the size of a hole into which threads will be tapped. If $N = 10$ threads per inch and $d = 0.65$ in., what is D, the outside diameter of the threading tool?

7. The formula $W = 2a + c$ is used to find the whole depth W of a spur gear tooth. Find c in terms of W and a.

8. Use the formula of exercise 7. If $W = 0.2157$ in. and $a = 0.1$ in., what is the value of c, the clearance?

9. In trigonometry, $\cos A = b/c$ is a formula used to solve for one angle of a right triangle. Solve the formula for b.

10. In the formula of exercise 9, if $\cos A = 0.866$ and $c = 6$ in., what is the value of b?

11. A formula for finding the number of teeth T in one of two mating gears is $T = ts/S$. Solve the formula for S.

12. Use the formula of exercise 11. If $t = 20$ teeth, $T = 30$ teeth, and $s = 150$ rpm, what is the rpm S of the other gear?

13. The formula for the area of triangle is $A = ab/2$. Solve this formula for a.

14. In Exercise 13, if $A = 1665$ sq mm and $b = 406.4$ mm, find the altitude a of the triangle.

15. The double depth D of a metric thread is found by the formula $D = P(1.299)/2$. Solve for pitch P.

16. Use the formula of exercise 15 to find the depth D of a 6.5-mm-pitch thread.

17. The area of a trapezoid is found by using the formula $A = a(b + b')/2$. Solve for a.

18. Use the formula of exercise 17. If $A = 1371.6$ sq mm, $b = 1219.2$ mm, and b′ $= 609.6$ mm, what is the value of the altitude a?

5-5   Square Root

OBJECTIVES
After completing this section, you should be able to:
1. Find the square root of a number by factoring.
2. Find the square root of a number by using a table of square roots.
3. Find the square root of a number by using the property of square roots.
4. Find the square root of a number by using a calculator.

Recall that *factors* of a number are numbers that, when multiplied together, yield the original number. The **square root** of a number is one of *two equal factors* whose product is the original number. A **perfect square** has a square root that is a whole number.

Example 5-17   4, 16, 9, and 144 are perfect squares.

$$2 \times 2 = 4, \quad 4 \times 4 = 16, \quad 3 \times 3 = 9,$$
$$12 \times 12 = 144$$

Not all numbers, however, are perfect squares. This will be evident as you learn to find square roots.

The **radical sign** signifies that a root of a number is to be extracted. It is written as $\sqrt{}$. Attached to this radical sign is a **vinculum**, which means "roof". Now the root sign looks like this: $\sqrt{}$. Finally, an **index**, or small number, is placed in the opening of the radical sign to indicate what root is to be taken. For example, $\sqrt[3]{27}$ is the cube root of 27. The fourth root of 625 would be written $\sqrt[4]{625}$. The index of the root is omitted, however, when the square root is indicated. Thus the square root of 9 is written $\sqrt{9}$ and not $\sqrt[2]{9}$.

It can be seen that $\sqrt{16} = 4$ because $4 \cdot 4 = 16$. Also, $\sqrt{4} = 2$, $\sqrt{9} = 3$, and $\sqrt{144} = 12$. If the number whose square root is desired is not a perfect square, a table can be used to find its square root. Using Table V in the appendix, you can see that $\sqrt{27} = 5.196$ and $\sqrt{90} = 9.487$. (Actually, these values are approximations rounded to three decimal places.)

Example 5-18   Find $\sqrt{38}$ using the table of square roots in the appendix.

   Solution $\sqrt{38} = 6.164$

If the number whose square root is required can be factored as a perfect square times another number such as 2, 3, 5, or 6, a short method of extracting the square root can be used. To apply this method, however, you need to use the property that the square root of a product is the product of the square roots.

RULE 5-4   PROPERTY OF SQUARE ROOTS

$\sqrt{ab} = \sqrt{a}\sqrt{b}$

where a and b are greater than zero.

In the following examples, the square roots of 2, 3, 5, and 6 will be used.

$\sqrt{2} = 1.414$, $\sqrt{5} = 2.236$,

$\sqrt{3} = 1.732$, $\sqrt{6} = 2.449$

Example 5-19   Find $\sqrt{50}$.

   Solution   The number 50 can be factored as 25×2. Using Rule 5-4, $\sqrt{50} = \sqrt{25 \times 2} = \sqrt{25}\sqrt{2}$. Thus

$$\sqrt{50} = \sqrt{25 \times 2} = \sqrt{25}\sqrt{2} = 5\sqrt{2} = 5 \times 1.414$$
$$= 7.07$$

Example 5-20   Find $\sqrt{75}$.

   Solution $\sqrt{75} = \sqrt{25 \times 3} = \sqrt{25}\sqrt{3} = 5\sqrt{3} = 5 \times 1.732$
   $$= 8.66$$

A calculator can also be used to find the square root of a number.

Example 5-21   Use a calculator to find $\sqrt{59}$.

Solution   Enter   Display

            [5] [9]   59

              [√x̄]   7.681145748

Round the solution to three decimal places.

$\sqrt{59} = 7.681$

The ability to find the square root can be used in solving many practical shop problems. If the length of two sides of a right triangle are known, the length of the third side can be found. A right triangle is a triangle with a 90° angle. Figure 5-2 shows three triangles. The one labeled B is a right triangle. The others are not.

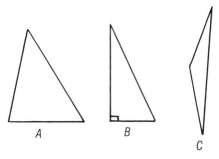

A   B

C

Figure 5-2

The sides of a right triangle are given names for identification. Figure 5-3 shows these names and the letters that are commonly substituted for them in formulas. To solve for the unknown side of a right triangle, use the following rule, known as the Pythagorean theorem.

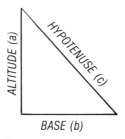

Figure 5-3 Right triangle.

RULE 5-5   PYTHAGOREAN THEOREM

The square of the hypotenuse is equal to the sum of the squares of the other two sides.

$$c^2 = a^2 + b^2$$

Figure 5-4 illustrates this rule.
Variations of the basic formula are

$$a^2 = c^2 - b^2 \quad \text{and} \quad b^2 = c^2 - a^2$$

These formulas give the squares of the sides. A machinist or metalworker, however, is more interested in finding the sides of a triangle than in finding the squares of the sides. To find the sides, take the square root of both sides of each formula.

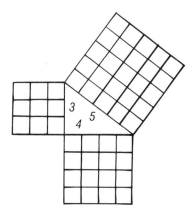

Figure 5-4

$$c^2 = a^2 + b^2 \quad \text{becomes} \quad c = \sqrt{a^2 + b^2}$$
$$a^2 = c^2 - b^2 \quad \text{becomes} \quad a = \sqrt{c^2 - b^2}$$
$$b^2 = c^2 - a^2 \quad \text{becomes} \quad b = \sqrt{c^2 - a^2}$$

(Note that $\sqrt{c^2} = c$, because $c \times c = c^2$, but the square root of $a^2 + b^2$ does not equal $a + b$, because $(a + b) \times (a + b)$ does not equal $a^2 + b^2$.)

All that remains to solving for one of the sides of the triangle is to substitute known values for the other two sides, using one of the three formulas given. Then solve the resulting equation to find the unknown side of the right triangle.

Example 5-22   Given the values shown, solve for c.

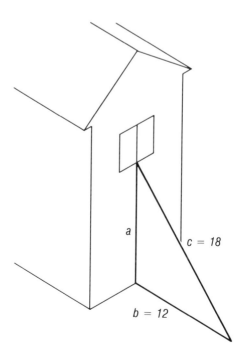

Solution $c = \sqrt{a^2 + b^2} = \sqrt{36 + 64} = \sqrt{100} = 10$

Example 5-23   Find b if $a = 7$ and $c = 11$.

Solution $b = \sqrt{c^2 - a^2} = \sqrt{121 - 49} = \sqrt{72} = 8.485$

Example 5-24   An 18-ft ladder is leaning against a house; the base of the ladder is 12 ft from the house. Will the ladder reach a window that is 15 ft from the ground?

Solution   Draw a picture, as shown here. Find a when $b = 12$ and $c = 18$.

$$a = \sqrt{c^2 - b^2} = \sqrt{324 - 144} = \sqrt{180} = 13.416$$

To two significant digits $a = 13$. The ladder will not reach the window.

Exercises 5-5

Give the square root of each perfect square.

| 1. 121 | 2. 625 | 3. 81 | 4. 196 | 5. 484 |
| 6. 900 | 7. 361 | 8. 2500 |

Find the square root of each number using Table V.

9. 52	10. 13	11. 28	12. 96	13. 98
14. 51	15. 63	16. 77	17. 85	18. 10
19. 19	20. 37

Find the square root of each number by factoring.

21. 8	22. 32	23. 27	24. 108	25. 72
26. 180	27. 54	28. 243	29. 200	30. 300
31. 486	32. 245	33. 320	34. 288	35. 363

36. The horizontal distance X between two holes must be moved by the table screw of a jig bore. Give the distance that must be traveled to the third decimal place. This problem is illustrated in Figure A.

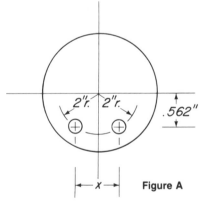

Figure A

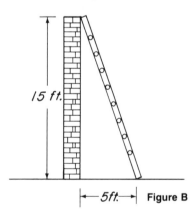

Figure B

37. A ladder is leaning against a wall as shown in Figure B. How long is the ladder?

38. The length of a shadow cast by a 50-ft-high pole is 30 ft measured from the base of the pole. How long a wire would be needed to stretch from the tip of the shadow to the top of the pole?

39. If the vertical length of a television picture tube is 14 in. and the length of a diagonal across its corners is 22.80 in., which of the following would come closest to being the horizontal length?

(a) 21 in.   (b) 18 in.   (c) 24 in.

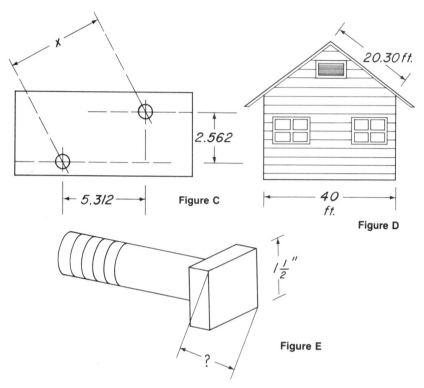

Figure C

Figure D

Figure E

40. Figure C illustrates a jig bore layout in toolmaking. If the two holes are bored in correct relationship to each other, that is, to the correct dimensions, what should be the distance X between the holes?

41. An end view of a house roof is shown in Figure D. How high is the roof?

42. What is the distance across the corners of a square-headed bolt that measures $1\frac{1}{2}$ in. across its flats? (See Figure E.)

5-6   Square Root Algorithm (Optional)

OBJECTIVE
After completing this section, you should be able to:

1. Find the square root of a number by using the square root algorithm.

The square root algorithm is a step-by-step procedure for finding the square root of a number. The following examples show how this procedure works. Notice that the basic arithmetic consists of a simple division, except for four additional steps.

RULE 5-6   SQUARE ROOT ALGORITHM

Step 1   Separate the number into groups of two, called periods, beginning at the right.

Step 2   Find the nearest perfect square of the first period that is less than the period.

Step 3   Bring down one period after each division.

Step 4   Double the quotient for each new division.

Example 5-25   Find $\sqrt{178929}$.

Solution

Step 1   Separate the number 178929 into periods of two figures each, starting at the right. This gives three periods, 17, 89, and 29.

$$\sqrt{17'89'29}$$

Step 2   Find the greatest perfect square that is less than the left-hand period, 17. The greatest perfect square less than 17 is 16, and the square root of 16 is 4. Place 16 under 17 and 4 in the answer.

$$\begin{array}{r} 4 \\ \sqrt{17'89'29} \\ 16 \end{array}$$

Step 3   Subtract 16 from 17, obtaining 1. Bring down the next period, 89, to obtain 189.

$$\begin{array}{r} 4 \\ \sqrt{17'89'29} \\ 16 \\ \overline{1\ 89} \end{array}$$

Step 4   Double the root already found and obtain the trial divisor, 8. Divide the trial divisor, 8, into the first two digits of 189; that is, divide 8 into 18. The quotient is 2 with a remainder. This 2 is the next figure in the answer. Place this 2 after 8, the trial divisor, and obtain 82, the complete divisor. Multiply the complete divisor by 2 to obtain 164, which is placed under 189.

$$\begin{array}{r} 4\ \ 2 \\ \sqrt{17'89'29} \\ 16 \\ 82 \quad \overline{1\ 89} \\ 1\ 64 \end{array}$$

Step 3   Subtract 164 from 189, obtaining 25. Bring down the last period, 29, to obtain 2529.

$$
\begin{array}{r}
\phantom{\sqrt{17'89'29}}4\ \ 2 \\
\sqrt{17'89'29} \\
16 \\
\hline
82\quad\ \ 1\ 89 \\
1\ 64 \\
\hline
25\ 29 \\
\end{array}
$$

Step 4   Double the root already found (that is, double the quotient 42) to obtain the trial divisor, 84. Divide the trial divisor, 84, into the first three digits of 2529; that is, divide 84 into 252. The quotient is 3, which is the last figure in the answer. Place 3 after 84, the trial divisor, and obtain 843, the complete divisor. Multiply the complete divisor by 3 and obtain 2529, which is placed under 2529.

$$
\begin{array}{r}
4\ \ 2\ \ 3 \\
\sqrt{17'89'29} \\
16 \\
\hline
82\quad\ \ 1\ 89 \\
1\ 64 \\
\hline
25\ 29 \\
843\quad\ 25\ 29 \\
\hline
\end{array}
$$

The square root of 178,929 is 423. The work can be checked by multiplying 423 by 423. It will be the perfect square 178,929.

Example 5-26   Find $\sqrt{625}$.

Solution

Step 1   Separate the number 625 into periods of two figures each, starting at the right. Since 625 contains three digits, it is separated as 6'25.

$$\sqrt{6'25}$$

Step 2   Find the greatest perfect square that is less than the left-hand period, 6. The greatest perfect square less than 6 is 4, and the square root of 4 is 2. Place 2 in the answer and 4 under 6.

$$
\begin{array}{r}
2 \\
\sqrt{6'25} \\
4 \\
\end{array}
$$

Step 3   Subtract 4 from 6, obtaining 2. Bring down the next period, 25, to obtain 225.

$$\begin{array}{r} 2 \\ \sqrt{6'25} \\ 4 \\ \hline 2\ 25 \end{array}$$

Step 4   Double the root already obtained and obtain the trial divisor, 4. Divide the trial divisor, 4, into the first two digits of 225; that is, divide 4 into 22. The quotient is 5 with a remainder. Place 5 in the answer. Place 5 after 4, the trial divisor, and obtain 45, the complete divisor. Multiply the complete divisor by 5 to obtain 225, which is placed under 225.

$$\begin{array}{r} 2\ \ 5 \\ \sqrt{6'25} \\ 4 \\ \hline 2\ 25 \\ 45 \quad\ \ 2\ 25 \end{array}$$

The square root of 625 is 25.

In the examples given so far, the square roots of whole numbers have been found. If the number is a decimal number greater than one, the periods must be marked off by beginning at the decimal point and marking off periods *in both directions*.

Example 5-27   Find $\sqrt{265.69}$.

Solution

$$\begin{array}{r} 1\ \ 6.\ \ 3 \\ \sqrt{2'65.'69} \\ 1 \\ \hline 1\ 65 \\ 26 \quad\ \ 1\ 56 \\ \hline 9\ 69 \\ 323 \quad\ \ 9\ 69 \end{array}$$

Recall that the first trial divisor is divided into 16 to obtain the complete divisor. Note that 2 divides 16 exactly 8 times. If 8 were used to obtain the complete divisor, the complete divisor would be 28, which is too great, since $28 \times 8 = 224$, which is greater than 165. Similarly, $27 \times 7 = 189$, also greater than 165. But 6×26 is less than 165, so 6 should be used to complete the divisor. The decimal point in the answer is placed directly above the decimal point in 265.69.

Sometimes it is necessary to find the square root of a number that is *not* a perfect square. The answer in such a case will not come out as a whole number, but will have a remainder. The number of decimal places to which the answer is to be carried determines how many pairs of periods must be added to the number after the decimal point. Zeros are used where needed to make up complete pairs.

Example 5-28   Find the square root of 56 to two decimal places.

Solution

$$
\begin{array}{r}
7 \,.\, 4 \quad 8 \quad 3 \\
\sqrt{56'.\ 00'\ 00'\ 00} \\
49 \\
\hline
7 \quad 00 \\
144 \qquad 5 \quad 76 \\
\hline
1 \quad 24 \quad 00 \\
1488 \qquad 1 \quad 19 \quad 04 \\
\hline
4 \quad 96 \quad 00 \\
14963 \qquad 4 \quad 48 \quad 89 \\
\hline
47 \quad 11
\end{array}
$$

Since you are required to obtain the square root of 56 to two decimal places, three periods of zeros are annexed to the right of 56. Observe that the answer is still not exact.

$\sqrt{56} = 7.48$, to two decimal places.

The least confusing method of handling fractions is to reduce them to decimal equivalents and take the square root of the resulting number.

Example 5-29   Find $\sqrt{\frac{1}{4}}$.

Solution   Write $\frac{1}{4}$ as 0.25 and find $\sqrt{0.25}$.

$$
\begin{array}{r}
0.\ 5 \\
\sqrt{0.25} \\
25 \\
\hline
\end{array}
$$

Exercises 5-6

Find the square root of each of the following numbers.

1. 25   **2.** 225   **3.** 1,296   **4.** 49,729   **5.** 2,704

6. 97,344   **7.** 5,929   **8.** 256   **9.** 2,274,064   **10.** 9,042,049

Find the square root of each of the following. Round your answer to two decimal places.

11. 2.25   **12.** 17.64   **13.** 70.56   **14.** 125.44

15. 2.89   **16.** 10.89   **17.** 7832.25   **18.** 3124.81

19. 27   **20.** 42   **21.** 125   **22.** 3,460

23. $\dfrac{1}{8}$ **24.** $\dfrac{1}{16}$ **25.** $\dfrac{1}{32}$ **26.** $\dfrac{5}{32}$

Unit 5 Review Exercises

Solve for the variable. (If necessary, round your answers to two decimal places.)

1. $3.12 + 5.12w = 740.5$ **2.** $\frac{3}{8}x + 23 = 64$

3. $\dfrac{A}{42.6} = 15.8$ **4.** $\frac{1}{3}(b + 142) = 24(b + 1)$

5. Calculate π from the formula $\pi = C/D$ if $C = 7.6875$ in. and $D = 2.4470$ in. (Round your answer to four decimal places.)

6. Find the perimeter of a rectangular sheet metal panel if the length is 13.2 cm and the width is 6.9 cm.

7. A 20-ft ladder just reaches a shelf 2 ft wide and 15 ft from the floor, as shown in Figure A. How far is the base of the ladder from the wall? (Round your answer to one decimal place.)

8. Find the square root of 150 by factoring.

9. Calculate the unknown dimension to complete the jig bore layout shown in Figure B. (Round your answer to three decimal places.)

10. Find the square root of 372 by factoring. (Round your answer to three decimal places.)

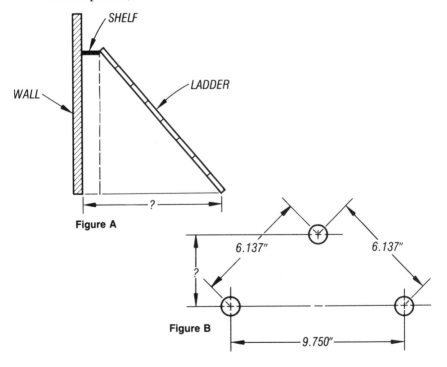

Figure A

Figure B

Ratio, Proportion, and Percent

Ratio and proportion are used in the shop to solve problems such as those involving simple gearing when cutting threads on a lathe. Proportion can also be used to solve percentage problems.

Percentage is a way of expressing a fractional part of something. The machinist's world is filled with percentages, in one form or another. If a machinist or toolmaker is given a piece of ferrous metal (containing iron) and asked to heat-treat it, one of the first things he or she must know is how much carbon it contains or, putting it another way, what the percentage of carbon is.

A piece of iron with more than 2.2 percent carbon content is so hard that it must be poured into a mold, rather than shaped by rollers or hammer blows. This is called *cast* iron. The machinist knows that cast iron cannot be machined as fast as other forms of iron because of this high percentage of carbon.

In this unit, you will learn about ratio, proportion, and percent and how to use them to solve practical shop problems.

6-1 Ratio

OBJECTIVES

After completing this section, you should be able to:

1. Write a ratio in simplest form.
2. Write a rate in simplest form.

A **ratio** is a relation between two quantities found by dividing the first quantity by the second quantity. The relation of the length of the two sides in Figure 6-1 is 6 inches to 3 inches. The ratio 6 to 3 is written

$$6 : 3 \quad \text{or} \quad \frac{6}{3}$$

A ratio can also be written as a decimal or percent.

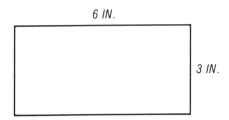

6 IN.

3 IN.

Figure 6-1

Example 6-1   Given two triangles with areas 11 sq m and 18 sq m, write two ratios comparing their areas.

Solution   The ratio of the area of the first triangle to that of the second is 11 : 18, or $\frac{11}{18}$.

The ratio of the area of the second triangle to that of the first is 18 : 11, or $\frac{18}{11}$.

Ratios are usually written in simplest form. Rule 6-1 tells how to do this.

RULE 6-1   WRITING A RATIO IN SIMPLEST FORM

Step 1   Write the ratio in the form $\frac{a}{b}$.

Step 2   If necessary, change any fractions in the numerator or denominator to whole numbers by multiplying the numerator and the denominator by the same number.

Step 3   Reduce the resulting fraction to lowest terms.

Example 6-2   Write $4\frac{1}{2}$: 3 in simplest form.

Solution

Step 1   Write the ratio in the form of a fraction.

$$4\frac{1}{2} : 3 \longrightarrow \frac{4\frac{1}{2}}{3}$$

Step 2   Multiply the numerator and the denominator by 2 to change the fraction in the numerator to a whole number.

$$\frac{4\frac{1}{2}}{3} = \frac{4\frac{1}{2} \times 2}{3 \times 2} = \frac{9}{6}$$

Step 3 Reduce to lowest terms.

$$\frac{9}{6} = \frac{9 \div 3}{6 \div 3} = \frac{3}{2}$$

Example 6-3 The ratio of the number of teeth in one gear to the number of teeth in another gear is 35 : 40. Write the ratio in simplest form.

Solution $35 : 40 = \dfrac{35}{40} = \dfrac{35 \div 5}{40 \div 5} = \dfrac{7}{8}$

The ratio is 7 : 8.

Sometimes it is necessary to compare two similar quantities expressed in different units of measure, such as two lengths or two weights. Before such a comparison can be written as a ratio, the quantities must be changed to the same unit of measure.

RULE 6-2 RATIOS AND MEASUREMENTS

When comparing two similar quantities with different units of measure, change the quantities to the same unit of measure before writing a ratio.

Example 6-4 Find the ratio of 2 qt to 3 gal.

Solution There are 4 qt in 1 gal, so multiply by 4 to change gallons to quarts.

$$3 \text{ gal} = 12 \text{ qt}$$

$$\frac{2 \text{ qt}}{3 \text{ gal}} = \frac{2 \text{ qt}}{12 \text{ qt}}$$

$$= \frac{2}{12}$$

$$= \frac{1}{6}$$

A ratio can also be used to express a rate. A **rate** is a comparison of two different kinds of measurements. For example, if a car travels 20 miles on 1 gallon of gasoline, the ratio of miles to gallons is 20 : 1. The rate is 20 miles per gallon. If 6 apples cost \$1, the rate is 6 apples per dollar. The ratio of apples to dollars is 6 : 1.

Example 6-5   Robin drove 400 mi in 8 h. Write a ratio in simplest form to express her rate in miles per hour.

Solution $\dfrac{400}{8} = \dfrac{400 \div 8}{8 \div 8} = \dfrac{50}{1}$, or 50 mi/h

---

Exercises 6-1

Write each ratio in the simplest terms.

1. 7 to 9   **2.** 3 to 2   **3.** 4 to 11   **4.** 5 to 3

5. 5 to 15   **6.** 20 to 15   **7.** 16 to 24   **8.** 9 to 3

9. 32 to 24   **10.** 30 to 12   **11.** 14 to 49   **12.** 12 to 16

13. 18 to 10   **14.** 12 to 20   **15.** 22 to 33   **16.** 70 to 84

Express each ratio in simplest terms.

17. 8 eggs to a dozen eggs   **18.** a nickel to a quarter

19. 3 ft to 4 yd   **20.** 45 s to 2 min

21. 8 ft 4 in. to 60 in.   **22.** 12 mo to 2 yr

23. 18 lb to 63 lb   **24.** 2 pt to 1 qt

25. 3 quarters to $2   **26.** 7 oz to 7 lb

27. 5¢ to 2 nickels   **28.** 50 cookies to 3 dozen cookies

Write each rate as a ratio in simplest terms.

29. 4200 revolutions in 5 min   **30.** 8 books for $16

31. 1750 mi in 4 h   **32.** 80 lb for 5 boxes

33. There are 20 nuts and 15 bolts in a box. Find the ratios asked for: (a) nuts to bolts, (b) bolts to nuts, (c) bolts to total number of nuts and bolts.

34. Each corner post Maria used in making her barbed-wire fence was 6 ft high and 4 in. in diameter. Find the ratio of height to diameter.

35. Richard works $2\frac{1}{2}$ h out of every 8 h to pay taxes. Express this as a ratio in simplest terms.

36. The pitch of a roof is the ratio of the rise to the span. What is the pitch of a roof with rise 8 ft and span 24 ft?

37. In a square thread screw, the depth of the thread is the ratio of 0.500 to the number of threads per inch. Find the depth if there are 36 threads per inch.

6-2 Solving Proportions

OBJECTIVES

After completing this section, you should be able to:

1. Identify a proportion.
2. Solve a proportion.

A statement that two ratios are equal is called a **proportion.** In the last section (Example 6-2), you saw that

$$\frac{4\frac{1}{2}}{3} \quad \text{and} \quad \frac{3}{2}$$

are equal ratios. Thus

$$\frac{4\frac{1}{2}}{3} = \frac{3}{2}$$

which can also be written $4\frac{1}{2} : 3 = 3 : 2$, is called a proportion. Both statements are read, "Four and one-half is to three as three is to two."

The numbers in a proportion are called **terms.**

$$\begin{array}{ll} \text{first term} \longrightarrow 4\frac{1}{2} \\ \text{second term} \longrightarrow 3 \end{array} = \begin{array}{ll} 3 \longleftarrow \text{third term} \\ 4 \longleftarrow \text{fourth term} \end{array}$$

The first and fourth terms are called the **extremes,** and the second and third terms are called the **means.**

RULE 6-3 EQUALITY OF CROSS PRODUCTS

In a proportion, the product of the means equals the product of the extremes.

Example 6-6 Is $4.8 : 1.6 = 6 : 2$ a proportion?

Solution Product of the means: $1.6 \times 6 = 9.6$

Product of the extremes: $4.8 \times 2 = 9.6$

Thus $4.8 : 1.6 = 6 : 2$ is a proportion.

Rule 6-3 can be used to find missing members in a proportion.

Example 6-7 Find n if $\dfrac{8}{10} = \dfrac{n}{15}$.

Solution
$$\frac{8}{10} = \frac{n}{15}$$

$10 \times n = 8 \times 15$ Product of means = product of extremes

$10n = 120$

$$n = \frac{120}{10} \quad \text{Property 4 of equations}$$

$n = 12$

Check $8 \times 5 = 120; \; 10 \times 12 = 120$

Example 6-8   The proportion

$$\frac{N_L}{N_C} = \frac{T_S}{T_L}$$

is used for lathe thread cutting computations involving simple gearing. The fixed stud gear and the spindle gear must have the same number of teeth. Find the number of threads per inch to be cut (N_C) where $N_L = 8$, $T_L = 42$, and $T_S = 28$.

Solution
$$\frac{N_L}{N_C} = \frac{T_S}{T_L}$$

$$\frac{8}{N_C} = \frac{28}{42}$$

$8 \times 42 = 28 \times N_C$

$336 = 28N_C$

$$\frac{336}{28} = N_C$$

$N_C = 12$

12 threads per inch are to be cut.

Check $28 \times 12 = 336; \; 8 \times 42 = 336$

Exercises 6-2

Write true *if the pair of numbers makes a proportion and* false *if it does not.*

1. $\dfrac{1}{2}, \dfrac{5}{11}$

2. $\dfrac{14}{16}, \dfrac{21}{24}$

3. $\dfrac{36}{200}, \dfrac{18}{100}$

4. $\dfrac{1}{5}, \dfrac{20}{100}$

5. $\dfrac{27}{50}, \dfrac{55}{100}$

6. $\dfrac{21}{52}, \dfrac{42}{104}$

7. $\dfrac{4\frac{1}{2}}{9}, \dfrac{25}{50}$ **8.** $\dfrac{6.3}{13.23}, \dfrac{18.07}{37.947}$

Find the missing term, and check your answer.

9. $\dfrac{5}{9} = \dfrac{n}{27}$ **10.** $\dfrac{4}{7} = \dfrac{28}{a}$ **11.** $\dfrac{n}{15} = \dfrac{1}{3}$

12. $\dfrac{n}{100} = \dfrac{28}{16}$ **13.** $\dfrac{18}{a} = \dfrac{22}{99}$ **14.** $\dfrac{7}{10} = \dfrac{70}{b}$

15. $\dfrac{n}{100} = \dfrac{36}{45}$ **16.** $\dfrac{6}{b} = \dfrac{4}{8}$ **17.** $\dfrac{3}{7} = \dfrac{n}{5}$

18. $\dfrac{16}{11} = \dfrac{96}{a}$ **19.** $\dfrac{5.1}{8} = \dfrac{n}{72}$ **20.** $\dfrac{2.6835}{y} = \dfrac{5}{1.4}$

Use a proportion to solve each problem. Check your answers.

21. A bottom-dump truck has a capacity of 100 tons. An electric shovel can fill it with ore in 3 scoops. How many scoops are required to fill 12 such trucks?

22. An airplane flew 1925 mi in 3.5 h. What is the rate in miles per hour?

23. A sampler x-ray station in a copper mill takes 4 readings per hour on the copper concentrate, tailings, and mill feed. How many readings are taken in $6\frac{1}{2}$ h?

24. The ratio of the circumference of a circle to its diameter is the same in every circle. If a circle of diameter 8 has circumference 25.12, find the diameter of a circle with circumference 18.84.

25. Use the formula given in Example 6-8 and find T_L if $N_L = 4$, $N_C = 8$, and $T_S = 32$.

6-3   Scale Drawings

OBJECTIVES
After completing this section, you should be able to:

1. Find the length needed on a scale drawing to represent a given dimension.
2. Find the length of a dimension given the scale and a scale drawing.

It is often not practical to work with a full-size drawing of an object. When the part or object to be drawn is very large or very small, a scale drawing can be used. A scale drawing shows the object accurately, but the drawing is a different size than the object.

A scale of 4 inches = 1 inch means the drawing is four times the size of the actual object. This scale can be expressed by a ratio of 4 : 1.

Figure 6-2 shows a scale drawing of a bolt. The scale shown on the drawing compares the lengths of the lines on the drawing to those of the bolt. This drawing is half the actual size of the bolt. The scale can be expressed by a ratio of 1 : 2.

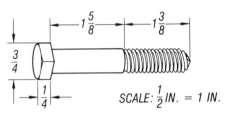

Figure 6-2

RULE 6-4   USING PROPORTION TO FIND DIMENSIONS

To find the actual size of a dimension from a scale drawing or the scale size of an actual dimension, use this proportion:

$$\frac{\text{First part of scale}}{\text{Second part of scale}} = \frac{\text{length on drawing}}{\text{actual length}}$$

Example 6-9   In a scale of 1 cm = 2 m, find the actual length of a dimension represented by 3.6 cm.

Solution $\dfrac{1 \text{ cm}}{2 \text{ m}} = \dfrac{3.6 \text{ cm}}{x}$

$1x = 2(3.6)$

$x = 7.2$

The actual length is 7.2 m.

Example 6-10   In a scale $\frac{1}{4}$ in. = 3 ft, find the length used to represent $4\frac{1}{2}$ ft.

Solution $\dfrac{\frac{1}{4}}{3} = \dfrac{x}{4\frac{1}{2}}$

$\dfrac{1}{4}\left(4\,\dfrac{1}{2}\right) = 3x$

$\dfrac{9}{8} = 3x$

$\dfrac{3}{8} = x$

$\frac{3}{8}$ in. is used to represent $4\frac{1}{2}$ ft.

Example 6-11   The world's largest truck tire is made for a 200-ton truck. Use the scale drawing to find out high the tire is.

Solution   Let h represent the actual height of the tire. The height in the scale drawing is $1\frac{1}{2}$ in.

$$\frac{1}{8} = \frac{1\frac{1}{2}}{h}$$

$$8 \times 1\frac{1}{2} = 1 \times h$$

$$12 = h$$

The tire is 12 ft high.

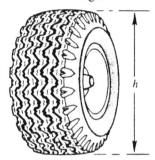

1 IN. = 8 FT

Example 6-12   The amoeba is drawn to a scale of 1 : 100. Find the actual width of the amoeba in inches.

Solution   Let w represent the actual width of the amoeba. The width in the scale drawing is 2 in.

AMOEBA × 100

$$\frac{100}{1} = \frac{2}{w}$$

$$100 \times w = 1 \times 2$$

$$\frac{100 \times w}{100} = \frac{2}{100}$$

$$w = 0.02$$

The amoeba is 0.02 in. wide.

Exercises 6-3

Use a scale of 3 in. = 10 ft. Find the actual distance represented by each scale length.

1. 1 in.   **2.** 15 in.   **3.** $2\frac{1}{2}$ in.   **4.** 14 in.

Use a scale of $\frac{1}{2}$ in. = 6 yd. Find how many scale inches represent each actual length.

5. 12 yd   **6.** 84 yd   **7.** 3 ft   **8.** 15 yd

9. Find the scale when the actual distance is 40 ft and the scale length is 5 in.

10. Find the length of the brick in Figure A. (Use a ruler to measure the scale length.)

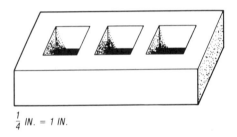

$\frac{1}{4}$ IN. = 1 IN.

Figure A

11. A tapered shim 10 in. long is drawn to a scale of 1 in. = 2 in. What will be the length of the line in the scale drawing that represents the length of the actual part?

12. A box 5 ft 4 in. long, 4 ft 8 in. wide, and 3 ft high is drawn to a scale of $\frac{1}{8}$ in. = 1 in. Calculate the length of the lines in the scale drawing that will represent the length, width, and height of the finished box.

13. What will be the produced size of the sine bar shown in Figure B? (Use a ruler to measure the scale length.)

14. Figure C shows a scale drawing of a vee block. What will be the produced width of the part?

15. Find the scale when the produced width of the vee block in Figure C is 1.33 in. (Use a ruler to measure the scale length.)

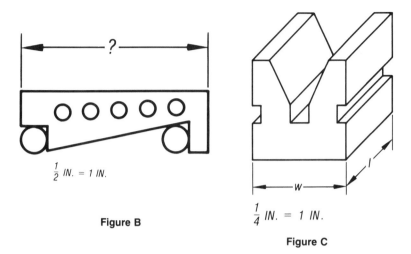

$\frac{1}{2}$ *IN. = 1 IN.*

Figure B

$\frac{1}{4}$ *IN. = 1 IN.*

Figure C

6-4   Variation

OBJECTIVES

After completing this section, you should be able to:

1. Tell whether two quantities vary directly or inversely.
2. Solve problems about direct and inverse variation.

If a machinist takes the same amount of time to produce each part he or she makes, then making more parts takes more time. In such a case, the amount of time varies **directly** as, or is directly proportional to, the number of parts made. The quotient of time to parts made is always constant.

RULE 6-5   DIRECT VARIATION

a varies directly as *b*, or is directly proportional to *b*, means that *a* /*b* has a constant value.

If it takes 10 minutes to make two parts you can use a proportion to find how many minutes it takes to make 12 parts. The ratio 10/2 will be the same as the ratio *t* /12, where *t* is the time required to make 12 parts. You can write a proportion and solve for *t*.

$$\text{first time} \longrightarrow \frac{10}{2} = \frac{t}{12} \longleftarrow \text{second time}$$

first number of parts $\longrightarrow$ ... $\longleftarrow$ second number of parts

$$2t = 120$$
$$t = 60$$

The machinist takes 60 minutes to make 12 parts.

Notice that an increase in a brings about an increase in b.

RULE 6-6   SOLVING DIRECT VARIATION PROBLEMS

Step 1   Set up a proportion in which both sides have a common ratio.

$$\frac{\text{First value of } a}{\text{First value of } b} = \frac{\text{second value of } a}{\text{second value of } b}$$

Step 2   Solve the proportion for the missing value.

Example 6-13   If a varies directly as b and $a = 15$ when $b = 2$, find a when $b = 5$.

Solution   Here $a = 15$ when $b = 2$, and $b = 5$ for some a.

$$\text{first } a \longrightarrow \frac{15}{2} = \frac{a}{5} \longleftarrow \text{second } a$$

first b $\longrightarrow$ ... $\longleftarrow$ second b

$$2a = 75$$
$$a = 37\frac{1}{2}$$

Check $37\frac{1}{2} : 5$ is the same as $75 : 10$, which reduces to $15 : 2$. Thus the pairs of quantities have a constant ratio.

Example 6-14   A card punch can punch 500 cards per minute. The number of cards punched is directly proportional to time. How many cards can it punch in an hour?

Solution   Let n be the number of cards punched in 1 h, which equals 60 min.

first number
of cards $\longrightarrow \dfrac{500}{1} = \dfrac{n}{60} \longleftarrow$ second number of cards
first time $\longrightarrow$ ... $\longleftarrow$ second time

$$n = 500(60)$$
$$n = 30{,}000$$

30,000 cards can be punched in 1 h.

Example 6-15   The No. 10 Brown and Sharpe taper has 0.516 in. taper per foot. Find the amount of taper in a 6-in. taper if the taper varies directly as the length.

Solution   Let t = the taper in a 6-in. taper. Write 1 ft as 12 in.

$$\frac{0.516}{12} = \frac{t}{6}$$

$$12t = 6(0.516)$$

$$t = \frac{6(0.516)}{12} = 0.258$$

Taper is 0.258 in. per 6 in.

If gas is held at a constant temperature and the pressure on the gas is increased, the volume will decrease. The volume of the gas varies **inversely** as, or is inversely proportional to, the pressure. Two quantities vary inversely if their product is constant.

RULE 6-7   INVERSE VARIATION

a varies inversely as *b*, or is inversely proportional to *b*, means that *ab* has a constant product.

Since the product is constant in inverse variation,

(First value of *a*) × (first value of *b*)
= (second value of *a*) × (second value of *b*)

This gives Rule 6-8 for solving inverse variation problems.

RULE 6-8   SOLVING INVERSE VARIATION PROBLEMS

Step 1   Write an equation in which both sides have a common product.

(First *a*) × (first *b*) = (second *a*) × (second *b*)

Step 2   Solve the equation for the missing value.

Example 6-16   If x varies inversely as y, and $x = 3$ when $y = 7$, find x when $y = 21$.

Solution   Here the first x is 3, the first y is 7, and the second y is 21.

first x ⟶ ⌐ second x
$$3 \cdot 7 = x \cdot 21$$
first y ⟶ ⌐ second y
$$21 = 21x$$
$$1 = x$$

As a check, $3 \times 7 = 21$ and $21 \times 1 = 21$, so the product is constant.

Example 6-17   The volume of a gas is 1500 cu cm at a pressure of 250 mm of mercury. Find the volume at 375 mm of mercury.

Solution   Let v = the volume at 375 mm of mercury.

first volume ⟶ ⌐ second volume
$$1500 \cdot 250 = v \cdot 375$$
first pressure ⟶ ⌐ second pressure
$$1500(250) = 375v$$
$$v = \frac{1500(250)}{375} = 1000$$

The volume is 1000 cu cm at 375 mm of mercury.

Example 6-18   For the given values of x and y, do x and y vary directly or inversely?

(a) $x = 2$ when $y = 5$, and $x = 4$ when $y = 2\frac{1}{2}$.

(b) $x = 1.5$ when $y = 3.3$, and $x = 4.5$ when $y = 9.9$.

(c) $x = 5$ when $y = 6$, and $x = 4$ when $y = 7$.

Solution   (a)   Since an increase in x brings about a decrease in y, check for inverse variation.

$$2 \times 5 \overset{?}{=} 4 \times 2\frac{1}{2}$$

$$10 = 10$$

These numbers have a constant product, so x and y are inversely proportional.

(b) Since an increase in x brings about an increase in y, check for direct variation.

$$\frac{1.5}{3.3} \overset{?}{=} \frac{4.5}{9.9}$$

$$1.5 \times 9.9 \overset{?}{=} 3.3 \times 4.5$$

$$14.85 = 14.85$$

The cross-products are equal, so x varies directly as y.

(c) An increase in x brings about a decrease in y, so check for inverse variation.

$$5 \times 6 \overset{?}{=} 4 \times 7$$

$$30 \neq 28$$

These numbers do not have a constant product, so they are not inversely proportional, nor are they directly proportional. Thus x varies neither directly nor inversely with y.

Exercises 6-4

Find the missing value if y varies directly as x.

1. If $x = 5$ when $y = 8$, find x when $y = 12$.
2. If $x = 2$ when $y = 7$, find x when $y = 28$.
3. If $x = 1.6$ when $y = 0.4$, find y when $x = 3.3$.
4. If $x = \frac{1}{2}$ when $y = 2$, find y when $x = 15$.
5. If $x = 1.2$ when $y = 0.8$, find x when $y = 4.32$.
6. If $x = 0.973$ when $y = 1.48$, find y when $x = 3.1136$.

Find the missing value if y varies inversely as x.

7. If $x = 8$ when $y = 15$, find x when $y = 5$.
8. If $x = 2.7$ when $y = 3.6$, find x when $y = 7.2$.
9. If $x = 7$ when $y = 4$, find x when $y = 28$.
10. If $x = 3.4$ when $y = 1.5$, find y when $x = 1.7$.
11. If $x = 21.79$ when $y = 13.32$, find x when $y = 3$.
12. If $x = 33.28$ when $y = 27.48$, find y when $x = 8.32$.

Solve the following problems using direct or inverse proportions.

13. A card reader can read 18,000 cards per minute. How long would it take to read 54,000 cards?

14. A Morse No. 0 taper has 0.0521 in. taper per inch. Find the taper in $2\frac{1}{2}$ in.

15. The pressure of a gas varies inversely as volume, at constant temperature. If the volume is 350 cu in. when the pressure is 15 lb/sq in., find the pressure when the volume is 70 cu in.

16. The weight at one end of a balanced lever varies inversely as its distance from the fulcrum. A 40-kg weight is 6 m from the fulcrum. Determine the weight needed at 15 m to balance the lever.

Do x and y vary directly, inversely, or not at all?

17. $x = 120$ when $y = 2$, and $x = 180$ when $y = 3$.

18. $x = 2$ when $y = 6$, and $x = 3$ when $y = 4$.

19. $x = 4.358$ when $y = 6.3$, and $x = 1.0895$ when $y = 25.2$.

20. $x = 35.98$ when $y = 15.225$, and $x = 82.754$ when $y = 35.02$.

6-5   Meaning of Percent

OBJECTIVES
After completing this section, you should be able to:

1. Write a percent as a decimal or a fraction.
2. Write a decimal or a fraction as a percent.

Percent means *per hundred.* The symbol for percent is %. One way to picture percent is by using money. There are 100 pennies, or cents, in one dollar. One-fourth of a dollar is 25 pennies, or 25% of a dollar. Written as a decimal, 25% becomes 0.25. Thus $\frac{1}{4} = 0.25 = 25\%$.

The comparison to pennies in a dollar is very easy to remember and expresses exactly the concept of percentage. One hundred pennies is 100% of a dollar. Fractional parts, or percentages, of that hundred are expressed decimally as if counting pennies.

The percentage sign represents two decimal places. Consequently, it is not used with the decimal point unless a fractional part of 1% is being expressed. Thus 2% = 0.02, but 0.2% = 0.002.

RULE 6-9   WRITING A PERCENT AS A DECIMAL

Move the decimal point two places to the left and drop the percent sign.

Using Rule 6-9, 5% = .05., or 0.05 as a decimal.

Example 6-19   10% = 0.10   80% = 0.80   17.6% = 0.176
                   25% = 0.25   75% = 0.75   0.2% = 0.002
                    8% = 0.08   2% = 0.02 $14\frac{1}{2}\% = 0.14\frac{1}{2} = 0.145$

A percent can also be written as a fraction by using the fact that percent means per hundred.

RULE 6-10   WRITING A PERCENT AS A FRACTION

Step 1   Drop the percent sign and write a ratio of the number to 100.

Step 2   Write the ratio as a fraction in lowest terms.

Thus, $50\% = \frac{50}{100} = \frac{1}{2}$.

Example 6-20

$$1\% = \frac{1}{100} \qquad\qquad 33\frac{1}{3}\% = \frac{33\frac{1}{3}}{100} = \frac{100}{300} = \frac{1}{3}$$

$$5\% = \frac{5}{100} = \frac{1}{20} \qquad\qquad 500\% = \frac{500}{100} = 5$$

$$20\% = \frac{20}{100} = \frac{1}{5} \qquad\qquad 625\% = \frac{625}{100} = 6\frac{25}{100} = 6\frac{1}{4}$$

$$75\% = \frac{75}{100} = \frac{3}{4} \qquad\qquad 87.5\% = \frac{87.5}{100} = \frac{875}{1000} = \frac{7}{8}$$

To write a decimal as a percent, use the following rule.

RULE 6-11   WRITING A DECIMAL AS A PERCENT

Move the decimal point two places to the right and add a percent sign.

Using Rule 6-11, 0.3 = 0.30, or 30%.

Example 6-21   0.45 = 45%   0.003 = 0.3%
                   0.67 = 67%   0.884 = 88.4%
                    1.1 = 110%   5 = 500%

RULE 6-12   WRITING A FRACTION AS A PERCENT

Write the fraction as a decimal and then use the rule for decimals.

Example 6-22   Write each fraction as a percent.

(a) $\dfrac{1}{4}$ (b) $\dfrac{1}{8}$ (c) $\dfrac{2}{3}$

Solution   (a)   First write $\frac{1}{4}$ as a decimal.

$$4\overline{)1.00} \quad 0.25$$

Then use Rule 6-11 to write 0.25 as a percent.

$$0.25 = 25\%$$

Thus $\frac{1}{4} = 0.25 = 25\%$.

(b) $8\overline{)1.000} \quad 0.125 \qquad \dfrac{1}{8} = 0.125 = 12.5\%$

(c) $3\overline{)2.00} \quad 0.66\frac{2}{3} \qquad \dfrac{2}{3} = 0.66\,\dfrac{2}{3} = 66\,\dfrac{2}{3}\%$

A decimal equivalent chart, such as the one in Figure 3-2, is also useful when changing a fraction to a percent. You can first read the decimal equivalent from the chart and then write the decimal as a percent.

Exercises 6-5

Write each percent as a decimal.

1. 21%   **2.** $19\dfrac{1}{2}\%$ **3.** 258%   **4.** 7%   **5.** $6\dfrac{1}{2}\%$

6. 43.8%   **7.** 0.62%   **8.** 188%   **9.** 5.3%   **10.** 48%

Write each percent as a fraction or mixed number in lowest terms.

11. 30%   **12.** 6%   **13.** 15%   **14.** 60%   **15.** 35%

16. $37\dfrac{1}{2}\%$ **17.** $187\dfrac{1}{2}\%$ **18.** 510%   **19.** 480%   **20.** $16\dfrac{2}{3}\%$

Write each decimal as a percent.

21. 0.67 **22.** 0.83 **23.** 0.233 **24.** 0.05 **25.** 0.092

26. $0.87\frac{1}{2}$ **27.** 1.22 **28.** 5.4 **29.** 8 **30.** 0.0007

Write each fraction as a percent.

31. $\frac{61}{100}$ **32.** $\frac{4}{100}$ **33.** $\frac{725}{100}$ **34.** $\frac{73\frac{2}{3}}{100}$ **35.** $\frac{3}{5}$

36. $\frac{1}{50}$ **37.** $\frac{1}{9}$ **38.** $\frac{7}{10}$ **39.** $\frac{5}{1000}$ **40.** $\frac{5}{6}$

Complete the following tables.

		Common Fraction	Decimal Fraction	Percent
41.	(a)	$\frac{1}{8}$	0.125	12.5%
	(b)	$\frac{1}{4}$		
	(c)	$\frac{3}{8}$		
	(d)	$\frac{1}{2}$		
	(e)	$\frac{5}{8}$		
	(f)	$\frac{3}{4}$		
	(g)	$\frac{7}{8}$		

		Common Fraction	Decimal Fraction	Percent
42.	(a)	$\frac{1}{6}$	$0.16\frac{2}{3}$	$16\frac{2}{3}\%$
	(b)	$\frac{1}{3}$		
	(c)	$\frac{1}{2}$		
	(d)	$\frac{2}{3}$		
	(e)	$\frac{5}{6}$		

	Common Fraction	Decimal Fraction	Percent
43. (a)	$\frac{1}{100}$		
(b)	$\frac{1}{50}$		
(c)	$\frac{1}{25}$		
(d)	$\frac{1}{20}$		
(e)	$\frac{1}{16}$		
(f)	$\frac{1}{12}$		

6-6   Using Proportion with Percent Problems

OBJECTIVE

After completing this section, you should be able to:

1. Use a proportion to solve a percent problem.

Proportions can be used to solve percent problems. To find 36% of 450, for example, recall that 36% can be written as the fraction, or ratio, $\frac{36}{100}$. To find 36% of 450, you need to find a number whose ratio to 450 is the same as the ratio 36 : 100.

Example 6-23   Find 36% of 450.

Solution   Let n be the number.

$$\frac{36}{100} = \frac{n}{450}$$ Find the number whose ratio to 450 is the same as $\frac{36}{100}$

$$100n = 36(450)$$
$$100n = 16,200$$
$$n = 162$$

Thus 36% of 450 is 162.

Proportion can also be used to find what percent one number is of another. For instance, if you want to know what percent 18 is of 36, you must find the number with the same ratio to 100 as the ratio 18 : 36.

Example 6-24 Find what percent 18 is of 36.

Solution Let n be the percent.

$$\frac{18}{36} = \frac{n}{100}$$

Find the number whose ratio to 100 is the same as $\frac{18}{36}$.

$$36n = 18 \times 100$$
$$36n = 1800$$
$$n = 50$$

18 is 50% of 36.

To answer a question such as "28.1 is $\frac{1}{4}$ percent of what number," find a number such that the ratio of 28.1 to that number is the same as the ratio of $\frac{1}{4}$ to 100.

Example 6-25 28.1 is $\frac{1}{4}$% of what number?

Solution Write $\frac{1}{4}$% as 0.25%. Let n be the number.

$$\frac{28.1}{n} = \frac{0.25}{100}$$

The ratio of 28.1 to the number is the same as $\frac{1}{4}$ (or 0.25) to 100.

$$0.25n = 28.1 \times 100$$
$$n = \frac{2810}{0.25}$$
$$= 11{,}240$$

28.1 is $\frac{1}{4}$% of 11,240

Exercises 6-6

Use a proportion to find each number.

1. 38% of 135 **2.** 5% of 90 **3.** 72% of $8000

4. 10% of 945 **5.** 12% of 63 **6.** 95% of 600

7. 170% of 58 **8.** 200% of 126 **9.** 450% of 39

10. 68.2% of 90 **11.** $120\frac{1}{2}$% of 9000 **12.** 7.2% of 638

13. 0.2% of 60 **14.** 0.31% of 900 **15.** $\frac{1}{2}$% of 80

Use a proportion to solve.

16. 27 is what percent of 40?

17. 75 is what percent of 150?

18. 8 is what percent of 80?

19. 7 is what percent of 8?

20. 50 is what percent of 60?

21. 2.4 is what percent of 40?

22. 0.48 is what percent of 3?

23. $\frac{1}{5}$ is what percent of 1?

24. 27 is what percent of 9?

25. 3 is what percent of 1.5?

26. 12 is what percent of 5?

27. 28 is what percent of 28?

28. 1.15 is what percent of 5.75?

29. 0.07 is what percent of 4?

30. 3.2 is what percent of 8?

Round each to the nearest tenth of a percent.

31. 70 is what percent of 90?

32. 8 is what percent of 4.8?

Use proportions to find the missing numbers.

33. 50% of what number is 10?

34. 7% of what number is 21?

35. 40% of what number is 200?

36. 90% of what number is 60?

37. 3% of what number is 56.13?

38. 12% of what number is 2.4?

39. 70% of what number is 49?

40. 55% of what number is 88?

41. $\frac{1}{2}$% of what number is 16?

42. $\frac{3}{4}$% of what number is 36?

43. 0.1% of what number is 73.5?

44. 150% of what number is 30?

45. 100% of what number is 68.5?

46. 500% of what number is 36.85?

47. 60% of what number is $84?

48. 10% of what number is $12.50?

6-7   Finding a Percent of a Number

OBJECTIVE
After completing this section, you should be able to:

1. Find a percent of a number using the percentage formula.

Proportions can be used to solve any percent problem; however, there is another way to solve these problems by using a formula. Using a proportion to find 25% of 20, you write

$$\frac{25}{100} = \frac{n}{20}$$

and solve for n, to find $n = 5$. If you let $r = \frac{25}{100}$ (the percent), $P = 5$ (the percentage, or part), and $b = 20$ (the base, or whole amount), you have the formula

$$r = \frac{P}{b}, \quad \text{or} \quad P = rb$$

RULE 6-13   USING A FORMULA TO FIND PERCENT

To find a percentage P of a number b, use the formula

$P = rb$

where r is the percent, written as a decimal or fraction.

Example 6-26   What is 23% of 600?

   Solution $P = rb$
   $= (0.23)600 = 138$

Example 6-27   What is 75% of 600?

   Solution $P = (0.75)600 = 450$

Example 6-28   Smith Tool and Die priced a shop machine at $125,000. Within a year, the price was 110% of that amount. Find the new price.

   Solution $P = rb$
   $= 1.1(125,000) = 137,500$

   The new price was $137,500.

Sometimes it is easier to use the fractional form.

Example 6-29   Find $33\frac{1}{3}\%$ of 375.

   Solution   Recall that $33\frac{1}{3}\% = \frac{1}{3}$.

   $$P = \frac{1}{3}(375) = 125$$

Exercises 6-7

Use the percentage formula to find each number.

1.	27% of 90	**2.**	85% of 69	**3.**	100% of 84
4.	50% of 60.8	**5.**	10% of 800	**6.**	140% of 86
7.	110% of 44	**8.**	$\frac{1}{2}$ % of 80	**9.**	60% of 9
10.	$37\frac{1}{2}$ % of 240	**11.**	8% of 621	**12.**	12% of $176
13.	$6\frac{1}{4}$ % of 108	**14.**	60% of 75	**15.**	0.7% of $8910
16.	$66\frac{2}{3}$ % of 18	**17.**	75% of $\frac{5}{9}$	**18.**	126.3% of 90,000

Solve using the percentage formula.

19. If a cubic foot of steel weighs 490 lb, how much does 33% of a cubic foot weigh?

20. The weight of a wood pattern used to make a casting is 7% of the 375-lb casting. How much does the pattern weigh?

21. A patternmaker buys 425 board feet of lumber. If $8\frac{1}{3}$ % of this amount is eventually scrapped, how many board feet remain for making patterns?

22. A precision measuring instrument costs an inspection department $84.00. If the price of this instrument increases 8% because of labor shortage, what is its new price in dollars and cents?

23. A brass casting contains 67% copper and 33% zinc. If the casting weighs 200 pounds, how many pounds each of copper and zinc are required for the heat (melting)?

6-8 Finding What Percent One Number Is of Another

OBJECTIVE

After completing this section, you should be able to:

1. Use the percentage formula to find what percent one number is of another.

As was seen in Section 6-7, the percentage formula can be used to find a percentage, or portion, of a whole quantity. If the whole quantity and a portion of it are already known, the percent of the whole quantity that the portion represents is found by using a variation of the basic percentage formula.

RULE 6-14   USING A FORMULA TO FIND WHAT PERCENT ONE NUMBER IS OF ANOTHER

To find the percent, divide the percentage by the base.

$$\text{Percent} = \frac{\text{percentage}}{\text{base}},$$

$$\text{or } r = \frac{P}{b}$$

Example 6-30   What percent of 300 is 60?

Solution $r = \dfrac{P}{b} = \dfrac{60}{300} = \dfrac{20}{100} = 20\%$

Example 6-31   What percent of 30 is 120?

Solution $r = \dfrac{P}{b} = \dfrac{120}{30} = \dfrac{4}{1} = 400\%$

Example 6-32   The desired taper per foot of a cut is $\frac{5}{16}$ in. The actual taper in 3 in. is 0.062 in. Find the percent of error.

Solution   The taper per inch is 0.312/12, or about 0.026 in. Thus the taper in 3 in. should be about 0.078. The error is 0.078 − 0.062 = 0.016.

$$r = \frac{P}{b} = \frac{0.016}{0.078}$$

$$= 0.205, \text{ or } 20.5\%, \text{ to the nearest tenth percent}$$

The percentage formula can also be used to find the base ($b = P/r$), but this form is not often used in shop math.

Exercises 6-8

Use the percentage formula to find what percent the first number is of the second.

1. 11, 25   **2.** 27, 45   **3.** 0.3, 1.2

4. 6, 5   **5.** 0.6, 120   **6.** $45, $50

7. $55, $1250   **8.** 60, 54   **9.** 7.2, 4

10. $\frac{1}{2}, \frac{2}{3}$ **11.** 21, 420   **12.** 15, 45

13. 4500, 1000   **14.** $75, $75   **15.** 0.26, 5.2

16. 46, 51   **17.** 14, 49   **18.** 11, 3

Solve using the percentage formula.

19. If a machinist works 36 h of a 40-h workweek, what percentage of her regular pay will she collect at the end of the week?

20. If a net profit of $960 is made on a job run in a machine shop and the contract price is $50,000, what percent profit has been realized?

21. The finished diameter of a shaft turned in an engine lathe measures 2.144 in. The blueprint calls for it to be 2.142 in. in diameter. What is the machinist's percent of error?

22. A slide rule is normally accurate to two decimal places. Any further accuracy must be estimated. The estimated answer to a problem solved by slide rule is 172,500. To check the accuracy of the slide rule, the problem is worked and an answer of 172,463 is obtained. What is the percent of error?

23. If a casting shrinks $\frac{1}{8}$ in. per linear foot in cooling, what is the percent of shrinkage?

24. A machine shop estimator must allocate the cost of making a customer's job as a percent of the total cost of the job. He charges the customer 15% above the cost of machining. If total machining requires 125 h at $12.00 per hour, what will he charge the customer?

25. It is company policy that a worker is allowed $2\frac{1}{2}$% of the working year for absences in a working year of 2080 h, whether excused or not. Any excess over this percent is grounds for possible discharge. A supervisor has six people with the following absentee records for one year:

> José: 50 h   George: 37 h
> Alice: 42 h   Maria: 65 h
> Vance: 54 h   Newton: 49 h

Which people could be discharged from their jobs?

26. Normalizing is a heat-treating process in which steel is heated above a critical temperature, held at that temperature until heated throughout, and then allowed to cool in still air. This relieves internal stresses and increases the strength of the steel approximately 20% above that of annealed steel. A machine part of tool steel with a range of strength of 125 to 145 thousand pounds per square inch (*PSI*) is normalized. What will be its new range of strength?

27. Classification of iron and steel is based on the percent of carbon and/or alloys present. The classification is as follows.

Wrought iron = traces to 0.08% carbon
Low-carbon steel = 0.10% to 0.30% carbon
Medium-carbon steel = 0.31% to 0.70% carbon
High-carbon steel = 0.71% to 2.2% carbon
Cast iron = 2.2% to 4.5% carbon

The following five pieces of material belong in one of the categories ranging from wrought iron to cast iron. Classify them according to the percent of carbon present.

	Total Weight	**Weight of Carbon**
Piece 1	50 lb	0.25 lb
Piece 2	18 lb	0.36 lb
Piece 3	35 lb	0.0245 lb
Piece 4	27 lb	0.81 lb
Piece 5	43 lb	0.086 lb

28. SAE 4140 is a type of alloyed steel much used in aircraft work. A 5-lb bar of this steel has the following chemical composition by percent.

Carbon = 0.40%
Phosphorus = 0.01%
Chromium = 1.5%
Manganese = 0.80%
Sulphur = 0.02%
Molybdenum = 0.20%
Iron = remaining percent

What percent of this bar is iron? What weight of each component will be found in a 5-lb bar?

Unit 6 Review Exercises

1. Write $7\frac{2}{3} : 15\frac{1}{3}$ in simplest form.

2. A machine produces 14,080 parts during an 8-h day. What is the machine's rate of output per hour?

3. Find 16.3% of 185.

4. Is $6.13/18.27 = 5.35/15.73$ a proportion?

5. If a circular swimming pool has a diameter of 18 ft and its circumference is 56.52 ft, what would be the circumference of a 24-ft-diameter pool?

6. A drawing of the new factory is scaled to $\frac{1}{4}$ in. = 1 ft. If the machine shop in the drawing measures 6.75 by 9.25 in., what are the actual machine shop dimensions?

7. Jim needs $4500 for expenses during the school year. Over the summer, Jim earned a total of $3750 from his temporary drafter's job, and he managed to save $2700 for school expenses. What percent of his school expenses do Jim's savings represent?

8. A group of employees decided to raise money to establish a credit union. After two months, $8736 had been collected, which was 58.25% of the required sum. What was the total amount the employees hoped to collect?

9. A 9-in.-diameter pulley driven by an electric motor turning at 1200 rpm is belted to a 3-in.-diameter pulley driving a drill chuck. How many revolutions per minute is the drill chuck turning?

10. A metal joint 23 ft long requires 115 rivets. How many rivets are required for a joint 9 ft long?

Precision Measuring Instruments

A human hair is approximately $2\frac{1}{2}$ thousandths of an inch in diameter. Skilled machinists and toolmakers can work to dimensions one-tenth the size of a human hair. Precision measuring instruments make this possible. Any tool capable of linear measure to 0.001 inch or smaller is a **precision instrument.** If an angle is to be measured, the protractor must be divided into minutes to be termed a precision tool.

Of all the precision measuring tools available to the ordinary machinist, the micrometer and vernier calipers are most used. Figure 7-1 illustrates one variation of the basic micrometer caliper, or *mike,* as it is often called in the shop. Manipulation of tools such as these is best learned in actual practice with test pieces of fixed sizes. Instructors can furnish gauge blocks, drill blanks, or hardened dowel pins for trial readings. However, it is equally important that the principle behind the reading be understood.

Figure 7-1 Micrometer caliper.
(Courtesy of The L. S. Starrett Company)

159

The ability to use precision measuring instruments is essential to the machinist, who must consistently produce parts to very exact specifications. The blueprints the machinist or toolmaker follows give precise dimensions and the amount of variation, or tolerance, that is allowed. Parts that do not meet the specified tolerance must be rejected, resulting in a waste of time, effort, and materials.

In this unit you will learn how to read precision instruments and how to apply precision measurements to meeting tolerances.

7-1 The Micrometer

OBJECTIVE
After completing this section, you should be able to:

1. Read a micrometer.

Suppose you wanted to make a micrometer. The first requirement is a hardened rod with one end flat and square to the axis of the rod. Its opposite end will have threads, 40 to the inch (see Figure 7-2). This rod is called a **spindle.**

Figure 7-2

Next, a **frame** with an **anvil** to oppose the spindle is needed. Inside the **barrel** of the frame is a fixed nut, called a **feed nut,** through which the spindle is threaded. This is shown in Figure 7-3.

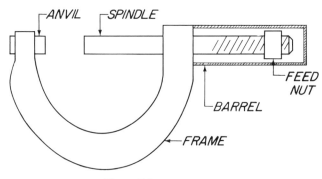

Figure 7-3

Now add graduations to the barrel, numbering every fourth one. Figure 7-4 shows this.

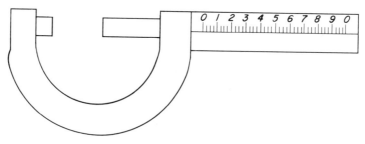

Figure 7-4

A **thimble** is attached to the spindle so that the spindle can be screwed in and out. There are 25 equally spaced graduations around the leading edge of the thimble (see Figure 7-5).

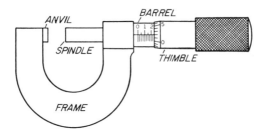

Figure 7-5

The micrometer is now ready for use. The heart of this tool is the spindle with its 40 threads per inch. Each complete turn of the thimble advances the spindle one-fortieth of an inch, or 0.025 inch.

To read an English micrometer, the total number of 0.025-inch graduations uncovered by the thimble is noted. To this total is added the number on the thimble closest to the *index line* on the barrel. Look at the micrometer reading in Figure 7-6.

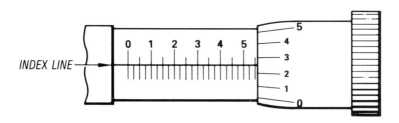

Figure 7-6

Example 7-1   Give the micrometer reading shown in Figure 7-6.

Solution   Numbered graduations exposed (5) = 0.500 in.
Additional unnumbered graduations (2) = 0.050 in.
Thimble reading (each graduation 0.001) = <u>0.003</u> in.
Total reading = 0.552 in.

When the index line falls between two thimble graduations, the graduation nearer to the index line is read. This would be only an approximation.

Exercises 7-1

Fill in the micrometer readings.

	Barrel Graduations Exposed	Thimble Reading	Micrometer Reading
1.	4	6	
2.	5	10	
3.	8	18	
4.	2	24	
5.	0	8	
6.	1	20	
7.	9	22	
8.	0	22	
9.	3	0	
10.	0	0	

11. What is the micrometer reading when the thimble makes five full turns, starting from zero on the barrel?

12. A micrometer thimble makes $\frac{2}{3}$ of a full turn. What linear travel of the spindle results?

13. A thimble is turned five times and then rotated in the opposite direction ten divisions on the thimble. How much total change in the gap between anvil and spindle has taken place?

14. How many full turns of the thimble are required to show a reading of 0.275 inch?

15. What is the micrometer reading when the edge of the thimble is between the 0.125 and 0.150 graduations, and graduation 18 on the thimble is the coinciding line?

16. Find the reading if the thimble is between 0.850 and 0.875, and graduation 3 on the thimble coincides.

17. What would be the reading on a micrometer if you can see three graduations from zero to the edge of the thimble, and thimble graduation 24 lines up with the index line?

18. Starting at zero, how many turns of the thimble would it take, and what number on the thimble would you line up with the index line, to get a reading of 0.269?

19. If the thimble is between zero and 0.025, and the coinciding graduation of the thimble is 8, what would be the micrometer reading?

20. The thimble is between 0.975 and 1.000, and the coinciding graduation on the thimble is 1. What is the reading?

7-2 The Vernier Scale

OBJECTIVE

After completing this unit, you should be able to:

1. Read a vernier.

A **vernier** is a secondary scale placed opposite an existing scale on a measuring tool to show fractional parts of the divisions of the primary scale. The vernier is based on the principle that while the human eye cannot measure the exact distance between two parallel lines, it *can* distinguish two opposing lines that align with each other. Figure 7-7 helps clarify this.

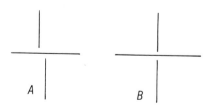

Figure 7-7

In (a), the opposing lines are separated from one another, although at what distance it cannot be determined. In (b), the two opposing lines coincide, or line up, with one another. Now refer to Figure 7-8.

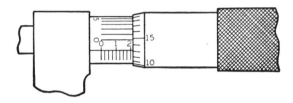

Figure 7-8

There are parallel lines marking off ten spaces on the micrometer barrel. Since one space on the thimble equals 0.001 inch, nine spaces equal 0.009 inch. If ten spaces on the vernier equal 0.009 inch, one vernier space equals one-tenth of 0.009 inch, or 0.0009 inch. The difference between a thimble and a vernier space is:

0.0010 (thimble)
−0.0009 (vernier)
0.0001 inch

Thus the value of each graduation on the vernier scale is 0.0001 inch.

The number of the graduation on the vernier that coincides with an opposing thimble graduation is added to the barrel and thimble reading to give a total reading.

Example 7-2   Give the reading shown in Figure 7-8.

Solution   Numbered graduations on barrel (2) = 0.200 in.
Additional graduations exposed (0) = 0.000 in.
Thimble reading = 0.013 in.
Vernier graduations (no. 4 coincides) = 0.0004 in.
Total reading = 0.2134 in.

One of the newer types of micrometers, shown in Figure 7-9, does not use a vernier on the barrel. A small white arrow on this model points to the reading in the window of the thimble. Tenths of an inch are read from the barrel, while thousandths and ten-thousandths are displayed in the window.

Figure 7-9 New micrometer.
(Photo courtesy of Brown & Sharpe Mfg. Co.)

Exercises 7-2

Fill in the complete readings for a ten-thousandths micrometer.

	Barrel Graduations Showing	Thimble Reading	Vernier Line Coinciding with Thimble Graduation	Complete Reading
1.	6	12	4	
2.	4	18	9	
3.	2	5	1	
4.	0	21	6	
5.	4	0	2	
6.	10	10	5	
7.	9	19	0	
8.	5	16	7	
9.	0	24	8	
10.	13	11	3	

11. A machinist measuring a piece of stock with a ten-thousandths micrometer got a reading as follows: the thimble between 0.125 and 0.150, index line between thimble graduations 16 and 17, coinciding vernier graduation 3. What dimension did the machinist have?

12. Give the four vernier micrometer readings in Figures A, B, C, and D.

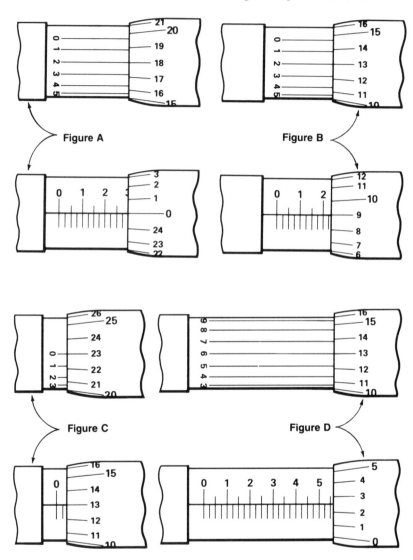

7-3 The Vernier Caliper

OBJECTIVE

After completing this section, you should be able to:

1. Read a vernier caliper.

The principle of the vernier on a caliper is identical to that of a ten-thousandths micrometer. A vernier caliper is shown in Figure 7-10.

Figure 7-10 Vernier caliper.   *(Courtesy of The L. S. Starrett Company)*

There are 25 spaces on the vernier scale, which are separated by 26 graduations. On the main, or primary, scale, there are 24 spaces in the same distance (see Figure 7-11).

Each division on the main scale = 0.025 inch
Each division on the vernier = 0.024 inch
Difference = 0.001 inch

To find the value of any vernier, divide one division of the main scale by the number of divisions on the vernier. The procedure for reading a vernier caliper is as follows:

1. Read the inches, tenths of an inch, and 0.025-inch marks up to the zero on the vernier.

2. To these, add the graduation on the vernier closest to a graduation on the main scale.

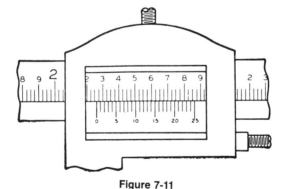

Figure 7-11

Example 7-3   Read the vernier caliper in Figure 7-11.

Solution   Number of inches (2) = 2.000 in.
Number of tenths of an inch (2) = 0.200 in.
Number of 0.025-in. marks (2) = 0.050 in.
Vernier graduation (no. 13 coincides) = 0.013 in.
Total reading = 2.263 in.

Figure 7-12 illustrates three types of calipers. The one at the top of the photograph has a direct-reading dial in thousandths of an inch. The caliper in the center is provided with a vernier for either thousandths of an inch or centimeters. At the bottom is a metric caliper with a vernier measuring to 150 millimeters by 0.02-millimeter increments.

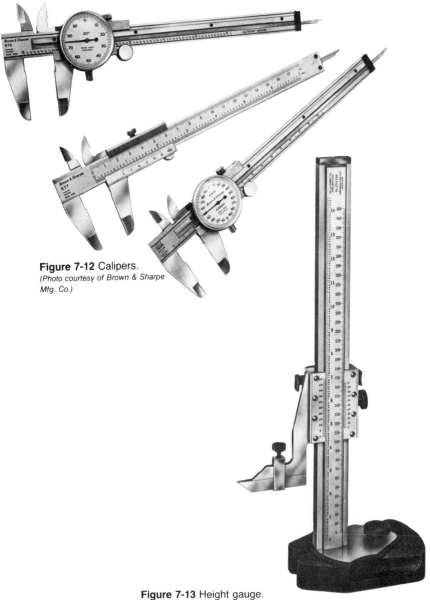

Figure 7-12 Calipers.
(Photo courtesy of Brown & Sharpe Mfg. Co.)

Figure 7-13 Height gauge.
(Courtesy of The L. S. Starrett Company)

Verniers may be placed on many varieties of measuring tools, thereby increasing the versatility of the tool. A height gauge, shown in Figure 7-13, has a vernier identical to those found on vernier calipers.

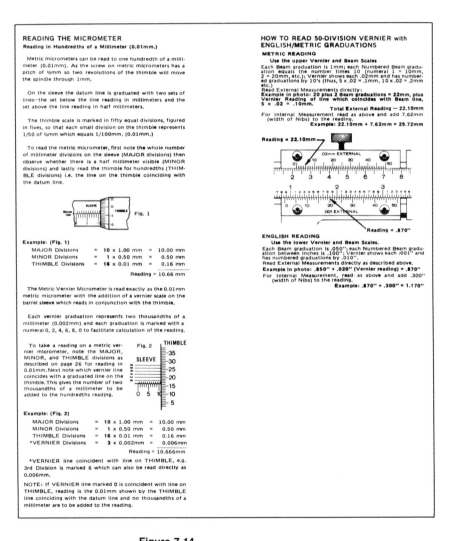

READING THE MICROMETER
Reading in Hundredths of a Millimeter (0.01mm.)

Metric micrometers can be read to one hundredth of a millimeter (0.01mm). As the screw on metric micrometers has a pitch of ½mm so two revolutions of the thimble will move the spindle through 1mm.

On the sleeve the datum line is graduated with two sets of lines—the set below the line reading in millimeters and the set above the line reading in half millimeters.

The thimble scale is marked in fifty equal divisions, figured in fives, so that each small division on the thimble represents 1/50 of ½mm which equals 1/100mm. (0.01mm.)

To read the metric micrometer, first note the whole number of millimeter divisions on the sleeve (MAJOR divisions) then observe whether there is a half millimeter visible (MINOR divisions) and lastly read the thimble for hundredths (THIMBLE divisions) i.e. the line on the thimble coinciding with the datum line.

Example: (Fig. 1)

MAJOR Divisions	= 10 x 1.00 mm	= 10.00 mm
MINOR Divisions	= 1 x 0.50 mm	= 0.50 mm
THIMBLE Divisions	= 16 x 0.01 mm	= 0.16 mm
	Reading =	10.66 mm

The Metric Vernier Micrometer is read exactly as the 0.01mm metric micrometer with the addition of a vernier scale on the barrel sleeve which reads in conjunction with the thimble.

Each vernier graduation represents two thousandths of a millimeter (0.002mm) and each graduation is marked with a numeral 0, 2, 4, 6, 8, 0 to facilitate calculation of the reading.

To take a reading on a metric vernier micrometer, note the MAJOR, MINOR, and THIMBLE divisions as described on page 26 for reading in 0.01mm. Next note which vernier line coincides with a graduated line on the thimble. This gives the number of two thousandths of a millimeter to be added to the hundredths reading.

Example: (Fig. 2)

MAJOR Divisions	= 10 x 1.00 mm	= 10.00 mm
MINOR Divisions	= 1 x 0.50 mm	= 0.50 mm
THIMBLE Divisions	= 16 x 0.01 mm	= 0.16 mm
*VERNIER Divisions	= 3 x 0.002mm	= 0.006mm
	Reading =	10.666mm

*VERNIER line coincident with line on THIMBLE, e.g. 3rd Division is marked 6 which can also be read directly as 0.006mm.

NOTE: If VERNIER line marked 0 is coincident with line on THIMBLE, reading is the 0.01mm shown by the THIMBLE line coinciding with the datum line and no thousandths of a millimeter are to be added to the reading.

HOW TO READ 50-DIVISION VERNIER with ENGLISH/METRIC GRADUATIONS
METRIC READING
Use the upper Vernier and Beam Scales
Each Beam graduation is 1mm; each Numbered Beam graduation equals the number times 10 (numeral 1 = 10mm, 2 = 20mm, etc.); Vernier shows each .02mm and has numbered graduations by 10's (thus, 5 x .02 = .1mm, 10 x .02 = .2mm etc.)
Read External Measurements directly:
Example in photo: 20 plus 2 Beam graduations = 22mm, plus Vernier Reading of line which coincides with Beam line, 5 x .02 = .10mm.
Total External Reading — 22.10mm
For Internal Measurement read as above and add 7.62mm (width of Nibs) to the reading.
Example: 22.10mm + 7.62mm = 29.72mm

Reading = 22.10mm

Reading = .870"

ENGLISH READING
Use the lower Vernier and Beam Scales.
Each Beam graduation is .050"; each Numbered Beam graduation between inches is .100"; Vernier shows each .001" and has numbered graduations by .010".
Read External Measurements directly as described above.
Example in photo: .850" + .020" (Vernier reading) = .870"
For Internal Measurement, read- as above and add .300" (width of Nibs) to the reading.
Example: .870" + .300" = 1.170"

Figure 7-14
(Courtesy of Brown & Sharpe Mfg. Co.)

Figure 7-14 illustrates how to read a metric or an English micrometer.

Exercises 7-3

The zero position of the vernier and the coinciding vernier line are given.
Fill in the vernier readings.

	Position of Zero on Vernier	Coinciding Vernier Line	Vernier Reading
1.	1 in. and 3 grad.	8	
2.	1 in. and 32 grad.	21	
3.	3 in. and 0 grad.	1	
4.	2 in. and 5 grad.	16	
5.	1 in. and 39 grad.	22	
6.	0 in. and 12 grad.	5	

	Position of Zero Between Lines	Coinciding Vernier Line	Vernier Reading
7.	0.125 and 0.150	16	
8.	0.250 and 0.275	18	
9.	0.000 and 0.025	6	
10.	0.050 and 0.075	24	
11.	0.350 and 0.375	14	

12. What would be the reading on a 6-in. vernier caliper if the zero were to the right of 1 in., $2\frac{1}{2}$ lines past the 8, and the coinciding vernier line were 11?

13. The index line (or zero) is between 0.275 in. and 0.300 in. on the main scale, and line 22 on the vernier coincides with a scale line. What is the reading?

14. Explain how to set a vernier caliper to read 3.629 in.

15. What would be the reading on a vernier caliper if the main scale read 4 in., the zero line were *between* the 4 and the first graduation, and the coinciding vernier line were 16?

16. If a machinist milled $\frac{7}{16}$ in. off a 1-in. part, how much material would be left? Explain the final reading on a vernier caliper if you had to measure the remaining material. Main scale reading? Vernier reading?

7-4 The Vernier Protractor

OBJECTIVE
After completing this section, you should be able to:

1. Read a vernier protractor.

A universal bevel protractor (Figure 7-15) is capable of measuring angles to the accuracy of 5 minutes. There are 60 minutes in each degree, so 5 minutes is one-twelfth of a degree.

The upper scale shows degrees. It is graduated in 1° increments, from 0 to 90 and back to 0 for each half of the dial. The lower scale is the vernier. It is graduated from 0 to 60 in both directions. Each graduation represents 5 minutes.

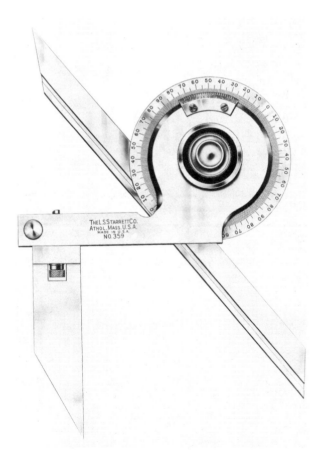

Figure 7-15 Universal bevel protractor.
(Courtesy of The L. S. Starrett Company)

If the numbers on the upper scale ascend to the left, as in Figure 7-16, vernier 0 lies to the left of the number of degrees.

To find the number of minutes, locate the vernier line to the left of vernier 0 that coincides with any line on the upper scale. On the vernier, count how many graduations the coinciding line is from 0. The number of graduations from 0 is multiplied by 5 to give the number of minutes.

Figure 7-16
(Courtesy of The L. S. Starrett Company)

Example 7-4   Read the vernier in Figure 7-16.

   Solution   Vernier 0 lies to the left of 50. $50° \, 0'$

The fourth vernier line to the left of vernier 0 coincides.
$4 \times 5 = 20$ $\underline{20'}$

Total reading $= 50° \, 20'$

If the twelfth vernier line coincides, the reading is not $60'$, but $0'$.

If the upper scale ascends to the right, vernier 0 lies to the right of the number of degrees, and the coinciding line is read to the right of vernier 0.

Exercises 7-4

Complete the protractor readings. The numbers on the upper scale ascend to the left.

	Vernier Zero Is to the Left of:	Coinciding Vernier Line to the Left of Zero Is:	Total Reading
1.	15°	fourth	
2.	56°	eleventh	
3.	23°	first	
4.	35°	first	
5.	89°	tenth	

Complete the protractor readings. The numbers on the upper scale ascend to the right.

	Vernier Zero Is to the Right of:	Coinciding Vernier Line to the Right of Zero Is:	Total Reading
6.	25°	second	
7.	5°	fifth	
8.	40°	ninth	
9.	36°	seventh	
10.	12°	third	

11. The vernier 0 coincides exactly with 27° on the upper scale. The twelfth line (60′) from vernier 0 coincides with a line on the upper scale. What is the reading?

12. Explain how to set a vernier protractor for an angle of 72° 45′ with the upper scale ascending to the right.

13. Give the Vernier protractor readings shown in Figures A, B, and C.

Figure A

Figure B

Figure C

(Courtesy of Scherr-Tumico)

7-5 Tolerances and Limits

OBJECTIVES

After completing this section, you should be able to:

1. Find the upper and lower limits of a measurement.
2. Determine the tolerance of a measurement.

One of the most important applications of addition and subtraction of decimals occurs when a designer must determine how two parts will work together as an assembly. The function of the assembly dictates the size that each of its mating parts will be.

Since it is very difficult to consistently manufacture precision parts that will be exactly the same size, a certain amount of variation from a basic, or desired, size is allowed. The amount that a given part is permitted to vary from the desired size is called **tolerance**. The machinist and toolmaker must understand how tolerance affects their work. Lack of knowledge in figuring tolerance can result in spoilage of machined parts that is both expensive and unnecessary.

On every blueprint, there is a title block in the lower right-hand corner. In this title block is a specification for tolerance (see Figure 7-17). Usual shop tolerance, unless otherwise specified, is $\pm \frac{1}{64}$ inch for fractional dimensions and ± 0.005 inch for decimal dimensions. This means that a shaft that is specified $1\frac{5}{8}$ inches in diameter can be as large as $1\frac{41}{64}$ inches or as small as $1\frac{39}{64}$ inches in diameter without being rejected as a bad part.

DO NOT SCALE	
TOLERANCE UNLESS OTHERWISE SPECIFIED: FRACTIONAL ± 1/64. DECIMAL ± .005. SHEET METAL FRACTIONAL ±1/16. DIMENSIONS ARE FINAL AND INCLUDE "FINISH" WHEN SPECIFIED.	FILE
	APR.
	CKD.
ASSEMBLY	DRW'N. BY
MAT'L	
DATE	QTY.REQ'D.
MACHINE	
NAME	
PART NO.	PATT. NO.
SCALE:	DWG. NO.

Figure 7-17

If the same shaft has a diameter of 1.625 inches, the diameter can be as large as 1.630 inches or as small as 1.620 inches. The *upper limit* of the tolerance is 1.630 inches and the *lower limit* of the tolerance is 1.620 inches. Every tolerance has three parts: the desired size, or **nominal size, the upper limit,** and the **lower limit.**

The designer assigns a specific nominal size and tolerance. If, for instance, a shaft is intended to fit into a precision bearing, the diameter of the shaft must be held to a *close,* or *tight,* size and tolerance. However, if the same shaft is merely to connect two moving parts, it may have an *open,* or *loose,* diameter and tolerance. Examples of both close and open tolerances are given in Figure 7-18.

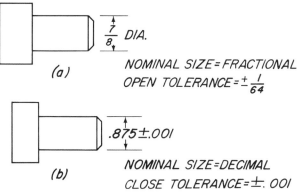

Figure 7-18

Sometimes the nominal size is also the upper or lower limit of the tolerance. The opening of the familiar "open-end" wrench used by all mechanics must be a minimum size or it will not fit on bolt heads of the size stamped on the wrench. The opening cannot be too large, on the other hand, or the wrench will not fit snugly. This will cause the corners of the bolt head to round off.

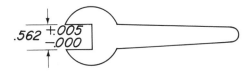

Figure 7-19

Figure 7-19 shows an open-end wrench with the dimension of its opening correctly stated. This is an example of *unilateral tolerance,* or tolerance in one direction only. The nominal size is also the lower limit of the tolerance, for reasons just discussed. The tolerance runs in a plus direction only. The lower limit (nominal size) is 0.562 inch. The upper limit is 0.567 inch. The complete tolerance allowed is the sum of both minus and plus limits—in this case, 0.005 inch.

If a hole to be bored in a workpiece is specified to be 1.500 ± 0.002 inches, this is called *bilateral tolerance.* Bilateral means running in both directions. The complete tolerance allowed is the sum of both plus and minus limits, or 0.002 + 0.002, which equals 0.004 inch. Therefore, the hole could vary in size from 1.498 inches to 1.502 inches, although 1.500 inches is nominal and is the size that the machinist attempts to bore.

Example 7-5 The finished dimension of a part is

$$1.500 \text{ cm } \begin{array}{c} + \ 0.002 \\ - \ 0.004 \end{array}$$

Find the nominal dimension, the upper limit, the lower limit, and the total tolerance.

Solution Nominal dimension: 1.500 cm
Upper limit: 1.500 + 0.002 = 1.502 cm
Lower limit: 1.500 − 0.004 = 1.496 cm
Total tolerance: 0.002 + 0.004 = 0.006 cm

Exercises 7-5

Complete the following table.

	Finished Dimension		Nominal Dimension	Upper Limit	Lower Limit	Total Tolerance
1.	1.150	+0.004 −0.000				
2.	1.625	+0.002 −0.000				
3.	0.875	+0.001 −0.002				
4.	1.750	±0.001				
5.	0.500	+0.0015 −0.0000				
6.	2.125	+0.003 −0.001				
7.	$\frac{5}{8}$	$\pm \frac{1}{64}$				
8.	1.312	+0.000 −0.001				
9.	0.8125	±0.0005				

Unit 7 Review Exercises

1. How many threads per inch are there on the micrometer spindle?

2. What part of an inch does each graduation on the thimble of the micrometer represent?

3. A machinist measuring the diameter of a part on the lathe reads the following on a micrometer: 5 graduations on the barrel are exposed and the thimble reading is 6. What is the diameter of the part?

4. The vernier scale on a micrometer shows the fractional parts of a division of the primary scale. What part of an inch does each vernier division represent?

5. What would be the reading on a ten-thousandths micrometer if 6 graduations were showing on the barrel, 12 graduations were showing on the thimble, and the vernier line coinciding with a thimble graduation were 4?

6. The sliding jaw of the vernier caliper contains the vernier scale. Each graduation on this scale represents what part of an inch?

7. What would be the reading on a vernier caliper if the zero were to the right of 5 in., $2\frac{1}{2}$ graduations past 2, and the coinciding line on the vernier scale were 16?

8. The upper scale on a vernier protractor ascends to the right. Vernier 0 lies to the right of 75. The fifth vernier line from 0 coincides with a line on the upper scale. What is the reading?

9. A machinist measured the angle of a cut and got a reading of 10° 30'. Where is the 0 on the vernier scale? What is the coinciding line?

10. A blueprint shows the following dimension for a part: 16.5 mm, +0.03 − 0.05. Find the nominal dimension, the lower limit, the upper limit, and the total tolerance.

Geometry

Geometry defines shapes and certain properties associated with those shapes. Machinists need this knowledge, since no machine is designed or mechanism constructed without the application of some of these properties.

Basic shapes include the triangle, circle, and parallelogram. Other shapes, such as cylinders, spheres, and pyramids, may be built from these. All shapes are a combination of straight or curved lines and flat or curved surfaces. The machinist needs to measure the angles formed by lines, the lengths of lines, and the areas enclosed by lines.

This unit will introduce some basic shapes and their properties.

8-1 The Angle

OBJECTIVES
After completing this section, you should be able to:
1. Find the measure of an included angle.
2. Subtract and divide angle measures.
3. Find the decimal equivalent of an angle measure.
4. Convert an angle measure expressed as a decimal to degrees, minutes, and seconds.

If two straight lines meet at a point, or **vertex,** they form an angle. The hands of a clock are a good example of this. The length of the hands does *not* determine the size of the angle formed by them. The amount of opening between the two joined lines fixes the size of an angle. Figure 8-1 illustrates this. The angles in Figure 8-1 are named by the points: angle *ACB*. The point at the vertex is in the center and is sometimes used alone to identify the angle. Thus, angle *ACB* is also called angle *C*.

179

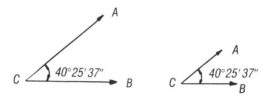

Figure 8-1 Angles.

Angles are commonly expressed in degrees or fractions of a degree. A degree is divided into 60 equal parts, called minutes. Each minute is divided into 60 seconds. This system of angular measurement, which is based on the number 60, comes from the ancient Chaldeans, who used it in astronomical studies. The following units are used in angular measurement.

1 degree (°) = 60 minutes(') = 3600 seconds (")

Note the symbols for degrees, minutes, and seconds. There are several types of angles of interest.

A **right** angle contains 90°.

A **straight** angle contains 180°.

An **acute** angle contains less than 90°.

An **obtuse** angle contains more than 90° but less than 180°.

Two angles are **complementary** if their sum equals 90°.

Two angles are **supplementary** if their sum equals 180°.

One of the most frequent applications of angles is found in the machining of tapered or sloped surfaces on workpieces in a lathe. The angle formed by the taper is designated on a blueprint as an **included** angle. An included angle is one equally divided by the center line, or central axis, of the workpiece. Figure 8-2 shows such a workpiece and the included angle as it might appear on a shop print.

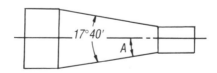

Figure 8-2 Included angle.

Example 8-1 Find the measure of angle A in Figure 8-2.

Solution Observe that A is half the included angle of $17°40'$.

$$\frac{17°}{2} = 8\tfrac{1}{2}° \text{ or } 8°30'$$

$$\frac{40'}{2} = 20'$$

$$\text{Angle } A = 8°30' + 20'$$
$$= 8°50'$$

If two angles are complementary, the measure of the unknown angle can be found by subtracting the measure of the given angle from $90°$. Similarly, if two angles are supplementary, the measure of the unknown angle can be found by subtracting the measure of the given angle from $180°$.

To subtract angles from $90°$ or $180°$, remember that $90°$ equals $89°60'$ or $89°59'60''$ and that $180°$ equals $179°60'$ or $179°59'60''$. Any other angle can be similarly renamed by rewriting 1 degree as 60 minutes and 1 minute as 60 seconds.

Example 8-2 Subtract $21°36'$ from $90°$.

Solution $90° = \quad 89°60'$
$\qquad\qquad\quad \underline{-21°36'}$
$\qquad\qquad\qquad 68°24'$

Example 8-3 Subtract $25°59'37''$ from $27°$.

Solution $27° = \quad 26°59'60''$
$\qquad\qquad\quad \underline{-25°59'37''}$
$\qquad\qquad\quad\ 1°\ 0'23'' \text{ or } 1°23''$

Example 8-4 Subtract $68°19'31''$ from $69°18'$.

Solution In this case, $60''$ is borrowed from $18'$, which then becomes $17'$. However, $19'$ cannot be subtracted from $17'$; therefore, $60'$ is borrowed from $69°$ and added to $17'$.

$\qquad\qquad 69°18' = \quad 68°77'60''$
$\qquad\qquad\qquad\qquad \underline{-68°19'31''}$
$\qquad\qquad\qquad\qquad\ 0°58'29'' \text{ or } 58'29''$

Fractional parts of a degree can be converted into decimal form by using 60 as a divisor.

Example 8-5   Convert 47′ to decimal form.

Solution $\dfrac{47'}{60} = 0.783°$

Example 8-6   Find the decimal equivalent of 18°19′.

Solution $18°19' = 18\dfrac{19°}{60} = 18.317°$

A decimal part of a degree can be changed into the equivalent number of minutes and seconds by multiplying by 60.

Example 8-7   Convert 6.53° to degrees, minutes, and seconds.

Solution $6.53 = 6° + (0.53 \times 60) = 6° + 31.8' = 6°32'$
$= 6° + 31' + (0.8 \times 60) = 6°31'48''$

Example 8-8   Change 11.7251° to degrees, minutes, and seconds.

Solution $11.7251° = 11° + (0.7251 \times 60) = 11° + 43.506'$
$= 11° + 43' + (0.506 \times 60) = 11°43'30''$

Many times, the machinist or toolmaker needs to divide an angle into several equal parts. Finding half the included angle of a tapered workpiece is a good example. The procedure is stated in Rule 8-1.

RULE 8-1   DIVIDING AN ANGLE

Step 1   Divide the angle by the desired number of divisions and carry the answer out to *four* decimal places.

Step 2   Convert the decimal into minutes and seconds with a multiplier of 60.

Example 8-9   Divide 27° into eight equal parts, to include minutes and seconds of a degree.

Solution $\dfrac{27°}{8} = 3.3750° = 3° + (0.375 \times 60) = 3°22.5'$
$= 3°22' + (0.5 \times 60) = 3°22'30''$

Exercises 8-1

Find the measure of each indicated angle.

1.   **2.**   **3.**

16°38'20"   10°39'47"   62°49'50"

Subtract each of the following measures from 90°.

4. 21°53'   **5.** 48°21'   **6.** 6°9'

7. 32°18'36"   **8.** 27°43'14"   **9.** 41°30'41"

10. 68°19'31"   **11.** 10°54'9"   **12.** 72°51'12"

Subtract the second measure from the first.

13. 180°; 37°28'   **14.** 180°; 118°16'

15. 180°; 97°48'   **16.** 180°; 70°36'15"

17. 29°; 18°37'24"   **18.** 56°; 10°50'8"

Express the following as degrees and decimal parts of a degree.

19. 25°45'   **20.** 17°15'   **21.** 16°10'   **22.** 40°50'

23. 36°12'   **24.** 21°20'   **25.** 27°36'   **26.** 19°18'

Convert the following to degrees, minutes, and seconds.

27. 16.337°   **28.** 89.117°   **29.** 31.755°

30. 11.212°   **31.** 54.545°   **32.** 1.111°

33. 19.625°   **34.** 13.483°   **35.** 67.812°

36. A circle of 360° divided into seven equal parts is shown in Figure A. What angle is contained in each part?

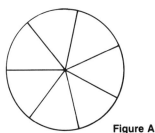

Figure A

37. If the included angle of a taper plug gauge is 9°57′, what is the angle with the center line in degrees, minutes, and seconds?

38. A lathe machinist turns a part at an angle with the center line of 28°14′17″. What is the included angle?

39. An apprentice toolmaker figures that one-half of $5\frac{2}{3}°$ is equal to 2°48′59″. Is he correct? Justify your answer.

40. What is one-twelfth of 32° in degrees, minutes, and seconds?

8-2   Lines and Angles

OBJECTIVES

After completing this section, you should be able to:

1. Identify vertical angles.
2. Find equal and supplementary angles in parallel lines cut by a transversal.

Both machinists and toolmakers must be able to analyze blueprints or working sketches to identify angles that may not be marked. Angles that have common sides formed by the same line are often equal in measure. An example of this is given in Figure 8-3.

Lines *GN* and *AB* are parallel. When parallel lines are cut by a line such as line RS, called a **transversal, alternate interior** angles are equal in measure. Angles *GPC* and *BCP* are alternate interior angles. They have the same measure. In Figure 8-3, these angles are identified as angles 3 and 5.

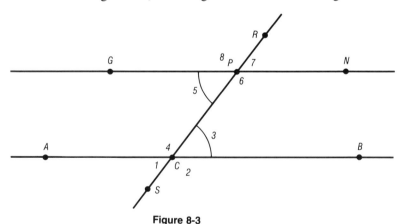

Figure 8-3

Example 8-10   Name two other equal angles in Figure 8-3.

Solution   Angles *NPC* and *ACP* are also alternate interior angles. Since lines *AB* and *GN* are parallel, angles *NPC* and *ACP* are equal.

There are other angle relationships that occur when parallel lines are cut by a transversal.

Corresponding angles on the same side of the transversal are equal in measure.

Alternate exterior angles are also equal in measure.

In Figure 8-3, angles 1 and 5 and angles 3 and 7 are corresponding angles. Angles 1 and 7 and angles 2 and 8 are alternate exterior angles.

Example 8-11   In Figure 8-3, find an angle equal to angle 4. Tell why they are equal.

Solution   Angle 6 is equal to angle 4 because when parallel lines are cut by a transversal, alternate interior angles are equal. Also, angle 8 is equal to angle 4 because when parallel lines are cut by a transversal, corresponding angles are equal.

Two intersecting lines form pairs of angles called **vertical** angles. In Figure 8-3, angles 5 and 7 are vertical angles, as are angles 8 and 6. Vertical angles are equal, so angles 5 and 7 are equal, as are angles 8 and 6.

Example 8-12   Angle 1 (in Figure 8-3) = 70°40'. What is the measure of angle 3?

Solution   Since angles 1 and 3 are vertical angles, they are equal. Angle 3 = 70°40'

Two angles such as 3 and 4 also have a special relationship. They have a side in common and their outer sides form a straight line. Angles 3 and 4 are therefore **supplementary,** which means their measures have a sum of 180°.

Example 8-13   Name other pairs of supplementary angles in Figure 8-3.

Solution   Angles 1 and 2   Angles 2 and 3
Angles 1 and 4   Angles 5 and 6
Angles 6 and 7   Angles 7 and 8
Angles 5 and 8

Example 8-14   Angle 1 (in Figure 8-3) = 70°40'. Find the measure of angle 4.

Solution   Subtract 70°40' from 180°.
$$180° = \quad 179°60'$$
$$\underline{- \quad 70°40'}$$
$$109°20' = \text{angle } 4$$

Exercises 8-2

Lines EF and GH are parallel. Lines AB and CD are also parallel. Complete the following groups of identical angles in Figure A.

1. Group 1: angle *GOB* = angle *RPS* = ?
2. Group 2: angle *BOX* = ?
3. Group 3: angle *JZY* = ?

4. Group 4: angle *DZJ* = ?
5. Group 5: angle *TSX* = ?

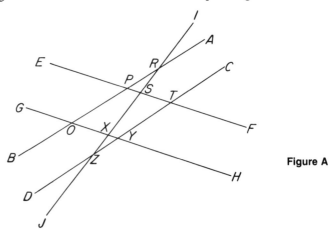

Figure A

Use Figure A for exercises 6–8.

6. If angle *TYH* = 48°20′, find angle *ZYO*.
7. If angle *CTF* = 48°20′, find angle *FTY*.
8. If angle *EPR* = 115°18′7″, find angle *EPO*.
9. In Figure B, if the shaper toolhead is swiveled 40° from the vertical position, how many degrees are in angle *A*?

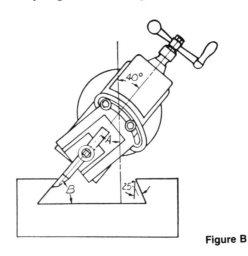

Figure B

10. To sharpen a drill, the machinist holds it against the grinding wheel, as shown in Figure C. How many degrees are in angle *C*? (*Hint:* Angle *C* = Angle *B*.)

11. In making the drill grinding gauge, Figure D, how many degrees are in angle *A*?

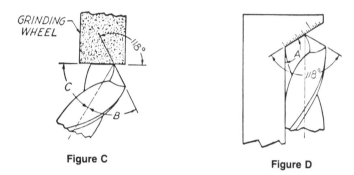

Figure C

Figure D

12. In Figure E, the machinist measures angle *A* to check the dovetail angle. What is angle *A*?

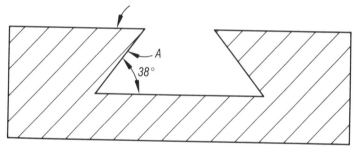

Figure E

13. In Figure F, find the bevel protractor reading for angle *B*.

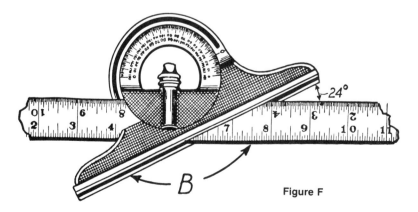

Figure F

14. In Figure G, how many degrees are in angle A? (*Hint:* $A + B + C = 180°$.)

15. In Figure H, gauge blocks are put under the discs of the sine bar until side b is parallel with the top of the surface plate. How many degrees are in angle A of the plug gauge?

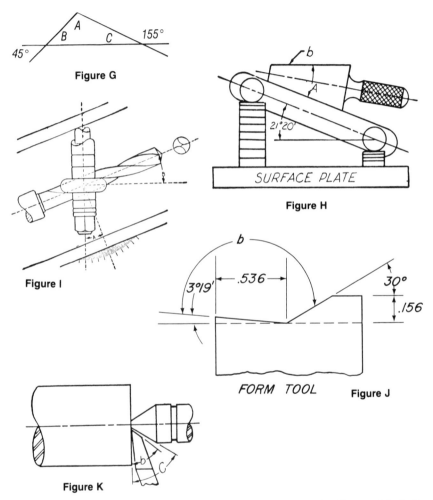

Figure G

Figure H

Figure I

Figure J

FORM TOOL

Figure K

16. In Figure I, angle B is the helix angle of a twist drill. To cut this helix on a universal milling machine, the toolmaker swivels the machine table until angle A reads $20°$ on the graduated scale at the base of the table. What is angle B?

17. In Figure J, what is angle b?

18. In Figure K, to what angle must the machinist grind the cutter bit so there will be $2°$ clearance between the side of the dead center and the side of the cutter bit? (The included angle of the dead center is $60°$.)

8-3 The Triangle

OBJECTIVES

After completing this section, you should be able to:

1. Identify the parts of a triangle.
2. Use the Pythagorean theorem.
3. Solve problems about triangles.

A **triangle** is a plane figure bounded by three straight lines. Every triangle contains three angles, the sum of which is 180°. If you know the value of two angles of a triangle, you can find the value of the third. Figure 8-4 shows a triangle with the value of each angle. Observe that their sum is 180°.

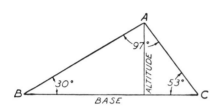

Figure 8-4 Triangle.

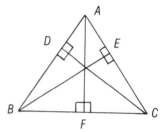

Figure 8-5 Altitudes of a triangle.

The **altitude** of a triangle is the perpendicular drawn from a vertex to the side opposite that vertex. A triangle has three sides and, therefore, three altitudes. In Figure 8-5, line *CD* is the altitude to side *AB*, line *BE* is the altitude to side *AC*, and line *AF* is the altitude to side *BC*.

An **isosceles triangle** has two equal sides (see Figure 8-6). In the isosceles triangle shown, side *AB* equals side *AC*. Furthermore, angle *B* is equal to angle *C*. The altitude drawn from the vertex to the third side of the isosceles triangle bisects the third side and forms two *equal right triangles*.

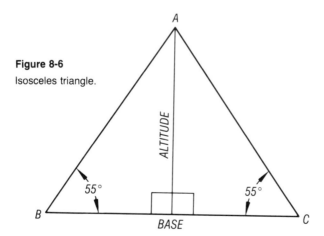

Figure 8-6

Isosceles triangle.

A **right triangle** contains one 90° angle. Figure 8-7 illustrates a right triangle. The side opposite the right angle is called the **hypotenuse** and is the longest side. Side *CB* is the *altitude*, and side *AC* is the *base*. If *BC* is the base, *AC* becomes the altitude.

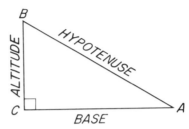

Figure 8-7 Right triangle.

The right triangle is one of the most important shapes in design and shop work. It is the basis for **rectangular coordinates,** which is a system for precision hole location in toolroom jig bore work. Figure 8-8 shows a set of rectangular coordinates utilizing the right triangle (shaded area).

The sum of the two acute angles of a right triangle equals 90°. Thus, if one acute angle of a given right triangle equals 40°, the other acute angle must equal 50°. Each acute angle is the *complement* of the other.

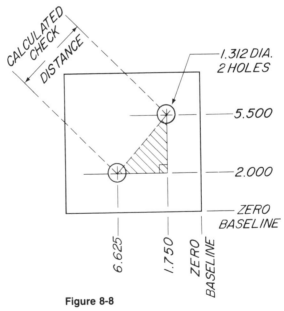

Figure 8-8

Unit 5 introduced the Pythagorean theorem. This theorem states the relationship between the three sides of a right triangle. If the values of two sides of a right triangle are known, this formula provides a way to find the value of the third side.

RULE 8-2   USING THE PYTHAGOREAN THEOREM TO SOLVE
RIGHT TRIANGLES

$c^2 = a^2 + b^2$ or (hypotenuse)^2 = (altitude)^2 + (base)^2

$c^2 = a^2 + b^2$ or $c = \sqrt{a^2 + b^2}$

$a^2 = c^2 - b^2$ or $a = \sqrt{c^2 - b^2}$

$b^2 = c^2 - a^2$ or $b = \sqrt{c^2 - a^2}$

Example 8-15   Find the value of the hypotenuse c.

Solution $c = \sqrt{a^2 + b^2}$

$= \sqrt{9^2 + 12^2}$

$= \sqrt{81 + 144}$

$= \sqrt{225} = 15$ ft

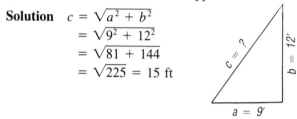

Example 8-16   Find the altitude of a right triangle whose hypotenuse is 13 ft and whose base is 5 ft.

Solution $a = \sqrt{c^2 - b^2}$

$= \sqrt{13^2 - 5^2}$

$= \sqrt{169 - 25}$

$= \sqrt{144} = 12$ ft

In an **equilateral triangle,** the three sides are equal and the three angles are equal. Since the sum of the three angles in any triangle equals 180°, *each angle of an equilateral triangle equals 60°.* The altitude drawn from any vertex of an equilateral triangle divides the triangle into two equal right triangles, and at the same time the altitude bisects the side to which it is drawn. In Figure 8-9, the altitude h divides the equilateral triangle into two equal right triangles. *CD* is equal to *DB*, and the angle *CAB* is bisected, thus making angle *CAD* and angle *BAD* each equal to 30°. These two equal right triangles are called *30°-60°-90°* triangles.

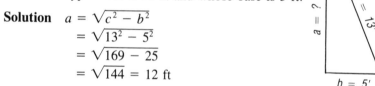

Figure 8-9 Equilateral triangle.

Example 8-17   Find the altitude of the triangle.

Solution   To find the altitude h, use the Pythagorean theorem.

$$h = a = \sqrt{c^2 - b^2}$$
$$= \sqrt{64 - 16} = \sqrt{48}$$
$$= 6.928 \text{ in.}$$

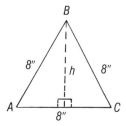

In a **scalene triangle,** no two sides are equal, and none of the angles of the triangle equals 90° (see Figure 8-10).

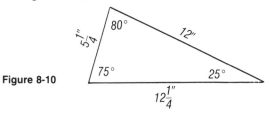

Figure 8-10

There are three *special right triangles.* In a **30°-60°-90° triangle,** the hypotenuse is *twice the length of the shortest side* (see Figure 8-11a). This property is used in the facing of work in an engine lathe to precise length. If the compound rest that holds the tool is set on 60° from the axis of the workpiece, the tool advances 0.0005 inch for each thousandth of an inch that the rest is fed toward the work. This is shown in Figure 8-12.

The second special triangle is the **45°-45°-90° triangle** (see Figure 8-11b). The two sides that include the 90° angle are equal. It is often called the *isosceles right triangle.*

The third special triangle is the **3-4-5 right triangle.** The three sides must be in the ratio of 3, 4, and 5. For example, 6 inches, 8 inches, and 10 inches are in the proper ratio. Check, using the Pythagorean theorem: $3^2 + 4^2 = 5^2$. This can be applied when it is required to square the corners of a plate or board. For example, in Figure 8-11c, it is required to square the plate at C. First, with C as a center, mark off 4 units on CD. With C as a center and a radius equal to 3 units, draw arc E. With A as a center and a radius equal to 5 units, draw arc F intersecting arc E at point B. Draw CB extended to the top, forming the 90° angle with edge CD.

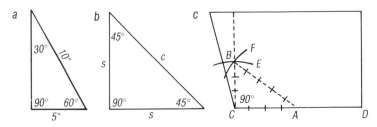

Figure 8-11

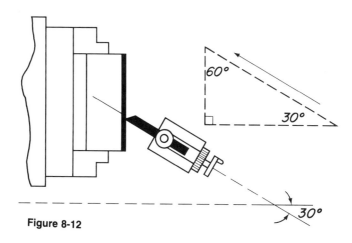

Figure 8-12

Exercises 8-3

Find the value of the third angle in each triangle.

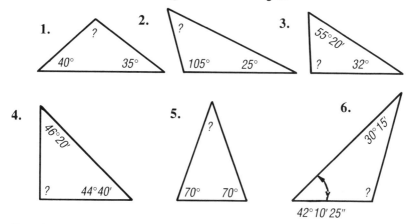

1. ? 40° 35°

2. ? 105° 25°

3. 55°20′ ? 32°

4. 46°20′ ? 44°40′

5. ? 70° 70°

6. 30°15′ 42°10′25″ ?

7. If one acute angle of a right triangle is 18°42′, what is the value of each of the other angles?

Find the unknown side of the right triangle, using the Pythagorean theorem.

8. $a = 6$; $b = 8$; find c.

9. $a = 9$; $b = 10$; find c.

10. $b = 13$; $c = 20$; find a.

11. $a = 6$; $b = 6$; find c.

12. $c = 8.4$; $a = 3$; find b.

13. $b = 5.3$; $c = 25$; find a.

14. $a = 2$; $c = 10$; find b.   **15.** $c = 8.484$; $a = 6$; find b.

16. $a = 9$; $b = 12$; find c.   **17.** $c = 14.2$; $b = 6.2$; find a.

18. In the isosceles triangle in Figure A, find the value of each unknown angle.

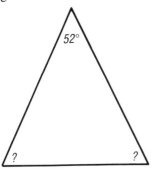

Figure A

19. Find the altitude of the isosceles triangle shown in Figure B.

20. In the isosceles triangle in Figure C, find the length of the base.

21. The base of an isosceles triangle is 15 in. and the altitude is 12 in. Find one of the equal sides.

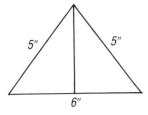 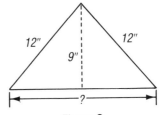

Figure B   Figure C

22. In the bevel gear drawing, Figure D, distance OA is called the *apex distance* and AB is called the *pitch diameter*. Find the pitch diameter if OA is $2\frac{13}{16}$ in. and OC is $1\frac{1}{4}$ in.

23. Compute the center distance between pulleys A and B in Figure E.

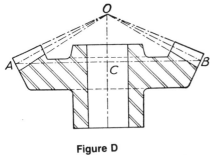

Figure D

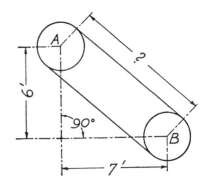

Figure E

24. Figure F illustrates a hole layout. What is distance *x*?

25. In Figure G, find the distance from hole *C* to hole *A*, center to center.

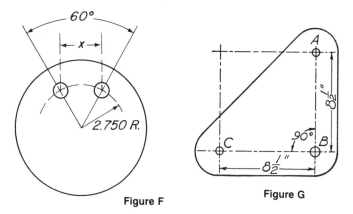

Figure F

Figure G

26. Compute the distance across the corners of the square-headed nut in Figure H.

27. Find (*a*) the length of the hypotenuse and (*b*) all the angles in the right triangle shown in Figure I.

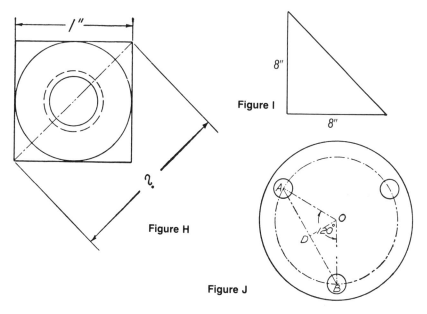

Figure I

Figure H

Figure J

28. The distance *0A* in Figure J is 6 in. What is the perpendicular distance *OD* to the center line of holes *A* and *B*?

29. A guy wire 60 ft in length is anchored to the ground 30 ft from the foot of the pole it supports. The ground is level from the foot of the pole to the end of the wire. How high is the pole?

30. In Figure K, the plate is to be burned off along line *AB,* which can be marked off by measuring *BC.* How long is *BC?*

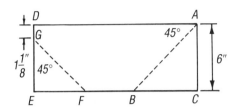

Figure K

31. In Figure K, find the distance *EF* to mark the plate so that the corner can be burned off along line *FG.*

32. In Figure K, what is the measurement *EG?*

33. In Figure L, the plate is to be burned off along line *AB.* What is the length of line *AB?*

34. In Figure L, if *AD* is to be 6 in., what must be the length of *AB* to maintain the 30° angle?

35. In Figure L, how many degrees are in angle *H?*

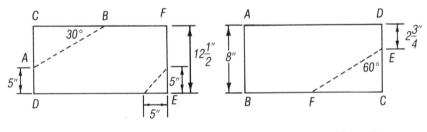

Figure L   Figure M

36. In Figure M, how many degrees are in angle *EFC?*

37. Figure N shows the section of a sharp V thread that is an equilateral triangle. If *p,* the pitch, equals ⅛ inch, find the thread depth *a*.

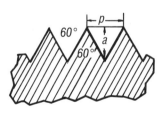

Figure N

8-4 Quadrilaterals and Hexagons

OBJECTIVES

After completing this section, you should be able to:

1. Solve problems about the angles and sides of quadrilaterals.
2. Find the distance across the flats of a regular hexagon.
3. Find the distance across the corners of a regular hexagon.

A **quadrilateral** is a plane figure bounded by four straight sides. The most common quadrilaterals found in machine shop work are the *parallelogram, rectangle,* and *square.* Another quadrilateral, the *trapezoid,* is useful in figuring the area of structural members.

A **parallelogram** is a quadrilateral whose opposite sides are parallel (see Figure 8-13). Opposite pairs of angles of a parallelogram are equal and the sum of all the angles is 360°. The altitude (*h*) of a parallelogram is the perpendicular distance between the base and the opposite side.

A **square** is parallelogram with four right angles and all sides equal in length. Figure 8-14 shows a square.

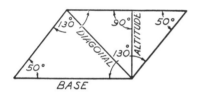

Figure 8-13 Parallelogram.

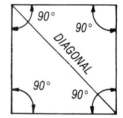

Figure 8-14 Square.

The formula for finding the diagonal of a square is the one most used by machinists and toolmakers for cutting square shapes from round workpieces. The diagonal of a square is, of course, the diameter of a circle that circumscribes, or just touches, each corner of the square (see Figure 8-15). Choosing a minimum-size bar enables the machinist to mill four flats that will form squares for a wrench or mating square hole.

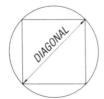

Figure 8-15

RULE 8-3 FINDING THE DIAGONAL OF A SQUARE

The diagonal of a square is approximately equal to 1.4142 × *s*, where *s* is the length of a side.

Example 8-18   What is the minimum-diameter stock that is necessary to mill a square 5 cm per side?

Solution   The diameter is equal to the length of the diagonal.

$$d = 1.4142 \times s = 1.4142 \times 5$$
$$= 7.071$$

The minimum diameter necessary is about 7.1 cm.

A **rectangle** is a parallelogram with four angles of 90°. All squares are rectangles, but not all rectangles are squares. Figure 8-16 shows a typical rectangle. A diagonal divides the rectangle into *two equal right triangles*. The diagonal of a rectangle is the hypotenuse common to both right triangles. Rule 8-4 gives a formula to find the diagonal.

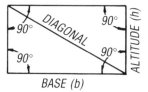

BASE (b)

Figure 8-16 Rectangle.

RULE 8-4   FINDING THE DIAGONAL OF A RECTANGLE

Diagonal = $\sqrt{\text{altitude}^2 + \text{base}^2}$, or $d = \sqrt{h^2 + b^2}$

Example 8-19   Find the diagonal of a rectangle with a base of 10 cm and an altitude of 7.8 cm.

Solution $d = \sqrt{h^2 + b^2}$
$$= \sqrt{(7.8)^2 + (10)^2} \quad = \sqrt{60.84 + 100} = \sqrt{160.84}$$
$$= 12.7$$

The diagonal is about 12.7 cm.

A **trapezoid** is a quadrilateral that has two sides parallel. The two parallel sides are called the *bases*. The two nonparallel sides are called the *legs*. The altitude of a trapezoid is the perpendicular distance between the bases.

The sum of the four angles of a trapezoid equals 360°. Likewise, the sum of the four angles of the quadrilaterals thus far discussed equals 360°. This is true of any quadrilateral.

There is a special trapezoid called the *isosceles trapezoid,* and it receives its name from the isosceles triangle. An isosceles trapezoid is one whose two bases are parallel and whose two nonparallel sides are equal. In Figure 8-17b, base *BC* is parallel to base *AD*, and side *AB* is equal to side *DC*.

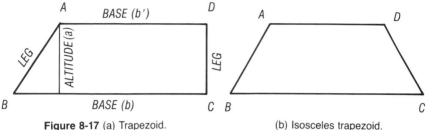

Figure 8-17 (a) Trapezoid.   (b) Isosceles trapezoid.

A **polygon** is any plane figure bounded by straight line segments joined at their end points. Triangles, squares, rectangles, parallelograms, and trapezoids are all examples of polygons. Other polygons are the hexagon (six sides) and the octagon (eight sides).

A **hexagon** is a six-sided figure. A *regular hexagon* contains six equal angles and six equal sides. Each angle of a regular hexagon contains 120°. The diagonals of a regular hexagon divide the hexagon into six equilateral triangles. (See Figure 8-18.)

Most shop people and mechanics are familiar with the *hex* or *Allen* wrenches for tightening socket-head cap screws. The sizes of these and the sockets they fit are measured across the flats of the hexagonal shape.

Any six-hole bolt circle forms a hexagon. The layout of six equally spaced holes on a given bolt circle is simple, since the distance between the holes is equal to the radius of the circle. Dividers are set equal to the radius and six places measured to be marked by a center punch. Such a layout is illustrated in Figure 8-19.

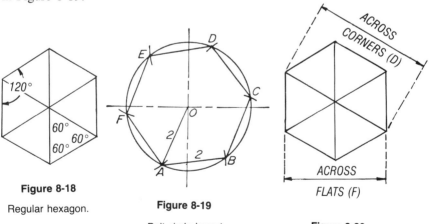

Figure 8-18

Regular hexagon.

Figure 8-19

Bolt circle layout.

Figure 8-20

The minimum-diameter stock needed to mill a hexagon is determined by the desired distance F across the flats of the hexagon. The distance across the corners (see Figure 8-20) is equal to $1.155 \times F$. The size of a hexagon is normally specified as the distance across its flats and will always so appear on blueprints. The largest hexagon that can be milled on a bar of diameter D has a distance across the flats equal to $D \times 0.866$.

RULE 8-5   FINDING THE DISTANCE ACROSS FLATS AND
CORNERS

If D is the distance across the corners of a regular hexagon
and F is the distance across the flats, then

$$D = 1.155 \times F \quad \text{and} \quad F = 0.866 \times D$$

Example 8-20   The distance across the corners of a regular hexagon is 8 cm.
Find the distance across the flats.

Solution $F = 0.866 \times D$
$= 0.866 \times 8 = 6.928$, or about 7 cm

Exercises 8-4

1. In Figure A, the number of degrees in angle B is 45°. How many
degrees are in angles A, C, and D?

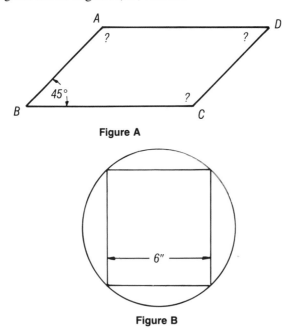

Figure A

Figure B

2. Find the length of the diagonal of a rectangle 18 ft by 14 ft.

3. If you are required to mill a square 6 in. per side, as shown in Figure
B, what is the minimum-diameter stock that would be necessary?

4. The distance across the flats of the square nut in Figure C is $2\frac{7}{8}$ in. What is the diagonal?

5. The square head of a bolt is milled from round stock $2\frac{3}{4}$ in. in diameter. What size wrench would fit this bolt, allowing $\frac{1}{32}$ in. for clearance?

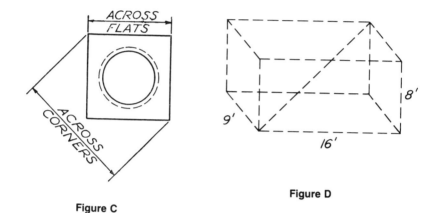

Figure D

Figure C

6. A rectangular room is 9 ft wide, 16 ft long, and 8 ft high. How long is a string stretched from the corner of the ceiling to the opposite corner of the floor? (See Figure D.)

Find the missing values of hexagons.

	Distance Across Corners	Distance Across Flats
7.	3 in.	
8.		17.32 in.
9.		20 in.
10.	3.27 in.	
11.	4 in.	
12.		1.625 in.
13.		1.031 in.
14.	$2\frac{3}{4}$ in.	
15.		$2\frac{3}{16}$ in.

8-5 The Circle

OBJECTIVES

After completing this section, you should be able to:

1. Find the diameter, given part of a circle.
2. Find the height of a circular segment.

A **circle** is a plane figure bounded by a curved line, every part of which is equidistant from a point called the **center**.

A whole class of machining is built around the generation of circular forms by a variety of cutting tools. The machinist and the toolmaker must have considerable knowledge of the circle's properties in order to carry out their work effectively.

The **circumference** of a circle is the length of that curved line which forms the circle. A line drawn through the center of a circle is called a **diameter.** A line drawn from the center of the circle to the circumference is called a **radius.** A radius is equal to one-half a diameter. In Figure 8-21, O is the center of the circle, AB is a diameter, OC is a radius, and the length of the closed, curved line is the circumference.

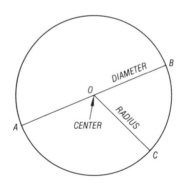

Figure 8-21 Parts of a circle

Sometimes a machinist is given a broken portion of a gear or other circular part and asked to find the diameter so that the workpiece can be reconstructed. Figure 8-22 shows such a part. A portion of a circle is called an **arc.** In Figure 8-22, $\overset{\frown}{AB}$ is an arc.

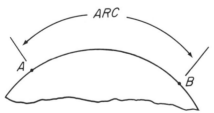

Figure 8-22

The machinist can place a straight edge on the workpiece, connecting A and B. This straight line connecting the two points on the circumference is a

chord. Figure 8-23 shows chord *AB*. The portion enclosed by chord *AB* and arc *AB* is called a **segment**.

Figure 8-23

The next step is for the machinist to measure the greatest height *h* from the arc to the midpoint of the chord, as shown in Figure 8-24.

If the length of the chord and the height *h* are known, the following formula may be used to find the missing radius *r*.

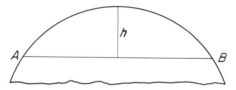

Figure 8-24

RULE 8-6   FINDING THE RADIUS OF A CIRCLE WHEN THE LENGTH OF A CHORD AND THE HEIGHT OF A SEGMENT ARE KNOWN

$$r = \frac{\left(\dfrac{l}{2}\right)^2 + h^2}{2h}$$

where *l* is the length of the chord and *h* is the height of the segment.

Example 8-21   A broken wheel is shown. A segment 2 ft high is cut off by a chord 8 ft in length. What is the complete diameter of this wheel?

Solution $r = \dfrac{\left(\dfrac{l}{2}\right)^2 + h^2}{2h} = \dfrac{(4)^2 + 4}{4} = \dfrac{20}{4} = 5 \text{ ft}$

Thus the diameter is 10 ft.

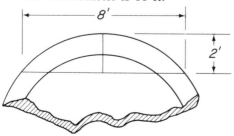

A Woodruff key is a circular half-moon used to key shafts to mating parts that mount on the shaft. A long keyway can be avoided by use of a Woodruff key. It has the advantage of being short in length but strong because of its greater depth compared with a square key.

In cutting the shaft slot for a Woodruff key, a circular cutter of the same radius as the key is plunged into the shaft. The cutter is sunk into the shaft to a depth measured from the sharp edge of the cut. Before a full width of cut results, the cutter must travel a distance equal to x shown in Figure 8-25. Ordinarily, to save time a machinist would consult a handbook that furnishes tables for the extra length.

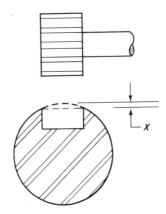

Figure 8-25

If a handbook is not available, there is a simple formula that gives this extra length to within 0.001-inch accuracy. You have probably recognized the distance x in Figure 8-25 as the *height of a circular segment*. If h stands for the height of the circular segment, D stands for the shaft diameter, and W stands for the length of the chord, the following formula can be used.

RULE 8-7   FINDING THE HEIGHT OF A CIRCULAR SEGMENT

$$h = \frac{W^2}{4D}$$ where W is the length of the chord
and D is the shaft diameter.

Example 8-22   A 3-in.-diameter shaft will have a $\frac{1}{4}$-in.-wide Woodruff keyway cut in it to dimensions given on the drawing. What *total* depth must the cutter be fed, once it touches the work?

.625

Solution $h = \dfrac{W^2}{4D} = \dfrac{(0.25)^2}{12} = \dfrac{0.0625}{12} = 0.005$

Total depth $= 0.005 + 0.625 = 0.630$ in.

This formula is also used to find the depth to feed a cutter when a flat surface is to be milled on a shaft, as in Figure 8-26.

Figure 8-26

Example 8-23   Find the depth to feed a cutter to produce a flat 2 in. wide on a 10-in.-diameter shaft.

Solution $h = \dfrac{W^2}{4D} = \dfrac{4}{40} = 0.100$ in.

Exercises 8-5

1. Find the radius of the arc in the casting drawn in Figure A.

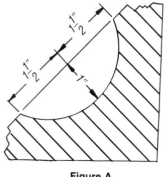

Figure A

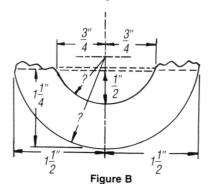

Figure B

2. What is the radius of a circle in which the midpoint of a chord 4 ft long is 1 ft from its arc?

3. A ring was broken and the part shown in Figure B was salvaged. To make a new ring, determine the inside and outside diameters of the original part.

4. The height and length of the chord of a segment in a circle are 2 ft 6 in. and 12 ft 3 in., respectively. What is the diameter of the circle?

5. A concrete arch is shown in Figure C. If *M* equals 24 ft and *h* is 6 ft, what is the radius of this arch?

6. In a No. 204 Woodruff key, shown in Figure D, the height *h* of the key is 0.203 in. and length *CD* is 0.4908 in. What is the key's radius?

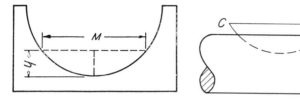

| Figure C | Figure D |

7. In Figure E, find the radius of the fillet if *W* is $\frac{1}{2}$ in. and *h* is $\frac{3}{16}$ in. Give your answer to the nearest sixty-fourth of an inch.

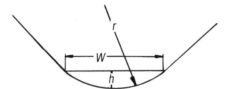

Figure E

8. A chord 203.2 mm long is drawn in a circle whose radius is 254 mm. What is the height of the segment cut by this chord?

9. A segment whose chord is 4 in. in length is cut from a 12-in.-diameter circle. What is the height of the segment?

10. The distance across the flat milled on a 152.4-mm shaft is 57.15 mm. Find the depth of the cutter must be fed, to the nearest thousandth of a millimeter.

11. Find the depth of the key seat shown in Figure F.

12. To what depth should a milling cutter be fed to cut the key seat shown in Figure G?

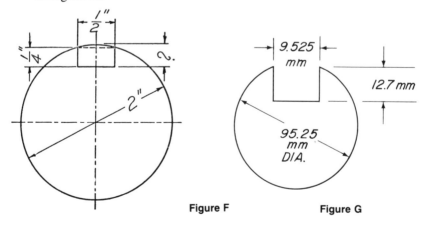

| Figure F | Figure G |

13. The key seat shown in Figure H is milled to a width of 11.112 mm in a shaft whose diameter is 41.275 mm. If the depth c of the key seat from the sharp corner is 4.762 mm, to what depth must the cutter be fed?

14. In Figure I, the 3-in. diameter shaft has a key seat cut to the depth shown. The height F of the key is equal to $\frac{21}{32}$ in. Find the depth to which the cutter must be fed into the shaft.

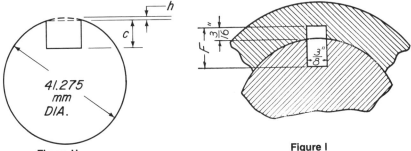

Figure H   **Figure I**

15. The key seat in a 101.6 mm shaft is to have a width of 15.875 mm. When the milling cutter grazes the top of the shaft, what additional distance must it be fed to cut a keyway whose side depth is 6.35 mm?

Unit 8 Review Exercises

1. Blueprint dimensions require $37\frac{1}{2}°$ spacing of holes, but your rotary table positions in decimal degrees. What is the decimal equivalent of $37\frac{1}{2}°$?

2. Find the difference between $43°42'53''$ and $38°45'37''$.

3. In Figure A, how many degrees are in angle B?

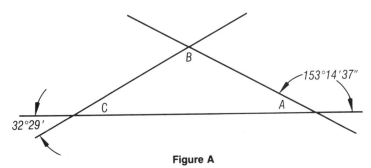

Figure A

4. What is the calculated check distance between the center lines of the two 1.312-diameter holes shown in Figure 8-8 (see page 190)?

5. Find the altitude of an equilateral triangle if one side measures 18.257 in.

6. Calculate the distance between the holes in the layout shown in Figure B.

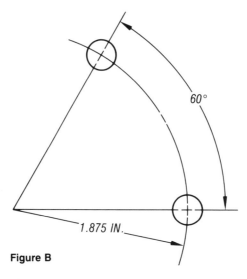

Figure B

7. What is the minimum-diameter stock required to mill a square 6.125 in. per side?

8. If you are required to hold a piece of square stock in a collet designed for round stock, what diameter collet is necessary, to the nearest $\frac{1}{8}$ in., for square stock measuring $\frac{51}{64}$ in. per side?

9. Find the depth to feed a cutter to machine a flat 0.875 in. wide on a 4.625-in.-diameter stock.

10. Find the bolt circle diameter, using the dimensions shown in Figure C.

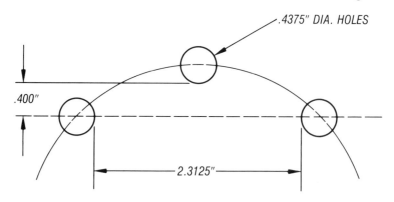

Figure C

Applied Geometry

The machinist will find many applications of basic geometry in shop work. The previous unit introduced basic shapes and some properties related to them. In this unit, you will learn how to solve problems about shape, such as finding perimeter, circumference, and area. This unit also introduces three-dimensional shapes and the formulas to find their volumes.

9-1 Perimeter and Circumference

OBJECTIVES
After completing this section, you should be able to:

1. Find the perimeter of a figure.
2. Find the circumference of a circle.

The **perimeter** of any plane figure is the distance around it. Thus, the perimeter of the polygon in Figure 9-1 is 38 m, the sum of the sides.

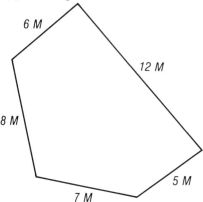

6 M

12 M

8 M

5 M

7 M

Figure 9-1

Because the opposite sides in a rectangle are the same length, the perimeter of a rectangle is always found by adding two widths and two lengths.

RULE 9-1   FINDING THE PERIMETER OF A RECTANGLE

$P = 2l + 2w$

where l is the length and w is the width.

Example 9-1   Find the perimeter of a rectangle whose length is 3.6 cm and whose width is 2.8 cm.

Solution $P = 2l + 2w$
$= 2 \times 3.6 + 2 \times 2.8$
$= 7.2 + 5.6 = 12.8$ cm

In a square, all four sides are the same length, so the perimeter of a square is found by multiplying the length of one side by 4.

RULE 9-2   FINDING THE PERIMETER OF A SQUARE

$P = 4s$

where s is the length of a side.

Example 9-2   Find the perimeter of a square whose sides are 90 ft in length.

Solution $P = 4s$
$= 4 \times 90$
$= 360$ ft

In a triangle, the perimeter is found by adding the lengths of the three sides.

RULE 9-3   FINDING THE PERIMETER OF A TRIANGLE

$P = a + b + c$

where a, b, and c are the lengths of the sides.

Example 9-3   Find the perimeter of a right triangle whose sides opposite the hypotenuse are 3 m and 4 m in length.

Solution   In a right triangle, $c^2 = a^2 + b^2$, where c is the length of the hypotenuse and a and b are the lengths of the sides.

$$c = \sqrt{3^2 + 4^2}$$
$$= \sqrt{25} = 5$$
$$P = a + b + c = 3 + 4 + 5 = 12 \text{ m}$$

Recall that the distance around a circle is called the circumference. The ratio between the diameter and the circumference of a circle is equal to a constant called *pi* (π). The value 3.1416 is used for π for ordinary shop purposes. A fractional approximation of π is $\frac{22}{7}$, which is close enough for woodworking but not metal cutting.

This ratio states that the *circumference of a circle is π, or about 3.1416, times the diameter.* Since the diameter of a circle is 2 times the radius, the ratio can also be stated in terms of the radius.

RULE 9-4   FINDING THE CIRCUMFERENCE OF A CIRCLE

$C = \pi D$ or $C = 2\pi r$

where D is the diameter and r is the radius of a circle.

From these two formulas, the following relations can be obtained:

$$D = \frac{C}{\pi} \qquad r = \frac{C}{2\pi}$$

The diameter of a circle can also be found by multiplying the circumference by 0.31831, but this requires memorizing another constant in addition to π. The important formula to remember is $C = \pi D$. The approximation for π, 3.1416, should be memorized, since machinists and toolmakers must use it to solve a variety of problems in their work. To demonstrate a nonshop, practical use for π, assume you have a very large tree in your yard. A quick way to find its diameter, without guessing, is to wrap a string around the trunk, straighten the string out, and measure its length. That length divided by π is the tree's diameter. If the string measures approximately $24\frac{9}{16}$ inches in length,

$$D = \frac{C}{\pi} = \frac{24.5625}{3.1416} = 7.8185, \text{ or about 7.8 in.}$$

Example 9-4   The diameter of a drill is 1.5 cm. What is the circumference of the hole it will drill?

Solution $C = \pi D$
$= 3.1416(1.5)$
$= 4.7124$, or about 4.7 cm

The following is a rather old problem called the "World with a Fence around It." This problem illustrates the constant ratio π that exists between the circumference and diameter of *any* circle.

Assume that the earth has a circumference of 25,000 miles at its equator. A fence consisting of a single strand of wire is strung on posts around the equator. The wire is 3 yards *longer* than the circumference of the earth. Can a person possibly get under this wire? Figure 9-2 illustrates the situation.

Figure 9-2

The answer is yes! There would be a space of 17 inches through which the person could crawl. The proof of this is left for you to work out. You will need to use the formula $r = C/2\pi$ plus a little logic.

Exercises 9-1

Find the perimeter or circumference of each figure.

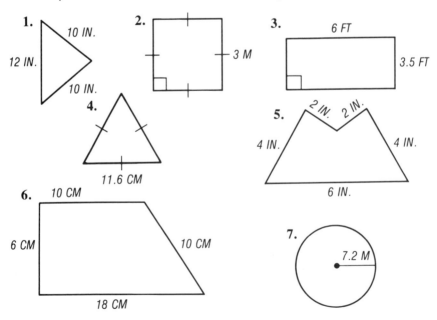

8. 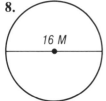   **9.**   (*Hint:* The curve in exercise 9 is a semicircle.)

Find the perimeter of each rectangle.

10. $l = 8.7$ in., $w = 5.2$ in.   **11.** $l = 9.7$ m, $w = 8.5$ m

12. $l = 17$ yd, $w = 11$ yd

Find the perimeter of each square.

13. $s = 16\frac{1}{2}$ ft   **14.** $s = 2.4$ cm

15.   Find the length of one side of a square with perimeter 8.4 yd.

Find the perimeter of each triangle.

16. $a = 7\frac{1}{2}$ ft, $b = 9\frac{2}{3}$ ft, $c = 8$ ft

17. $a = b = 10$ in., $c = 11\frac{1}{2}$ in.

18. $a = b = c = 15.9$ cm

Find the circumference of each circle.

19. $r = 21$ mm   **20.** $D = 18$ in.

21.   Figure A shows a conveyor belt passing over two wheels, each 12 in. in diameter. How far will an object on the belt travel in three turns of the wheels?

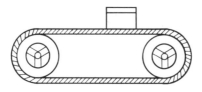

Figure A

22.   The diameter of a drill is $\frac{1}{2}$ in. What is the circumference of the hole it will drill?

23.   What length of wire is required to make a hoop for a barrel that has a $2\frac{1}{2}$-ft diameter, if 8 in. are allowed for fastening the ends of the wire?

24.   A 28-in.-diameter wheel on an auto makes $720\frac{1}{3}$ turns per minute. How fast is the auto traveling?

25.   Assume the earth's diameter to be 8000 mi. If the diameter were increased by 1 ft, what would be the increase in the circumference?

26. In calculating the length of steel needed for bending, the length of an imaginary line drawn through the center of the stock is the *developed length required*. Find the developed length of round stock required to bend the ring shown in Figure B if the stock is 1 in. in diameter.

27. Find the length of stock required to bend the piece shown in Figure C if the bent part is a quarter of a circle whose radius is shown.

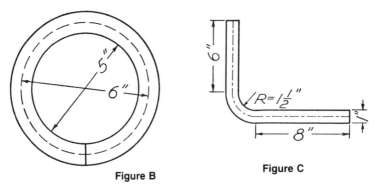

Figure B

Figure C

28. What length of stock is required for the U shown in Figure D?

29. Calculate the developed length of $\frac{3}{8}$-in.-diameter round stock required to forge the link shown in Figure E.

30. A ring whose outside diameter is 9 in. is to be forged from round stock 1 in. in diameter. What length of stock is required? (Refer to Figure F.)

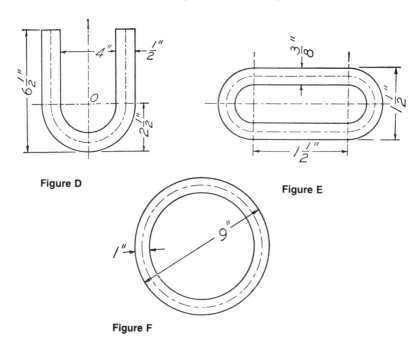

Figure D

Figure E

Figure F

9-2 Area: Triangles and Quadrilaterals

OBJECTIVES

After completing this section, you should be able to:

1. Find the area of a triangle.
2. Find the area of a rectangle or parallelogram.
3. Find the area of a square.
4. Find the area of a trapezoid.

The area of a region is the number of square units it contains. Thus, the area of the rectangular region in Figure 9-3a is 12 square centimeters, and the area of the triangular region in Figure 9-3b is 8 square inches.

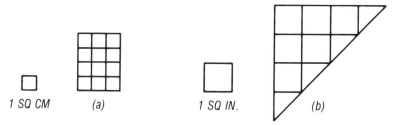

1 SQ CM   (a)   1 SQ IN.   (b)

Figure 9-3

When there is no chance of confusion, the expression "area of a triangle" or "area of a rectangle" will be used rather than "area of a triangular region" or "area of a rectangular region." There are many formulas to help you find the area of regions of various shapes. Rule 9-5 gives a formula to find the area of a triangle. Remember that in a triangle, any side can be used as a base. The altitude h is then the length of a perpendicular line from that base to the opposite vertex.

RULE 9-5 FINDING THE AREA OF A TRIANGLE

$$\text{Area} = \frac{\text{base} \times \text{altitude}}{2}, \quad \text{or} \quad A = \frac{bh}{2}$$

Example 9-5 The base of a triangle is 8 in. and its altitude is 6 in. Find its area.

Solution $b = 8$ and $h = 6$

$$A = \frac{bh}{2}$$

$$= \frac{8 \times 6}{2} = \frac{48}{2} = 24$$

The area is 24 sq. in.

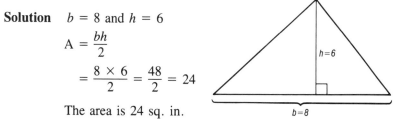

If the altitude of an equilateral (all sides equal) triangle is not known, the area can still be found. An equilateral triangle has three 60° angles. The altitude drawn from any vertex divides the triangle into two triangles with angles 30°, 60°, and 90°.

Example 9-6 Find the area of the equilateral triangle.

Solution In triangle ABD, side AD is one-half of side AB.

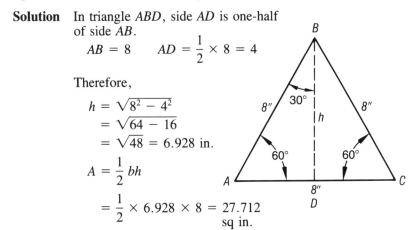

$$AB = 8 \qquad AD = \frac{1}{2} \times 8 = 4$$

Therefore,

$$h = \sqrt{8^2 - 4^2}$$
$$= \sqrt{64 - 16}$$
$$= \sqrt{48} = 6.928 \text{ in.}$$

$$A = \frac{1}{2} bh$$
$$= \frac{1}{2} \times 6.928 \times 8 = 27.712 \text{ sq in.}$$

In a right triangle, you can always find the altitude (or base) when given the hypotenuse and one side, by using the Pythagorean theorem.

For scalene or oblique triangles (triangles without a 90° angle), where the altitude is not given, the area of the triangle can still be found if its three sides are known. The procedure consists of the three steps given in Rule 9-6.

RULE 9-6 FINDING THE AREA OF A TRIANGLE THAT IS NOT A RIGHT TRIANGLE

Step 1 Find half the sum of the sides and find the difference between this sum and each side.

Step 2 Multiply the differences by each other and by half the sum.

Step 3 Extract the square root of the resulting product.

$$A = \sqrt{s(s - a)(s - b)(s - c)}$$

where $s = \dfrac{a + b + c}{2}$ and a, b, and c are the lengths of the sides.

Example 9-7 Find the area of the triangle.

Solution $s = \dfrac{9 + 15 + 20}{2} = 22$

$A = \sqrt{s(s - a)(s - b)(s - c)}$
$= \sqrt{22(22 - 9)(22 - 15)(22 - 20)}$
$= \sqrt{22(13)(7)(2)}$
$= \sqrt{4004}$
$= 63.28$ sq in.

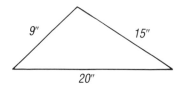

The formula for finding the area of a rectangle and the area of a parallelogram is the same.

RULE 9-7   FINDING THE AREA OF A RECTANGLE OR PARALLELOGRAM

Area = base × altitude   or $A = bh$

Example 9-8 Find the area of a parallelogram with base 3 ft 4 in. and altitude 2 ft 6 in.

Solution First write both base and altitude as feet:

$$3 \text{ ft } 4 \text{ in.} = 3\,\frac{4}{12} \text{ ft} = 3\,\frac{1}{3} \text{ ft}$$

$$2 \text{ ft } 6 \text{ in.} = 2\,\frac{6}{12} \text{ ft} = 2\,\frac{1}{2} \text{ ft}$$

$A = bh$

$$= 3\,\frac{1}{3} \times 2\,\frac{1}{2} = \frac{10}{3} \times \frac{5}{2} = \frac{25}{3} = 8\,\frac{1}{3} \text{ sq ft}$$

The same formula can also be used to find the altitude, given the area and the base.

Example 9-9   Find the altitude of a rectangle with base 8 cm and area 32.8 sq cm.

Solution   First solve the area formula for h.

$$A = bh$$

$$h = \frac{A}{b}$$

$$= \frac{32.8}{8}$$

$$= 4.1$$

The altitude is 4.1 cm.

The area of a square is used by engineers to calculate the torsional (twisting) load of a bar for strength. A sheet metal worker or die layout person would also be interested in the amount of material contained in a given sheet or strip of metal. And a machine shop estimator is concerned with area in figuring weight of material.

The area of any square is equal to one side squared.

RULE 9-8   FINDING THE AREA OF A SQUARE

Area = side squared   or $A = s^2$

Example 9-10   Find the area of a square with side 8 cm.

Solution $A = s^2 = 8 \times 8 = 64$ sq cm

If the diagonal of the square is known, another formula for area can also be used:

$$A = \frac{d^2}{2}$$

where d is the diagonal. To find the approximate value of the diagonal when it is not known, the following formula can be used:

$$d = 1.4142s$$

where s is the side.

Since the trapezoid is often a part of the area of a region, it is important to be able to find the area of a trapezoid. The formula for the area of a trapezoid is given in Rule 9-9.

RULE 9-9   FINDING THE AREA OF A TRAPEZOID

$$A = \frac{1}{2} h (b + b')$$

where h is the altitude, b is the bottom base, and b' is the top base.

Example 9-11   Find the area of the trapezoid.

Solution $A = \frac{1}{2} h (b + b')$

$$= \frac{1}{2} \times \frac{3}{2} (4 + 2)$$

$$= \frac{3}{4} \times \frac{6}{1} = 4.5 \text{ sq ft}$$

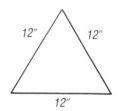

$b' = 2\ FT$
$h = 1\frac{1}{2}\ FT$
$b = 4\ FT$

Exercises 9-2

Find the area of each triangle.

1.   **2.**   **3.**   **4.**

6.928″

8.4″

10″

10 cm

6.42 cm

12″   12″

12″

5.   **6.**   **7.**   **8.**

60°   12⅛″   90°   30°

5′   5′

4′-6″

4′-6″   4′-6″

4′

4′   3′-6″

3′-6″

9.   **10.**   **11.**   **12.**

11″   12″

8″

10′   10′

10′

11′-6″   9′-6″

10′-6″

21″   17″

16″

13. Find the area of a parallelogram whose base is 2 ft 9 in. and whose altitude is 5 ft 4 in.

14. If the area of a parallelogram is 63.5 sq in. and its altitude is 1 ft 3 in., what is its base, to the nearest hundredth of an inch?

Find the area of each rectangle.

15. $b = 8$ in., $h = 5$ in.

16. $b = 3.4$ m, $h = 2.9$ m

17. $b = 15$ cm, $h = 1.3$ cm

Find the area of each square.

18. $s = 15$ ft   **19.** $s = 3.26$ m   **20.** $s = 1\frac{1}{2}$ yd

21. In the isosceles triangle in Figure A, find the length of the base and the area.

22. Find the altitude of the isosceles triangle shown in Figure B. Find its area.

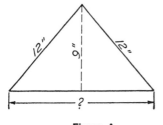

Figure A

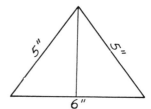

Figure B

23. What is the largest square that will fit on a 33-in.-diameter magnetic chuck or table of a vertical grinder without the corners of the workpiece projecting over the edge of the table?

24. The area of a rectangle is 108 sq ft and its base is $6\frac{1}{2}$ ft. Find the value of its altitude to the nearest thirty-second of an inch.

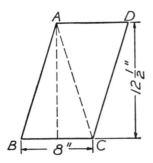

Figure C

25. If the diagonal of a rectangle is 125 ft and its altitude is 75 ft, find its base and its area.

26. Given the parallelogram in Figure C, with base of 8 in. and altitude of $12\frac{1}{2}$ in., find the number of square inches in triangle *ADC*.

Find the area of the following trapezoids whose bases and altitudes are as follows:

27. Bottom base 6 ft; top base 4 ft; altitude 3 ft.

28. Bottom base 5.5 ft; top base 3.2 ft; altitude 1.4 ft.

29. Bottom base 3 yd; top base 1.75 yd; altitude $1\frac{3}{4}$ yd.

30. Bottom base 2 ft 9 in.; top base 1 ft 8 in.; altitude 8 in.

31. Bottom base 37 in.; top base 23 in.; altitude 14 in.

Solve for the area of the isosceles trapezoids.

32.   **33.**   **34.**

35.   **36.**   **37.**

38. Find the area of the cross section of the I beam shown in Figure D.

39. Find the area of the cross section of the channel shown in Figure E.

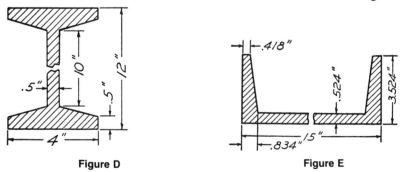

Figure D   Figure E

9-3   Area of a Circle; Central Angles and Arc Length

OBJECTIVES

After completing this section, you should be able to:

1. Find the area of a circle.
2. Find the length of the arc intercepted by a given central angle.
3. Find the measure of a central angle of a circle.

A circle may be part of a larger geometric shape, such as a cylinder, piston, or cross section of a solid, round object. The total area or portions of the area of a circle may be needed to solve such problems as those listed below.

Pressure on the head of a piston

Strength of a bar of steel

Weight of metal objects

Capacity of tanks

Displacement of material for counterbalancing

Cubic measurement of solid objects

There are two general formulas for finding the area of a circle. They are given in Rule 9-10.

RULE 9-10   FINDING THE AREA OF A CIRCLE

$A = \pi r^2$, where r is the radius.

$A = \dfrac{\pi D^2}{4}$, where D is the diameter.

If the formula $A = \pi D^2/4$ is chosen, an extra step of dividing by 4 must be performed. Therefore, the formula $A = \pi r^2$ will be used here exclusively. The examples that follow illustrate the use of this formula.

Example 9-12   Find the area of a circle whose radius is 10 in.

Solution $A = \pi r^2 = 3.1416(10^2) = 314.16$ sq in.

Example 9-13   A circle has a 6-ft diameter. Find its area.

Solution $A = \pi r^2 = 3.1416(3^2) = 28.27$ sq ft

If the area of a circle is given, the diameter can be found with a variation of the same formula.

RULE 9-11   FINDING THE DIAMETER OF A CIRCLE

$$D = 2\sqrt{\frac{A}{\pi}}$$

Example 9-14   Find the diameter of a circle whose area is 78.54 sq ft.

Solution $D = 2\sqrt{\dfrac{A}{\pi}} = 2\sqrt{\dfrac{78.54}{3.1416}} = 2\sqrt{25} = 10$ ft

The toolmaker or experimental machinist is occasionally asked to construct a fixture or a prototype mechanism that specifies that a certain length of arc be used. The length of arc AB is shown in Figure 9-4a.

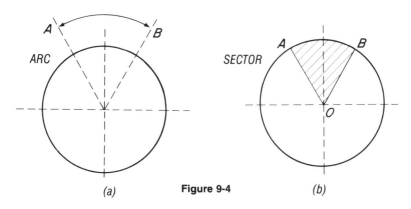

(a)   **Figure 9-4**   (b)

If radii are connected to the limits of the arc, the area enclosed is called a **sector** of a circle. The radii form a **central angle** AOB. The central angle contains the same number of degrees as the arc that it intercepts. If, in Figure 9-4b, arc AB contains 60°, so does the central angle AOB. In general, the length of the arc is measured in linear units rather than degrees. The important thing to remember is that the central angle cuts a part of the circumference proportional to the ratio of the angle to 360°. Thus, if angle AOB is 60°, arc AB is $\frac{60}{360}$ or $\frac{1}{6}$ of the circumference.

RULE 9-12   FINDING THE LENGTH OF AN ARC

$$\text{Length of arc} = \frac{\text{central angle} \times \text{circumference}}{360}$$

Example 9-15   Find the length of the arc AB intercepted by a central angle of 40° on a 12-in.-diameter circle.

Solution   Length of arc $= \dfrac{\text{central angle} \times \text{circumference}}{360}$

$$= \dfrac{40(37.699)}{360} = 4.189 \text{ in.}$$

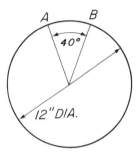

To find the central angle, knowing the length of the arc, the formula can be used if solved for the central angle, as given in Rule 9-13.

RULE 9-13   FINDING THE CENTRAL ANGLE

Central angle $= \dfrac{\text{(length of arc)(360)}}{\text{circumference}}$

Example 9-16   Use the arc length found in Example 9-15 to find the central angle.

Solution   Central angle $= \dfrac{\text{(length of arc)(360)}}{\text{circumference}} = \dfrac{(4.189)(360)}{(3.1416)(12)}$

$$= \dfrac{1508.04}{37.699} = 40°$$

Exercises 9-3

Fill in the missing values.

	Area	Diameter
1.		6 in.
2.		12 ft

3.		$6\frac{1}{4}$ ft
4.	14 sq in.	
5.	752 sq ft	
6.	16.32 sq in.	

7. A detent is to be milled either on a rotary table or in a dividing head to the shape and specifications given in Figure A. Through what angle should the part be rotated to produce the arc on this part?

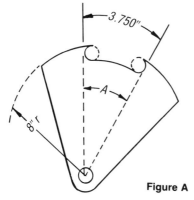

Figure A

8. Figure B shows another detent. What is the circular distance between the centers of the slots at the circumference?

9. What is the length of the arc *AB* in Figure C?

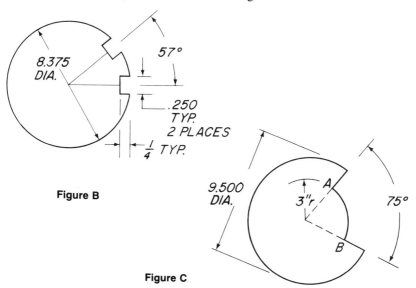

Figure B

Figure C

Find the length of arc, given the following conditions.

10. A 55° sector of a 20-in.-diameter circle.

11. A 10° sector of a 225.425-mm-diameter circle.

12. A $27\frac{1}{2}°$ sector of a 12-in.-diameter circle.

13. The radius of a circle is $2\frac{1}{2}$ in. and a sector of the circle has an angle of 120°. What is the length of arc formed by the central angle at the circumference?

14. The diameter of a piston is 6.421 in. What is the total pressure on this piston, if the steam pressure is 75 lb/sq in.?

15. The diameter of a water valve is 1.75 in. What is its area?

16. A steel shaft has a $2\frac{3}{4}$-in. diameter. How many square inches are in its cross section?

17. Steel expands 0.0000065 in. per inch of length for each degree Fahrenheit that the temperature is raised. What will be the new diameter of a 1-in. standard plug gauge if the temperature is raised 50° F?

18. Figure D shows the largest possible circular disc that can be cut from a piece of sheet steel that measures 304.8 mm by 355.6 mm. How many square millimeters of waste will result?

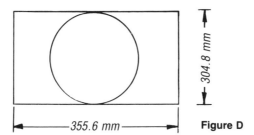

304.8 mm

355.6 mm **Figure D**

9-4 Volume

OBJECTIVES

After completing this section, you should be able to:

1. Find the volume of a right prism.
2. Find the volume of a right cylinder.
3. Find the volume of a right pyramid or cone.
4. Find the volume of a sphere.

A **prism** is a solid whose bases are parallel polygons and whose faces are parallelograms. Since a polygon has three or more sides, the bases of a prism can have three or more sides. The sides of a prism, which are parallelograms, are called the **faces** of the prism. If the edges of the faces of the prism are perpendicular to the bases, the prism is called a **right prism:** if the edges of the faces of a prism are not perpendicular to the bases, the prism is called an **oblique prism.**

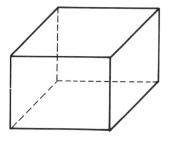

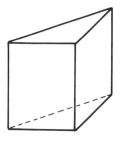

Figure 9-5 (a) Rectangular prism. (b) Right triangular prism.

A **rectangular prism** is a right prism whose bases are rectangles. A brick is a rectangular prism (see Figure 9-5a).

A **cube** is a rectangular prism whose six faces are equal squares.

A **right triangular prism** is one whose edges are perpendicular to the bases and whose bases are triangles (see Figure 9-5b).

The volume of a right prism is equal to the area of the base multiplied by the altitude. This is stated as a formula in Rule 9-14.

RULE 9-14   FINDING THE VOLUME V OF A RIGHT PRISM

$V = A \times h$ where A is the area of the base
and h is the altitude.

Example 9-17   Find the volume of a right prism whose altitude is 9 in. and whose base is a rectangle 12 in. long and 6 in. wide.

Solution $V = A \times h$
$= 12 \times 6 \times 9 = 648$ cu in.

Example 9-18   The base of a right prism is a square 10 in. on a side and the altitude is 15 in. Find the volume.

Solution $V = A \times h$
$= 10 \times 10 \times 15 = 1500$ cu in.

A **right cylinder** is illustrated in Figure 9-6. Its bases are circles and the altitude is perpendicular to each of the bases.

When a drill or other rotary cutting tool penetrates a workpiece, the hole it produces is in the shape of a cylinder. If you know how to figure the volume of a cylinder and know the weight of metal being removed, the total weight removed from the workpiece by a rotary tool can be figured.

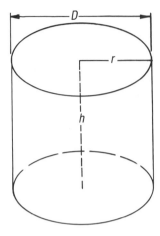

Figure 9-6 Right cylinder.

This provides a good way to balance a pulley or other device that must revolve at high speed without vibration. This process is called *dynamic balancing* and must be performed on flywheels and shafts that could be dangerous if unbalanced.

RULE 9-15   FINDING THE VOLUME OF A RIGHT CYLINDER

Volume = area of base × altitude   or $V = \pi r^2 h$

where r is the radius and h is the altitude.

Example 9-19   A hole 16 in. in diameter is cut through a 2-in.-thick metal plate. How many cubic inches of metal have been removed?

Solution $V = \pi r^2 h$

$$= 3.1416(8^2)(2) = 402\frac{1}{8} \text{ cu in.}$$

Example 9-20   If a $\frac{3}{4}$-in.-diameter hole is drilled through a 3-in.-thick cast-iron plate, what weight of metal has been removed? Cast iron weighs 0.26 lb/cu in.

Solution $V = \pi r^2 h = 3.1416\left(\frac{3}{8}\right)^2 (3) = 1.33 \text{ cu in.}$

Weight of cast iron equals 0.26 lb/cu in.

$0.26 \times 1.33 = 0.35 \text{ lb}$

A **right pyramid** is a solid whose base is a regular polygon of any number of sides and whose faces are equal isosceles triangles whose vertices meet at a point, called the **apex** of the pyramid.

The **altitude** of a right pyramid is the line dropped from the apex perpendicular to the base. The altitude of a right pyramid falls on the center of the base of the pyramid.

The line drawn from the vertex of one of the equal isosceles triangles comprising the faces of the pyramid to the midpoint of the base of the triangle is called the **slant height.** In a right pyramid the slant height is the same for all of the equal triangles comprising the faces.

The intersection of any two faces is a straight line called the **lateral edge** of the pyramid.

In a right pyramid, it is important to remember that the foot of the altitude is the center of the base. In addition, since the base of a right pyramid is a regular polygon, the sides of the base are equal. Figure 9-7 shows a right pyramid whose base is a square.

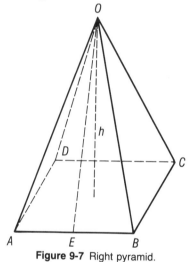

Figure 9-7 Right pyramid.

Point O in Figure 9-7 is the apex of the pyramid, h the altitude, OE a slant height, and OB a lateral edge. The base of this right pyramid is the regular polygon, $ABCD$. This regular polygon is a square.

In geometry, the proof is given for a theorem that states that the *volume of a right pyramid is one-third the volume of a right prism provided they have equivalent bases and altitudes.* Since the volume of a right prism is given by the formula $V = Ah$, the volume of a right pyramid is one-third of Ah.

RULE 9-16   FINDING THE VOLUME OF A RIGHT PYRAMID

$$V = \frac{1}{3} Ah$$

where A is the area of the base and h is the altitude.

Example 9-21   Find the volume of a right pyramid whose altitude is 14 in. and whose base is 8 in. on a side.

Solution   The base of this right pyramid is an equilateral triangle. To find the area of the base of the pyramid, it is necessary to find the area of an equilateral triangle 8 in. on a side. Use the Pythagorean theorem to find x:

$$x = \sqrt{8^2 - 4^2}$$
$$= \sqrt{64 - 16} = \sqrt{48} = 6.928$$

Then find A, the area of the base:

$$A = \frac{1}{2} bh$$

$$= \frac{1}{2} \times 8 \times 6.928$$

$$= 27.712$$

Finally, find V:

$$V = \frac{1}{3} Ah$$

$$= \frac{1}{3} \times 27.712 \times 14 = 129.323 \text{ cu in.}$$

If a right triangle is revolved about its altitude as an axis, the solid cutout will be a **right circular cone.** The base of a cone is a circle. The lateral surface tapers to a point called the **apex.** The **altitude** is a perpendicular from the apex to the base, and it falls on the base at its center. The **slant height** is a line drawn from the apex to the circumference of the base.

Figure 9-8 shows a right circular cone. It should be noted that the altitude, radius of the base, and slant height form a right triangle. Given the altitude and radius of the base, the slant height can be found by use of the Pythagorean theorem. If the radius is 6 inches and altitude is 10 inches, then the slant height is

$$\sqrt{6^2 + 10^2} = \sqrt{136} = 11.6619 \text{ in.}$$

Figure 9-8 Right circular cone.

RULE 9-17   FINDING THE VOLUME OF A RIGHT CIRCULAR
             CONE

$$V = \frac{1}{3} Ah = \frac{1}{3} \pi r^2 h$$

where A is the area of the base, h is the altitude,
and r is the radius of the base.

Since $r = D/2$ and $r^2 = D^2/4$, the above formula can also be written

$$V = \frac{\pi D^2 h}{12}$$

Example 9-22   Find the volume of a cone whose altitude is 10 in. and whose
diameter of the base is 5 in.

Solution $V = \dfrac{1}{3} Ah = \dfrac{1}{3} \times \dfrac{\pi D^2}{4} \times h$

$$= \frac{3.1416 \times 25}{12} \times \frac{10}{1} = 65.45 \text{ cu in.}$$

A **sphere** is a solid bounded by a curved surface every point of which
is equally distant from a point called the **center** (O in Figure 9-9). The area
of the curved surface is called the **surface area** of the sphere. The distance
from the curved surface to the center of the sphere is the **radius.** The
diameter is a line that passes through the center and terminates in the
surface.

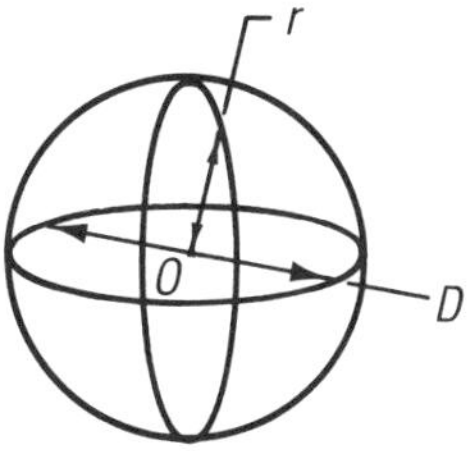

Figure 9-9 Sphere.

RULE 9-18   FINDING THE SURFACE AREA AND VOLUME OF
             A SPHERE

$$S = 4\pi r^2 = \pi D^2 \qquad V = \frac{4}{3} \pi r^3 = \frac{1}{6} \pi D^3$$

where S is the surface area, r is the radius, D is
the diameter, and V is the volume.

Example 9-23   What is the surface area of a sphere whose radius is 8 in.?

Solution $S = 4\pi r^2$

$S = 4 \times 3.1416 \times (8 \times 8) = 804.25$ sq in.

Example 9-24   Find the volume of a sphere whose diameter is 4 in.

Solution $V = \dfrac{4}{3} \pi r^3$

$V = \dfrac{4}{3} \times \dfrac{3.1416}{1} \times \dfrac{2 \times 2 \times 2}{1} = 33.51$ cu in.

Exercises 9-4

1.   A concrete pillar 15 ft high has a cross section that is a regular hexagon 1 ft 6 in. on a side. How many cubic yards of concrete does the pillar contain? (*Hint:* A regular hexagon can be divided into six equal equilateral triangles.)

2.   If a tank whose shape is a rectangular prism is 12 ft long, 9 ft wide, and 6 ft deep, how many gallons of water will it contain when it is two-thirds full? (1 gal = 231 cu in.)

3.   A triangular right prism has an altitude of 15 ft. If the base is an equilateral triangle 3 ft 3 in. on a side, what is the volume of the prism?

4.   Cast iron shrinks $\frac{1}{8}$ in. per linear foot in cooling. If a casting whose shape is a cube is desired, what will be the decrease in volume if the cube is 1 ft 6 in. on an edge after cooling?

5.   A bar of steel 12 ft long has a cross section that is a regular hexagon. If the distance across the flats is 2 in., what is the volume?

6.   Find the volume of 6 ft of length in an I beam whose section is shown in Figure A.

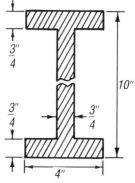

Figure A

7. In a cylinder, $D = 6$ in.; $h = 15$ in. Find the volume.

8. In a cylinder, $D = 3$ ft 6 in.; $h = 9$ ft 8 in. Find the volume.

9. In a cylinder, $r = 1$ ft; $h = 3$ ft 6 in. Find the volume.

10. In a cylinder, the volume equals 1536 cu in. and the area of the base is 235 sq in. What is the altitude of the cylinder?

11. How many cubic yards of concrete will be required to fill a cylindrical column 18 ft long and 6 ft 5 in. in diameter?

12. If concrete weighs 154 lb/cu ft, what is the weight of the column in exercise 11?

13. Find the number of cubic inches of metal in a pipe 16 ft long if the inside diameter is 3 in. and the thickness of the pipe is $\frac{3}{16}$ in.

14. A $\frac{1}{2}$-in.-diameter drill cuts a hole through a casting 4 in. thick. If cast iron weighs 0.26 lb/cu in., what is the weight of the metal that was removed?

15. A cast-iron flywheel has a flat-bottomed hole sunk into its rim by an end mill for balancing purposes. The end mill is 1 in. in diameter. The flywheel is 2 in. thick and the amount of metal to be removed is $\frac{1}{4}$ lb. How deep is the end mill plunged into the rim of the flywheel?

16. A cylindrical tank is required to hold 8000 gal. If its altitude is 16 ft, what diameter must it have? (1 gal = 231 cu in.)

17. You are asked to design and construct a sheet metal container for 5 gal of oil. The inside diameter must be 18 in. How high will this container be?

18. Five holes, each $\frac{5}{16}$ in. in diameter, are drilled in a steel disc. The disc is $8\frac{1}{4}$ in. in diameter and $\frac{1}{2}$ in. thick. Steel weighs approximately 0.2833 lb/cu in. If the disc has a weight of 7.572 lb before drilling, what is its weight after drilling?

19. A rectangular strip of steel as shown in Figure B, has four holes drilled through the strip, each $1\frac{1}{2}$ in. in diameter. Find the weight of the steel strip in pounds, after drilling. (See exercise 18.)

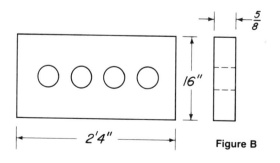

Figure B

20. Calculate the weight of the casting shown in Figure C. The material is cast iron with a 2-in. thickness. (See exercise 14.)

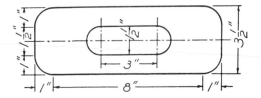

Figure C

21. Figure D is the side view of a cast iron cylinder with a hole drilled through it. Find its weight. (See exercise 14.)

22. Figure E is a cross-section view of a round steel shaft with a square hole through the middle. The length of the shaft is 12 in. What is its weight? (See exercise 18.)

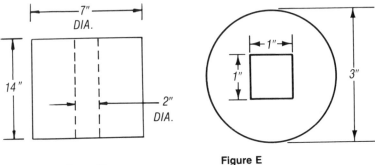

Figure E

Figure D

23. What is the volume of a pyramid whose altitude is 1 ft 6 in. and whose base is a square 2 ft 3 in. on a side?

24. Find the volume of a pyramid whose altitude is 3.25 ft and whose base is a hexagon 2.25 ft on a side.

25. What is the volume of a pyramid whose base is a square 4 in. on a side and whose altitude equals 8 in.?

26. Find the volume of a pyramid with a square base 8 in. on a side and an altitude equal to $6\frac{1}{2}$ in.

27. Find the volume of a cone whose altitude is 25 in. and whose diameter of base is 10 in.

28. Find the volume of a cone if the radius of the base is 4 in. and the altitude is 15 in.

29. The slant height of a cone makes an angle of 60° with the base. If the slant height is 20 in., what is the volume?

30. The area of the base of a cone is 152 sq in. If the altitude is 12 in., what is the volume?

31. A pile of sand is in the shape of a right cone. The circumference of the base is measured and found to be 36 ft. The slant height is found to make an angle of 45° with the base. Find the number of cubic yards in the pile.

32. If the base of a pyramid is an equilateral triangle 6.125 in. on a side and the altitude is 8.250 in., what is its volume?

33. Find the surface of a sphere whose radius is 9 in.

34. Find the volume of a sphere whose radius is 12 in.

35. A hollow sphere is contained in a solid cube 10 in. on an edge. Find the volume of the sphere and the volume of the solid cube.

36. Find volume and surface area of a hemisphere whose radius is 2.3 in.

37. A cylindrical tank 12 ft in diameter has ends that are hemispheres. If the total length of the tank is 36 ft, find the volume.

38. The surface of a sphere contains 236 sq in. Find the radius.

39. Find the volume of metal in a hollow sphere if the metal is 1 in. thick and the inside diameter of the hollow sphere is 5 in.

40. Find the volume of a sphere whose diameter is 7 in.

9-5   Similar Figures

OBJECTIVE

After completing this section, you should be able to:

1. Solve problems involving similar figures.

Figures that are alike in shape are known as **similar** figures. Any square is similar to any other square and any circle is similar to any other circle. Two triangles may not be similar, since they may have different shapes. A ratio exists between the corresponding parts of similar figures. A comparison between quantities of the same thing is a ratio. The two squares A and B in Figure 9-10 illustrate this.

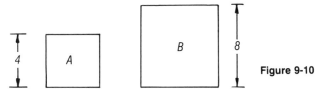

Figure 9-10

A side of square A is half the length of a side of square B. This can be expressed in a practical form as

$$\frac{\text{side } A}{\text{side } B} = \frac{4}{8} = \frac{1}{2}$$

If the length of the sides have ratio $\frac{1}{2}$, their areas will have ratio $(\frac{1}{2})^2$ or $\frac{1}{4}$. This is because area A = (side $A)^2$ and area B = (side $B)^2$. Thus,

$$\frac{\text{Area } A}{\text{Area } B} = \frac{(\text{side } A)^2}{(\text{side } B)^2} = \left(\frac{\text{side } A}{\text{side } B}\right)^2 = \left(\frac{1}{2}\right)^2 = \frac{1}{4}$$

In general,

$$\frac{(\text{Side } A)^2}{(\text{Side } B)^2} = \frac{\text{area } A}{\text{area } B}$$

Example 9-25 If the side of one square is 4 in. and the side of a second square is 12 in., how do their areas compare?

Solution $\dfrac{12^2}{4^2} = \dfrac{144}{16} = \dfrac{9}{1}$

By the same reasoning, the areas of circles are directly proportional to the squares of their diameters.

$$\frac{\text{Area large circle}}{\text{Area small circle}} = \frac{\text{Diameter}^2}{\text{diameter}^2} \quad \text{or} \quad \frac{A}{a} = \frac{D^2}{d^2}$$

Example 9-26 Compare the areas of two circles whose diameters are 6 in. and 12 in.

Solution $\dfrac{12^2}{6^2} = \dfrac{144}{36} = \dfrac{4}{1}$

Example 9-27 A 4-in. pipe can carry how many times as much water as a 2-in. pipe?

Solution $\dfrac{4^2}{2^2} = \dfrac{16}{4} = \dfrac{4}{1}$ The 4-in. pipe carries four times as much water as the 2-in. pipe.

In the above examples, notice that when the diameter is doubled, the area of the circle increases four times. Doubling the radius will achieve the same result. The same is true of the areas of squares and their sides. Rule 9-19 states these relationships.

RULE 9-19   RATIO OF DIAMETER AND RADIUS TO THE AREA OF A CIRCLE AND OF LENGTH OF SIDE TO THE AREA OF A SQUARE

- Doubling the diameter or radius of a circle increases its area four times.
- Doubling the length of the side of a square increases its area four times.

Similar triangles can be used to find heights of objects and distances between points.

Example 9-28   A pole 10 ft high is placed in a vertical position 60 ft from the foot of a smokestack. A line sighted along the top of the pole and smokestack touches the ground 15 ft from the foot of the pole. Find the length of the smokestack.

Solution   Let *h* equal the height of the smokestack. Then, from similar triangles:

$$\frac{15}{75} = \frac{10}{h}$$

$$15h = (10)(75)$$

$$h = 50 \text{ ft}$$

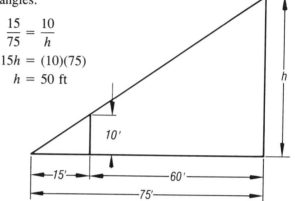

Exercises 9-5

Find the ratio of the area of the first figure to that of the second.

1. A square 4 m on a side; a square 8 m on a side.

2. A circle 4.2 cm in radius; a circle 6.3 cm in radius.

3. A rectangle 4 ft 7 in. by 6 in.; a rectangle 6 in. by 9 ft 2 in.

4. A circle 3.6 in. in diameter; a circle 1.8 in. in radius.

5. A circle 5.77 m in radius; a circle 11.54 m in radius.

6. A 6-in. steam pipe discharges into two smaller pipes of equal diameter. What are the diameters of the smaller pipes?

7. What is the diameter of a circle whose area is equal to the sum of the areas of two circles 6 in. and 8 in. in diameter?

8. How many 4-in.-diameter pipes carry the same volume as a pipe with an 8-in. inside diameter?

9. The flow of water in a 12-in. pipe is carried off by some number of 4-in. pipes. How many?

10. From a point 100 ft from the base of a building, a person sights the ridge of the roof on a line with the top of a 12-ft pole. If the pole is 6.096 m from the person, whose eye is 1.524 m above the ground, how high is the roof of the building?

Unit 9 Review Exercises

1. Find the area of the trapezoid shown in Figure A.

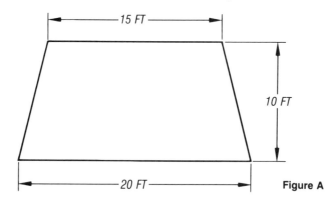

Figure A

2. Determine the combined area of the four holes in the metal plate in Figure B.

3. Determine the amount of metal in the plate in Figure B.

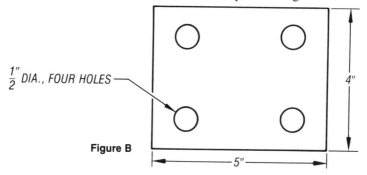

$\frac{1''}{2}$ DIA., FOUR HOLES

Figure B

4. The circumference of a pulley is 28.260 in. What is its diameter?

5. Find the area of a ring with an outside diameter of 4.000 in. and an inside diameter of 1.000 in. (Figure C.)

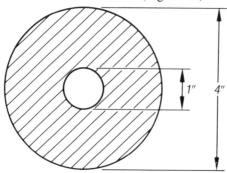

Figure C

6. Find the area formed by the two rafters and joist in Figure D, assuming the rafters are each 5 ft long and the rise is 3 ft. (*Note:* The rise is perpendicular to the joist.)

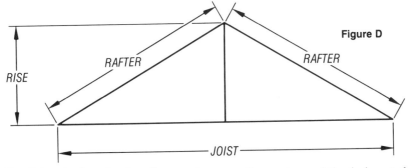

Figure D

7. What is the circular distance between the centers of the holes at the radius given in Figure E?

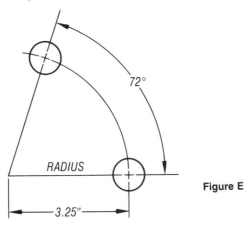

Figure E

8. Find the height of a tree whose shadow measures 28 ft if a 6-ft person casts a shadow 4 ft long at the same time of day. (See Figure F.)

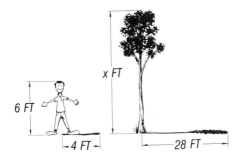

Figure F

9. Find the volume of material removed from the part in Figure G by making a hole $\frac{5}{16}$ in. in diameter by 0.800 in. deep.

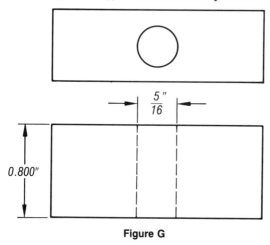

Figure G

10. Find the volume of a right-triangular-shaped aluminum extrusion having equal sides of 5 in. and containing a similar triangular opening with 2 in. sides, if the extrusion is 32. in. long. (See Figure H.)

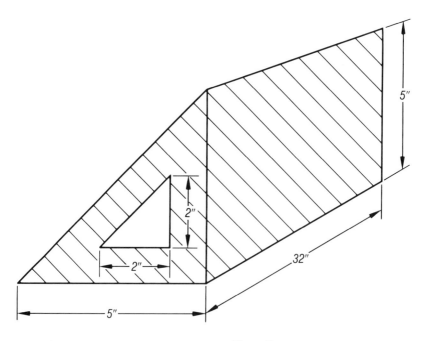

Figure H

Shop Trigonometry

Trigonometry is the study of angles and their functions. A toolmaker must be proficient in trigonometry to complete many routine assignments. Good toolmakers are recruited from the ranks of machinists who took the trouble to learn trigonometry. There is no reason why any machinist cannot learn to use a table of trigonometric functions and solve shop problems.

It often happens that the dimensions required to make a workpiece from a blueprint are not obvious on that print or perhaps not even directly stated, since adding every possible dimension would clutter a blueprint to the point of confusion. Some dimensions must be calculated indirectly. The draftsperson must insert enough information for the part to be made.

Figure 10-1 is an example in which a minimum amount of information is given on a print. It is not necessary to specify that the six equally spaced holes of the bolt circle should be 60° apart, since any machinist can divide 6 into 360° and come up with a value of 60°.

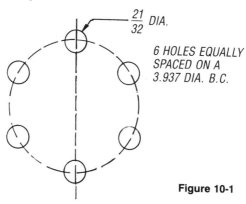

$\frac{21}{32}$ DIA.

6 HOLES EQUALLY
SPACED ON A
3.937 DIA. B.C.

Figure 10-1

Right-angle trigonometry deals with finding unknown values of parts of a right triangle. These can be sides, the hypotenuse, or angles. Sometimes it is necessary to solve shop problems involving triangles that are not right triangles. Triangles that do not contain a right angle are called *oblique* triangles. This unit introduces techniques to solve right triangles and oblique triangles.

10-1   Trigonometric Functions

OBJECTIVES
After completing this section, you should be able to:

1. Name the trigonometric functions of an angle.
2. Use cofunctions.

As the acute angles of a right triangle change in size, the sides opposite these angles increase or decrease in a predictable ratio. This ratio is a *function* of any given angle. That is, if an angle of a right triangle is changed, so are the sides that depend on that angle. If an angle increases, so does the side opposite that angle; if the angle decreases, so does the side.

Consider the right triangle constructed in a semicircle in Figure 10-2. This is an easy way to show results of changes in sides and angles. Since both acute angles are the same value, the sides opposite them are also equal. We can express the relationship of sides *a* and *b* as

$$\frac{a}{b} = \frac{1}{1}$$

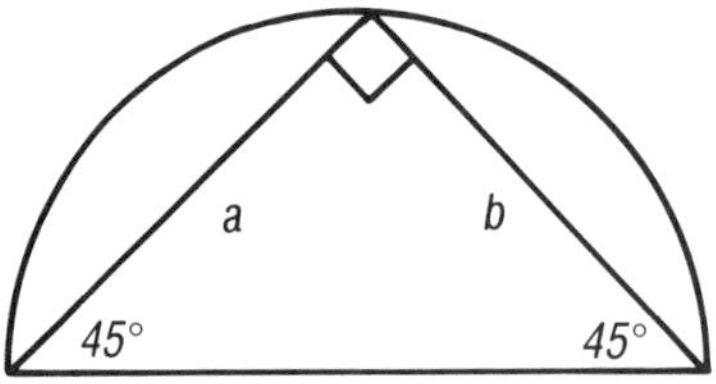

Figure 10-2

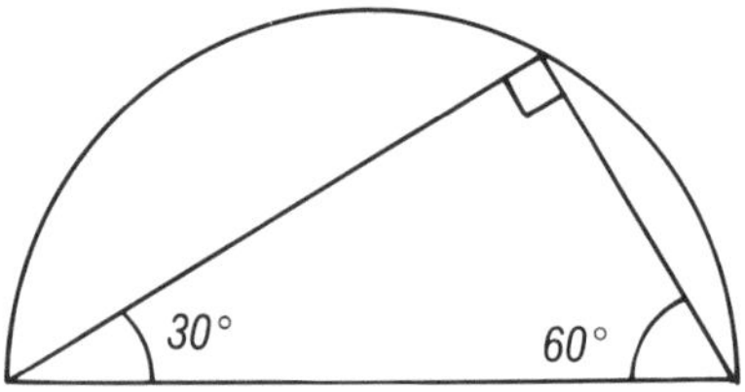

Figure 10-3

Figure 10-3 shows another triangle with different acute angles. In this triangle, one acute angle is half the value of the other. Observe that the side opposite the 60° angle is now longer than the side opposite the 30° angle. When a given acute angle changes in value, the ratio of the sides to the hypotenuse changes. This ratio is the quotient resulting from division of one side by the other or of the side by the hypotenuse.

Every angle forms its own ratio of sides and hypotenuse. This ratio is called a *function* of the angle. There are four trigonometric functions commonly used in the shop. They are listed in Table 10-1.

Function	Abbreviation
sine	sin
cosine	cos
tangent	tan
cotangent	cot

Table 10-1

Cosine is a contraction of *complement of the sine* and *cotangent* is a similar contraction of *complement of the tangent*.

Figure 10-4 is a conventional drawing of a right triangle with all its parts labeled. This figure will be used in defining these functions.

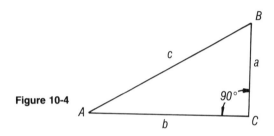

Figure 10-4

$$\text{sine of angle } BAC = \frac{\text{side opposite}}{\text{hypotenuse}} = \frac{a}{c}$$

$$\text{cosine of angle } BAC = \frac{\text{side adjacent}}{\text{hypotenuse}} = \frac{b}{c}$$

$$\text{tangent of angle } BAC = \frac{\text{side opposite}}{\text{side adjacent}} = \frac{a}{b}$$

$$\text{cotangent of angle } BAC = \frac{\text{side adjacent}}{\text{side opposite}} = \frac{b}{a}$$

When the functions of the other acute angle, angle *ABC*, are listed, something interesting can be observed.

$$\text{sine of angle } ABC = \frac{b}{c}$$

$$\text{cosine of angle } ABC = \frac{a}{c}$$

$$\text{tangent of angle } ABC = \frac{b}{a}$$

$$\text{cotangent of angle } ABC = \frac{a}{b}$$

There is a definite relationship between the functions of the two acute angles of a right triangle.

The sine of one acute angle equals the cosine of the other acute angle.
The cosine of one acute angle equals the sine of the other acute angle.
The tangent of one acute angle equals the contangent of the other acute angle.
The cotangent of one acute angle equals the tangent of the other acute angle.

The sine and *co*sine are said to be *co*functions of each other, as are the tangent and *co*tangent.

RULE 10-1   COFUNCTIONS

In a right triangle, any trigonometric function of one acute angle is equal to the corresponding cofunction of the other acute angle.

For example, if angle *BAC* (Figure 10-4) is equal to 30°, its complementary angle *ABC* is equal to 60°. Therefore:

$$\sin 30° = \cos 60° \quad \tan 30° = \cot 60°$$
$$\cos 30° = \sin 60° \quad \cot 30° = \tan 60°$$

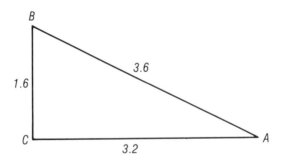

Example 10-1   Find the value of the trigonometric functions sine, cosine, tangent, and cotangent for angle *A*.

Solution $\sin A = \dfrac{1.6}{3.6} = 0.44 \qquad \tan A = \dfrac{1.6}{3.2} = 0.5$

$\cos A = \dfrac{3.2}{3.6} = 0.89 \qquad \cot A = \dfrac{3.2}{1.6} = 2$

Example 10-2   Use Example 10-1 to find cot B.

  Solution   cot B = tan A = 0.5

The following trigonometric functions should be memorized so that you can identify any function of an angle from a drawing of the triangle.

$$\sin A = \frac{\text{side opposite } A}{\text{hypotenuse}}$$ $$\tan A = \frac{\text{side opposite } A}{\text{side adjacent to } A}$$

$$\cos A = \frac{\text{side adjacent to } A}{\text{hypotenuse}}$$ $$\cot A = \frac{\text{side adjacent to } A}{\text{side opposite } A}$$

Exercises 10-1

Complete each sentence.

1. sin 46° = cos _____
2. cot 52° = tan _____
3. cos 3° = sin _____
4. tan 37° = cot _____
5. _____ 17° = tan 73°
6. sin 25° = _____ 65°
7. cot 89° = _____ 1°
8. cos 56° = _____ 34°

Use Figure A to find the value of the specified function.

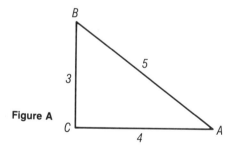

Figure A

9. sin A 10. cos A
11. tan A 12. cot A
13. sin B 14. cos B
15. tan B 16. cot B

17. If cos 82°31′ = 0.12447, find A when sin A = 0.12447.
18. If cot 31°25′ = 0.16372, find B when tan B = 0.16372.
19. If a = 8.17 and b = 13.2, find tan A.
20. If a = 7.1 and b = 3.8, find sin B.

10-2   The Table of Natural Functions

OBJECTIVES

After completing this section, you should be able to:

1. Use a table of trigonometric functions to read the value of a function of an angle.
2. Use a table and interpolation to find the value of a function of an angle.
3. Use a table and interpolation to find the value of an angle given a function.

The number of degrees in the acute angles of a right triangle can vary from 0° to 90°; the values of the trigonometric functions expressed as ratios will change as the size of the acute angle changes.

Tables have been worked out that give the values of the four trigonometric functions most used in shop work. The values of the ratios are approximate numbers commonly expressed to five decimal places. This type of table is called a **table of natural trigonometric functions.**

The tables in the back of this book contain values of the sine, cosine, tangent, and cotangent functions. Most tables list the angles from 0° to 45° at the top of the page. This table lists the functions from 0° to 44° at the top and 45° to 89° at the bottom of the page. To find an angle 45° or greater, begin at the lower right-hand corner of the table and read *upwards* for values.

Example 10-3   Find the tangent of 36° 50′.

Solution   Find 36°50′ in the upper left-hand corner, since this angle is *less* than 45°. Opposite 36° 50′ in the column headed *tan* at the top, read 0.74900, which is the tangent of 36° 50′.

Example 10-4   Find the value of cosine 77° 20′

Solution   When the value is *greater* than 44°, find the name of the function at the bottom of the page and read upward. The cosine of 77° 20′ is found to be 0.21928.

Most handbooks contain tables giving trigonometric functions for every 10 minutes (rather than for every minute). If such a table is not available, a procedure known as **interpolation** can be used to find approximations of such values.

Example 10-5   Find the sine of 15° 23′.

Solution   The sine of 15° 23′ is between sin 15° 20′ (0.26443) and sin 15° 30′ (0.26724). The difference between these two numbers is 0.00281.

$$10\left\{3\begin{cases}\sin 15°\,20' = 0.26443\\\sin 15°\,23' = \quad ?\\\sin 15°\,30' = 0.26724\end{cases}\right\}0.00281$$

The difference between sin 15° 23′ and sin 15° 20′ should be $\frac{3}{10}$ × 0.00281, or about 0.00084, since 23′ is $\frac{3}{10}$ of the way between 20′ and 30′. Therefore, add the difference of 0.00084 to the sine of 15° 20′.

0.26443
+0.00084
0.26527

Therefore, sin 15° 23′ = 0.26527.

Example 10-6 Find the cosine of 82° 51′.

Solution

$$10\left\{1\begin{cases}\cos 82°\,50' = 0.12476\\\cos 82°\,51' = \quad ?\\\cos 83° \quad\;\; = 0.12187\end{cases}\right\}\text{The difference is } 0.00289.$$

$\frac{1}{10}$ × 0.00289 = 0.000289, or about 0.00029

Subtract 0.00029 from cosine 82° 50′, since the cosine is decreasing as the angle increases.

0.12476
−0.00029
0.12447

Therefore, cos 82° 51′ = 0.12447.

It is possible to use interpolation to find the value of an angle in degrees and minutes when the value of the given function falls between two values in the table.

Example 10-7 Find the value of angle A if cot A = 1.6372.

Solution In a column headed *cot* at the top, find 1.6319 and 1.6426.

$$10'\left\{\begin{cases}\cot 31°\,20' = 1.6426\\\cot A \quad\;\;\; = 1.6372\\\cot 31°\,30' = 1.6319\end{cases}0.0054\right\}0.0107$$

$\frac{0.0054}{0.0107}$ × 10′ = 5.047, or 5′

Add 5′ to 31° 20′, so A = 31° 25′.

Exercises 10-2

Fill in the missing values. Do not use the table at the back of the book.

1.	$\sin 13°$ $= 0.22495$	$\sin 13° \, 10' = 0.22778$	$\sin 13° \, 8' =$?
2.	$\tan 28° \, 20' = 0.53920$	$\tan 28° \, 30' = 0.54296$	$\tan$? $= 0.54107$
3.	$\tan 75° \, 20' = 3.82083$	$\tan 75° \, 30' = 3.86671$	$\tan 75° \, 25' =$?
4.	$\sin 42°$ $= 0.66913$	$\sin 42° \, 10' = 0.67129$	$\sin$? $= 0.66978$
5.	$\cos 22° \, 40' = 0.92276$	$\cos 22° \, 50' = 0.92164$	$\cos$? $= 0.92231$

Find the indicated function of each angle.

6. $\sin 30°$ **7.** $\cos 28° 13'$ **8.** $\tan 37° 16'$

9. $\cot 18°$ **10.** $\sin 74° 59'$ **11.** $\cos 81° 48'$

12. $\tan 62°$ **13.** $\cot 49° 25'$ **14.** $\sin 48° 50'$

15. $\cos 78° 40'$ **16.** $\tan 36° 20'$ **17.** $\cot 12° 50'$

18. $\sin 32° 10'$ **19.** $\cos 27° 20'$ **20.** $\tan 89° 50'$

21. $\cot 47° 40'$ **22.** $\sin 11° 18'$ **23.** $\cos \ 2° 40'$

24. $\tan 35° 27'$ **25.** $\cot 21° 46'$ **26.** $\sin 53° 52'$

27. $\cos 72° 22'$

Find the indicated angle.

28. $\sin A = 0.98163$ **29.** $\cos A = 0.89879$ **30.** $\tan B = 0.59691$

31. $\cot C = 0.56962$ **32.** $\sin B = 0.78980$ **33.** $\cos B = 0.74703$

34. $\tan A = 0.26795$ **35.** $\cot A = 2.35585$ **36.** $\sin A = 0.97100$

37. $\cos A = 0.87036$ **38.** $\tan A = 0.98441$ **39.** $\cot A = 0.10275$

40. $\sin B = 0.09874$ **41.** $\cos C = 0.77347$ **42.** $\sin W = 0.76791$

43. $\tan B = 3.75388$ **44.** $\cot A = 0.68642$ **45.** $\tan D = 0.24008$

10-3 Solution of Right Triangles

OBJECTIVE

After completing this section, you should be able to:

1. Solve right triangles.

The rules listed below are for use in finding one side of a right triangle when a *side* and an acute *angle* are known. You can choose the most suitable operation, depending on the information given.

RULE 10-2   SOLVING RIGHT TRIANGLES

To Find	Operation:
Length of hypotenuse	$\dfrac{\text{Side opposite}}{\text{Sine}}$, or $\dfrac{\text{side adjacent}}{\text{cosine}}$
Length of side opposite	Hypotenuse × sine, or side adjacent × tangent
Length of side adjacent	Hypotenuse × cosine, or $\dfrac{\text{side opposite}}{\text{tangent}}$

Example 10-8   Find sides a and b.

Solution a = length of side opposite = hypotenuse × sin 41° 10′
              = 10 × 0.65825 = 6.583 in.

b = length of side adjacent = hypotenuse × cos 41° 10′
   = 10 × 0.75280 = 7.528 in.

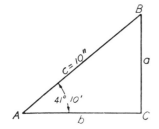

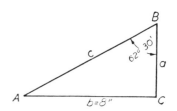

Example 10-9   Find the hypotenuse.

Solution c = length of hypotenuse = $\dfrac{\text{side opposite}}{\sin 62° 30'}$

$$= \frac{8}{0.88701} = 9.019 \text{ in.}$$

The following are examples of right triangles that have only the value of two sides or one side and the hypotenuse given. If one acute angle can be found, the missing value for the hypotenuse or side can be found in terms of an angle and side.

Example 10-10 Find angles *A* and *B* and side *a*.

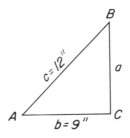

Solution First find angle *A*.

$$\cos A = \frac{\text{side adjacent}}{\text{hypotenuse}} = \frac{9}{12} = 0.75000$$

Angle $A = 41°\,25'$

Angle B = angle C − angle $A = 89°\,60' - 41°\,25'$
$$= 48°\,35'$$

Now, side *a* can be found.

$a = \text{hypotenuse} \times \sin A = 12 \times 0.66153$
$$= 7.938 \text{ in.}$$

Example 10-11 Find angles *A* and *B* and hypotenuse *c*.

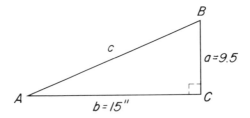

Solution First find angle *A*.

$$\tan A = \frac{\text{side opposite}}{\text{side adjacent}} = \frac{9.5}{15} = 0.63333$$

Angle $A = 32°\,21'$

Angle B = angle C − angle $A = 89°\,60' - 32°\,21'$
$$= 57°\,39'$$

Now find the hypotenuse c.

$$\text{Hypotenuse} = \frac{\text{side adjacent}}{\cos A}$$

$$= \frac{15}{0.84480}$$

$$= 17.756 \text{ in.}$$

Exercises 10-3

1. In Figure A, find dimension c.

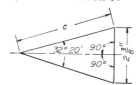

Figure A

2. Figure B shows three holes on the ear of a part being drilled by a numerically controlled drill. Find distances a and b to move from point 1 to point 2.

3. In Figure B, find distances x and y to move from point 2 to point 3.

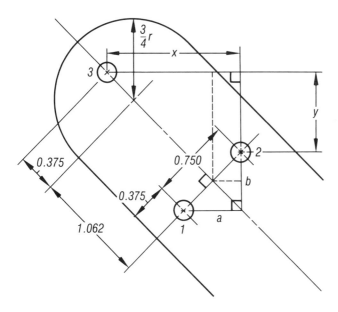

Figure B

4. Normally, the depth of a hole is given for the full diameter only, excluding the point of a drill. In Figure C, the depth of full diameter is $1\frac{5}{16}$ in. A question arises about the amount of metal that will remain between the drill point and the outside of the workpiece. What is distance T if the drill is $\frac{7}{8}$ in. in diameter?

5. If the drill in Figure C is 28.575 mm in diameter, what is distance T?

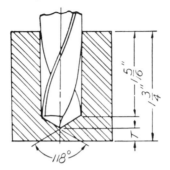

Figure C

6. In Figure D, to cut the thread to depth h, 0.0542 in., how many thousandths of an inch must the tool be advanced in the direction of the feed?

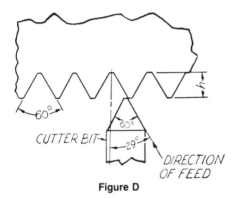

Figure D

7. In Figure E, an angled jaw for a cut-off saw is shown. It will clamp channel iron so that a cut of $1°\,20'$ will result. What must be the length of the small end to produce $1°\,20'$ slope on this jaw?

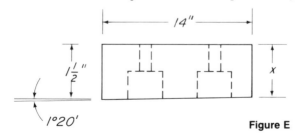

Figure E

8. Figure F shows the compound rest of a lathe swiveled to take a very light cut. If it is swiveled 30° from the center line of the lathe and the handle turns the graduated dial through five graduations, what depth of cut *d* will be taken?

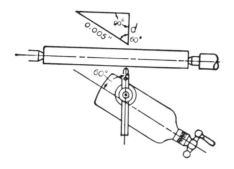

Figure F

9. In exercise 8, the compound rest is swiveled 84° and the graduated dial is in increments of 0.1 mm. What depth of cut will be taken in 25 graduations?

10. Figure G is a diagram of a hole layout for jig boring. What is the included angle *Y* that the center lines of the holes form with the center of the disc?

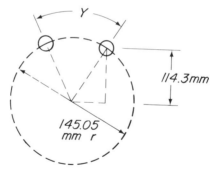

Figure G

11. In Figure H, find angle *A* and hypotenuse *c*.

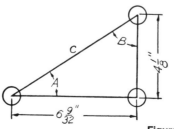

Figure H

12. Figure I illustrates a two-diameter workpiece with a chamfered shoulder. Although this method of designating a chamfer is poor drafting practice, the draftsperson has given enough information for a toolmaker to find angle *A*. A form tool ground to angle *A*, when advanced straight into the work, will produce angle *B*. You are the toolmaker who will grind the form tool. Find angle *A*.

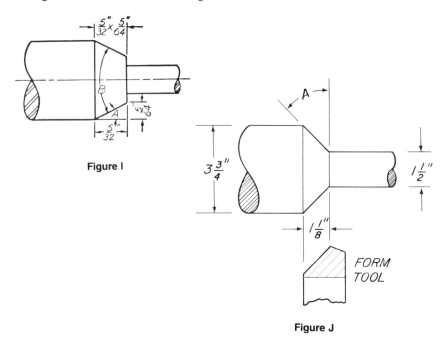

Figure I

Figure J

13. A turning tool with its leading edge ground on angle *A* (Figure J) will produce the chamfer on the workpiece illustrated. What should angle *A* be?

14. In Figure K, a milling machine with tilting head is set up to cut angle *A* with the side of an end mill. Find angle *A* to the nearest degree.

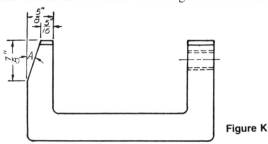

Figure K

15. An oil filtration line for a machine must slope 3 ft in every 100 ft. What will be its angle with the horizontal?

10-4   The Sine Bar

OBJECTIVE

After completing this section, you should be able to:

1.   Use a sine bar.

The sine bar is a precision instrument for layout, inspection, and setting of accurate angles in conjunction with gauge blocks. Figure 10-5 illustrates a basic sine bar.

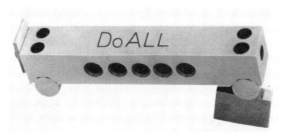

Figure 10-5   *(Photo courtesy of DoALL Company)*

In construction, the sine bar is a hardened and ground steel bar whose ends are fitted securely with hardened and ground discs of identical diameter, usually 1.000 inch. The centers of these discs are in a line parallel with the top surface of the bar, and the distance between their centers is either 5.000 or 10.000 inches.

Figure 10-6 is a drawing of a sine bar. It can be seen immediately that when one end is elevated, the bar forms the hypotenuse of a right triangle.

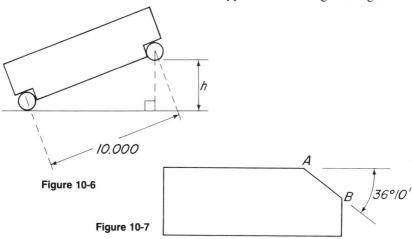

Figure 10-6

Figure 10-7

The effective use of a sine bar is best illustrated by a sample problem in inspection. Figure 10-7 shows a workpiece with a sloped surface bounded by points *A* and *B*. The sloped surface is to form an angle of 36° 10′. Checking procedure is as follows: The workpiece is rested on the top surface of a sine bar and clamped against an angle plate as in Figure 10-8. The 10-inch sine bar has a combination of gauge blocks under one disc sufficient to tilt the bar and workpiece 36° 10′. A test indicator mounted on a height gauge is traversed from one end to the other. If there is no variation in height, the indicator remains at zero reading and the surface is angled correctly.

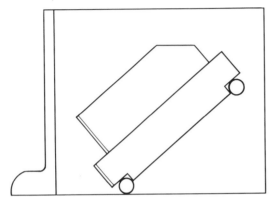

Figure 10-8

Example 10-12   Find the combination of gauge blocks needed to tilt the sine bar 36° 10′.

Solution   Basically, there is a triangle with the dimensions shown below. The height h is the altitude of the triangle.

$$\sin 36° 10′ = \frac{h}{\text{hypotenuse}} = \frac{h}{10}$$

$$h = 10(0.59014) = 5.9014 \text{ in.}$$

A combination of gauge blocks that equals 5.9014 in. must be inserted under one disc of the sine bar.

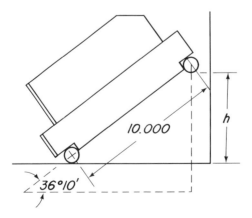

Most handbooks contain tables for the difference in height of the discs on either a 5-inch or 10-inch base. However, a table of sines will give the sine value, and the height can be found if the function is multiplied by the length of the sine bar. The sine bar is invaluable for checking tapers (see Unit 11).

Exercises 10-4

1. At what angle is the sine bar in Figure A tilted from the horizontal?
2. What is the value of angle *A* in Figure B?

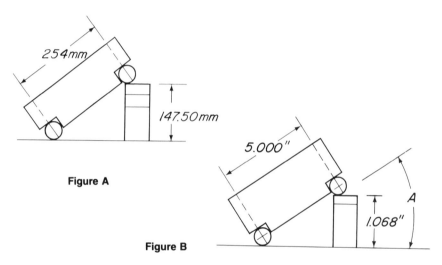

Figure A

Figure B

3. What total combination of gauge blocks is needed to tilt a 5-in. sine bar 19° 5'?

4. In Figure C, what is the angle *A* at which the sine bar is tilted?

5. In Figure C, what would be the angle *A* at which the sine bar was tilted if a 5.000-in. sine bar were used?

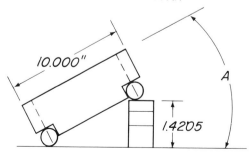

Figure C

10-5   Length of a Chord

OBJECTIVE

After completing this section, you should be able to:

1. Find the length of a chord by using trigonometry.

In Unit 9, a method was discussed for finding the length of a circular arc cut by a chord. There are times when the *straight* distance between holes on a circle must be established. This involves the application of right-angle trigonometry. In Figure 10-9, it is required to position eight equally spaced holes on a bolt circle with a 10-inch diameter.

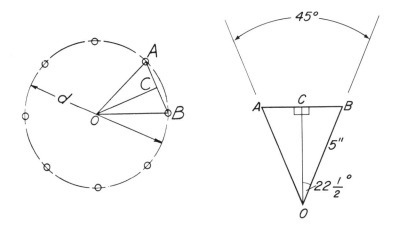

Figure 10-9

The eight holes divide the circle into eight sectors with central angles of 45°. If OC is drawn perpendicular to chord AB, it divides both the central 45° angle and the chord into equal parts. Two right triangles result. Solve for side CB or CA, then double the result. This will be the length of the chord AB.

Example 10-13   Find the length of chord AB.

Solution   In triangle OCB,

$$\sin 22\tfrac{1}{2}° = \frac{CB}{5}$$

$$5(0.38268) = CB = 1.9134$$

Therefore, chord $AB = 3.8268$ in.

Exercises 10-5

Fill in the chord lengths.

	No. of Equally Spaced Holes	Bolt Circle	Chord Distance Between Holes
1.	4	12	
2.	6	6	
3.	10	15	
4.	16	11.125	
5.	51	21	
6.	3	8	

7. In Figure A, find the center-to-center distance on a straight line between the two holes.

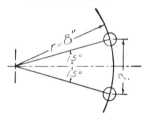

Figure A

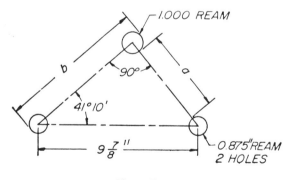

Figure B

8. In Figure B, the toolmaker must check dimensions *a* and *b* after the holes are reamed. Find *a* and *b*.

9. The workpiece shown in Figure C has been clamped to a jig bore table and hole *B* bored. Now the table is to be repositioned a distance + *x* units to the right and − *y* units toward the operator to locate point A under the machine spindle. Find the + *x* and − *y* distances.

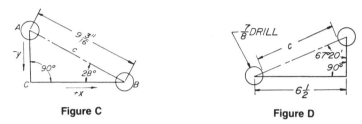

Figure C   **Figure D**

10. After the two holes are drilled, in Figure D, the toolmaker uses dimension *c* to check the accuracy. What is dimension *c*?

11. Figure E is a layout for three holes. What is center-to-center distance *b*?

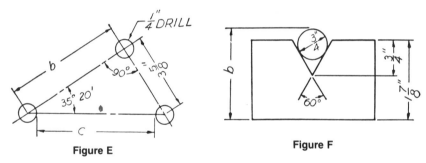

Figure E   **Figure F**

12. In Figure F, find measurement *b*.

13. In Figure G, to what diameter is the plug turned to measure the accuracy of the angular cut if the plug is flush with *DE*?

14. The diameter of each plug in Figure H is $\frac{3}{4}$ in. The toolmaker measures the distance over the plugs, dimension *C*, with a vernier caliper. What is dimension *C* to three decimal places?

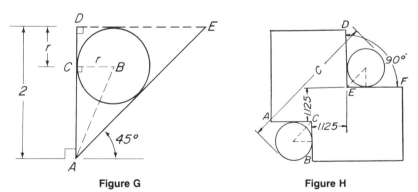

Figure G   **Figure H**

15. In Figure I, the edges of the casting are milled after the 1-in.-diameter holes are drilled, so that the distances marked *A* are equal. Find distance *A*.

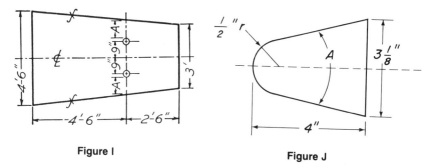

Figure I

Figure J

16. In Figure J, find the value of angle *A* in degrees and minutes.

17. Find the distance *x* in Figure K.

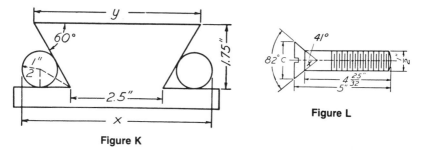

Figure K

Figure L

18. A flat-head bolt, Figure L, is made to the dimensions shown. Find the diameter *c* to the nearest thirty-second of an inch.

19. In Figure M, find the number of degrees and minutes in angle *A*.

20. Figure N shows a 0.875-in.-diameter end mill machining the outside perimeter of a part. Find the offset value *x*. (*Note:* Line *AC* bisects angle *DAB*.)

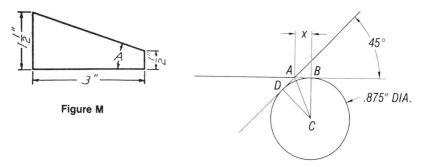

Figure M

Figure N

21. In Figure O, find the dimensions x and y. (*Note:* Line AB bisects the angle formed by the tangents on the end mill.)

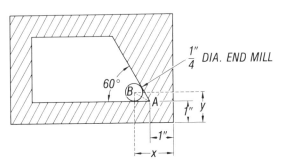

Figure O

10-6   Law of Sines

OBJECTIVE

After completing this section, you should be able to:

1. Use the law of sines to solve oblique triangles.

An **oblique** triangle does not contain a right angle. This type of triangle has many practical applications. It is invaluable for surveyors in the calculation of civil engineering projects, such as construction of tunnels, grading of roads, and establishment of distances to objects that are difficult to approach.

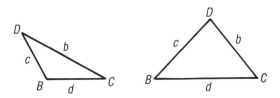

Figure 10-10

Toolmakers frequently make use of the laws of oblique triangles to locate holes accurately. Typical oblique triangles encountered in toolroom work are shown in Figure 10-10. The triangle on the left contains one obtuse angle and two acute angles. The triangle on the right has three acute angles. It is good practice to label the angles of an oblique triangle with letters B, C, and D to avoid confusion with the standard A, B, and C labels used for right triangles.

Three parts of an oblique triangle must be known to solve for the other unknown angles or sides. At least one of the three unknown parts must be a side. It can be proved for any triangle BCD that $\dfrac{b}{\sin B} = \dfrac{c}{\sin C} = \dfrac{d}{\sin D}$. This proportion can be written as three separate proportions, as in Rule 10-3. If any three parts in any of the proportions are known, the fourth part can be found.

RULE 10-3 LAW OF SINES

In any triangle, the ratios of the sides to the sines of the angles opposite them are equal.

$$\frac{c}{\sin C} = \frac{d}{\sin D} \qquad \frac{b}{\sin B} = \frac{c}{\sin C} \qquad \frac{b}{\sin B} = \frac{d}{\sin D}$$

Example 10-14   Find c.

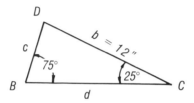

Solution $\dfrac{b}{\sin B} = \dfrac{c}{\sin C}$ or $\dfrac{12}{\sin 75°} = \dfrac{c}{\sin 25°}$

By cross multiplication,

$$\frac{12 \sin 25°}{\sin 75°} = c, \quad \text{or} \quad c = \frac{12(0.42262)}{0.96593} = 5.25 \text{ in.}$$

RULE 10-4   APPLYING THE LAW OF SINES TO SOLVE
OBLIQUE TRIANGLES

Apply the law of sines when:
Two angles and a side are given.
Two sides and an angle opposite one of them is given.

Example 10-14 illustrates the first case, given two angles and a side. The triangle involved three *acute* angles. It sometimes happens in the solution of an oblique triangle that one of the angles is an *obtuse* angle. This is an angle containing more than 90° but less than 180°. If the law of sines is applied to an oblique triangle containing an obtuse angle, it is necessary to find the value of the *sine of an angle greater than 90° and less than 180°*, provided the side opposite the obtuse angle is not given. Referring to Example 10-14 once again, if angle B were 130°, cos 40° would be used in the ratio $\dfrac{b}{\sin B}$ instead of sin 130°.

The sine of an angle between 90° and 180° can be written as $\sin(90° + A)$, where A represents an *acute* angle. It can be proved that $\sin(90° + A) = \cos A$. Thus, $\sin 130° = \sin(90° + A) = \sin(90° + 40°) = \cos 40°$, which is equal to 0.76604. To find the value of $\sin 120°$, look for $\cos 30°$ in the table; to find the value of $\sin 140°$, look for $\cos 50°$; and so on.

RULE 10-5   FINDING THE FUNCTION OF AN OBTUSE ANGLE

$\sin(90° + A) = \cos A$

Example 10-15   Find the value of b.

Solution $\dfrac{b}{\sin B} = \dfrac{c}{\sin C}$

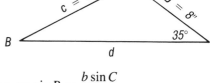

$\dfrac{b}{\sin 125°} = \dfrac{8}{\sin 40°}$

$b = \dfrac{8 \sin 125°}{\sin 40°} = \dfrac{8 \cos 35°}{\sin 40°}$

$= \dfrac{8(0.81915)}{0.64279} = 10.19$ m

Example 10-16   Find B, D, and d.

Solution   For angle B

$\dfrac{b}{\sin B} = \dfrac{c}{\sin C}$, so $\sin B = \dfrac{b \sin C}{c}$

$\sin B = \dfrac{8(\sin 35°)}{10} = \dfrac{8(0.57358)}{10} = 0.45886$

Angle $B = 27° 19'$

For angle D

$D = 180° - (35° + 27° 19') = 117° 41'$

$\dfrac{d}{\sin D} = \dfrac{c}{\sin C}$, so $d = \dfrac{c(\sin D)}{\sin C}$

For d

$d = \dfrac{10(\sin 117° 41')}{\sin 35°} = \dfrac{10(\cos 27° 41')}{\sin 35°}$

$= \dfrac{10(0.88550)}{0.57358} = 15.44$ in.

Exercises 10-6

In the following exercises, where a triangle is not shown, draw one and label all sides and angles. Then find the missing parts.

1. If $B = 25°$, $D = 80°$, and $b = 15$ in., find C, c, and d.
2. If $C = 27°$, $D = 72°$, and $c = 8.4$ in., find B, b, and d.
3. If $B = 118°27'$, $C = 19°18'$, and $d = 4$ mm, find D, b, and c.
4. Figure A is a diagram of three holes in a tooling plate. You, as a toolmaker, are asked to find the center-to-center distance BC (side d). Solve for this distance.

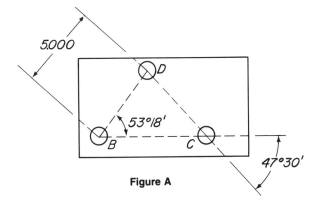

Figure A

5. Figure B illustrates two steel blocks that, when bolted together, form a special locating angle X. What is the value of angle X?

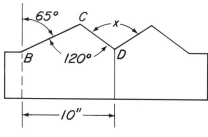

Figure B

6. In Figure B, find the length of side d.
7. Angle C in an oblique triangle equals $60°$ and sides b and c equal 11 in. and 12 in., respectively. Find angle B and angle D.

8. In Figure C, find the length of side d.

9. In Figure C, what is angle D?

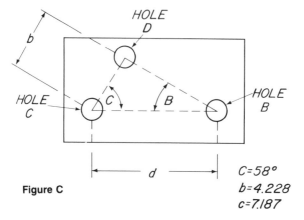

Figure C

$C=58°$
$b=4.228$
$c=7.187$

10. Find the distance in Figure D from point B across a river to point D, if length BC is 283 m, angle BCD is 66° 18′, and angle CBD is 38°.

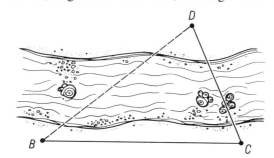

Figure D

10-7   Law of Cosines

OBJECTIVE

After completing this section, you will be able to

1. Use the law of cosines to solve oblique triangles.

The **law of cosines** is another valuable tool in the solution of oblique triangles when one of two conditions exist.

RULE 10-6   APPLYING THE LAW OF COSINES TO SOLVE OBLIQUE TRIANGLES

Apply the law of cosines when:
Two sides and the included angle are given.
Three sides are given.

The law of cosines states that in any triangle, the *square of any side equals the sum of the squares of the other two sides diminished by twice the product of these two sides multiplied by the cosine of the included angle.* This is a case where a formula expresses the statement better than words could do.

RULE 10-7   LAW OF COSINES

$$b^2 = c^2 + d^2 - 2cd \cos B$$
$$c^2 = b^2 + d^2 - 2bd \cos C$$
$$d^2 = b^2 + c^2 - 2bc \cos D$$

Figure 10-11 illustrates the notation in the above formulas.

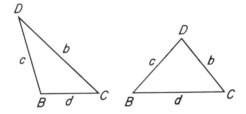

Figure 10-11

Although the law of cosines can be used to solve any triangle, it is usually employed to solve oblique triangles when two sides and the included angle or three sides are given. Given a right triangle where the right angle is considered the included angle (see Figure 10-12), $c^2 = b^2 + d^2 - 2bd \cos C$. Since angle C equals $90°$, $\cos C$ equals $\cos 90°$, which is zero. Multiplying $2bd$ by $\cos 90°$ is thus the same as $2bd$ times zero, which is zero. Therefore, in the case of a right triangle, the formula $c^2 = b^2 + d^2 - 2bd \cos C$ becomes $c^2 = b^2 + d^2$. This is the Pythagorean theorem. It is not good practice to attempt a solution for one side of a right triangle by employing the law of cosines when a shorter solution is offered by the Pythagorean theorem.

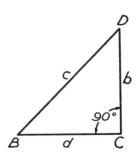

Figure 10-12

The solution of oblique triangles by the law of cosines presents the two following cases:

1. The included angle is acute.
2. The included angle is obtuse.

You will usually encounter the first case.

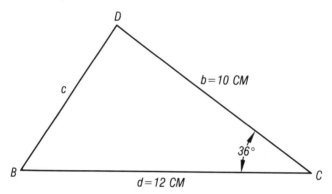

Example 10-17   Find c, D, and B.

  Solution   Use the law of cosines to find c:

$$c^2 = b^2 + d^2 - 2bd\cos C$$
$$c^2 = 10^2 + 12^2 - 240(0.80902)$$
$$c^2 = 100 + 144 - 194.165 = 49.835$$
$$c = \sqrt{49.835} = 7.059 \text{ cm}$$

Use the law of sines to find B:

$$\frac{b}{\sin B} = \frac{c}{\sin C}, \text{ so } \sin B = \frac{b\sin C}{c}$$

$$\sin B = \frac{10(0.58779)}{7.059} = 0.83268$$

$$B = 56° 23'$$

Find D:

$$D = 180° - (36° + 56° 23') = 87° 37'$$

The angles of an oblique triangle can be found using the law of cosines if *three sides* are given. The following three formulas, which are derived from the law of cosines, are used to find the value of the three angles when three sides are given.

RULE 10-8 FINDING THE ANGLES WHEN THREE SIDES ARE
GIVEN

$$\cos B = \frac{c^2 + d^2 - b^2}{2cd}$$

$$\cos C = \frac{b^2 + d^2 - c^2}{2bd}$$

$$\cos D = \frac{b^2 + c^2 - d^2}{2bc}$$

Example 10-18 Find B, C, and D.

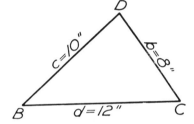

Solution $\cos B = \dfrac{c^2 + d^2 - b^2}{2cd} = \dfrac{10^2 + 12^2 - 8^2}{240} = 0.75000$

$B = 41° 25'$

$\cos C = \dfrac{b^2 + d^2 - c^2}{2bd} = \dfrac{8^2 + 12^2 - 10^2}{192} = 0.56250$

$C = 55° 46'$

$\cos D = \dfrac{b^2 + c^2 - d^2}{2bc} = \dfrac{8^2 + 10^2 - 12^2}{160} = 0.12500$

$D = 82° 49'$

Exercises 10-7

In the following exercises, solve for the third side of the oblique triangle by using the law of cosines; then the second and third angles can be found using the law of sines.

1. If $B = 30°$, $c = 10$ in., and $d = 15$ in., find b, C, and D.

2. If $B = 42°$, $c = 7$ m, and $d = 5$ m, find b, C, and D.

3. If $C = 27°$, $b = 11$ cm, and $d = 9$ cm, find c, B, and D.

4. If $C = 23°$, $d = 13$ in., and $b = 19$ in., find c, B, and D.

5. If $D = 42°40'$, $c = 14$ in., and $b = 12$ in., find d, B, and C.

6. If $D = 67°50'$, $b = 13$ mm, and $c = 10$ mm, find d, B, and C.

Solve the following oblique triangles, given three sides.

7. If $b = 8$ in., $c = 15$ in., and $d = 14$ in., find B, C, and D.

8. If $b = 4.5$ m, $c = 8$ m, and $d = 7.3$ m, find B, C, and D.

9. If $b = 12$ cm, $c = 14$ cm, and $d = 8.5$ cm, find B, C, and D.

10. If $b = 6$ ft, $c = 7$ ft, and $d = 9$ ft, find B, C, and D.

Unit 10 Review Exercises

---

1. In Figure A, the circle with radius r contains a right triangle with angle X and sides x and y. Using these terms, write formulas to find $\sin X$, $\cos X$, $\tan X$, and $\cot X$.

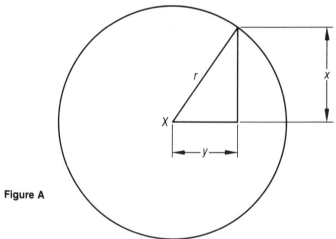

Figure A

2. In Figure A, if X is $47°38'$, what is $\cos X$?

3. Interpolate:
$\sin 29° = 0.48481$
$\sin 29°5' = \quad ?$
$\sin 29°10' = 0.48735$

4. Interpolate:
$\cos 6°10' = 0.99421$
$\cos \quad ? \quad = 0.99399$
$\cos 6°20' = 0.99390$

5. In Figure B, find sides *c* and *b*.

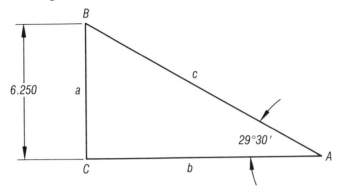

Figure B

6. Find the value of *h* in Figure C.

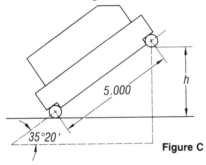

Figure C

7. In Figure C, find the value of *h* if the angle is 15°.

8. In the eight-hole bolt circle shown in Figure D, what is the distance between the centers of the holes? Round your answer to four decimal places.

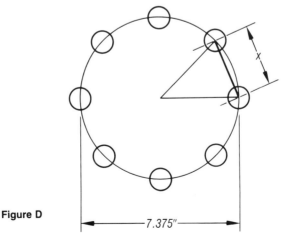

Figure D

9. Find the length along edge *CB* of the blade shown in Figure E. Round your answer to three decimal places.

10. Given three holes spaced center to center as shown in Figure F, find angles *B*, *C*, and *D*.

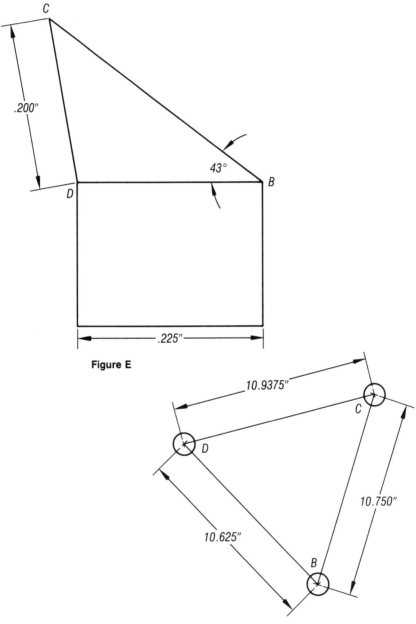

Figure E

Figure F

Tapers

When a round or flat object changes in size in a given distance, it is said to have **taper,** or to be **tapered.** Figure 11-1 illustrates taper on a wedge and on a pin.

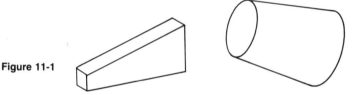

Figure 11-1

11-1 Types of Tapers

OBJECTIVES

After completing this section, you should be able to:

1. Identify common types of tapers.
2. Find the diameter of the ends of a taper pin and find the amount of taper.

Taper per foot is the unit of measurement for taper in countries using the English system of measurement. If, for example, the diameters of a workpiece 1 foot long are 3 inches and $2\frac{1}{2}$ inches, the piece would have a taper of $\frac{1}{2}$ inch per foot. Often the workpiece is less than a foot in length. In such cases, the taper per foot is divided by 12 to give **taper per inch.**

The metric system designates amount of taper in degrees from a base line or the central axis of the work. This would correspond to one-half the included angle in American shop practice. This, incidentally, points out one of the weaknesses of the metric system: There is no convenient large unit, such as our foot, to use in measuring taper until the meter is reached. Anything less must be in centimeters or millimeters. The decimeter is rarely used in industrial practice for machined work.

The common tapers found in every machine shop fall into one of two categories: *fast* or *slow* tapers. A **slow taper** is one with an included angle of not more than 3°. Tapers of this category result in a wedging action when forced into a mating taper. These are valuable for holding work or tools securely.

A **fast taper** has an included angle too large to be effective for wedging, but serves as a locator in a tapered hole. This type of taper has the advantage of quick assembly and disassembly. The best way to identify fast and slow tapers is by name, amount of taper, and use.

Morse Taper

The **Morse taper** is used on rotary cutting tools (drills, reamers, counterbores, and the like) that will fit into the spindle of a drill press or tailstock spindle of an American lathe. Lathes have a Morse taper in their *tailstock* spindles but do not necessarily have the same type of taper in the *headstock* spindles. The latter may be a Jarno, Brown & Sharpe, or any other kind of taper that the manufacturer decides is proper for a particular lathe.

The Morse taper is *generally* $\frac{5}{8}$ inch per foot. This is a *slow* taper. Table 11-1 lists the essential data needed to identify a Morse taper of any particular number. In this table and those that follow, *T.P.F.* and *T.P.I.* are abbreviations for *taper per foot* and *taper per inch,* respectively.

MORSE TAPER			
No.	T.P.F.	T.P.I.	Diameter at Small End
0	0.6246	0.0521	0.252
1	0.5986	0.0499	0.369
2	0.5994	0.0499	0.572
3	0.6023	0.0512	0.778
4	0.6233	0.0519	1.020
5	0.6312	0.0526	1.475
6	0.6256	0.0521	2.116
7	0.6240	0.0520	2.750

Table 11–1.

Brown & Sharpe Taper

The **Brown & Sharpe taper** is $\frac{1}{2}$ inch per foot, except for the No. 10, which has 0.516 inch taper per foot. Brown & Sharpe tapers are found in some grinding machines and the spindle holes of most dividing heads. Any tapered-shank end mill or other milling cutter will usually have a Brown & Sharpe shank. This is a holdover from an earlier time when the spindles of milling machines had a Brown & Sharpe taper, requiring tools to be secured by a drawbar reaching through the machine spindle to screw into the rear end of the tool.

Brown & Sharpe tapers are *slow* tapers. Care must be exercised in their use, since it is easy for a new worker to mistake a Brown & Sharpe for a Morse taper. If a Brown & Sharpe tapered tool is inserted into a Morse drill sleeve, for example, it will appear to fit but will come loose under sideways pressure. Tapers *must* be of the same type to mate together. Table 11-2 gives the most frequently used Brown & Sharpe tapers, together with the diameter at the small end, for easy identification.

BROWN & SHARPE TAPER	
No.	Diameter at Small End
1	0.200
2	0.250
3	0.312
4	0.350
5	0.450
6	0.500
7	0.600
8	0.750
9	0.900
10	1.045
11	1.250
12	1.500

Table 11–2.

Taper Pin

One of the *slow,* or self-holding, tapers used in machine assembly is the **taper pin,** which is ¼ inch per foot. The size is given by the diameter at the large end. The sizes of taper pins overlap, as Figure 11-2 illustrates. Table 11-3 gives the number of the taper and corresponding large diameter.

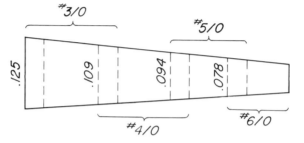

Figure 11-2

TAPER PIN TAPER	
No. of Taper	Diameter of Large End
6/0	0.078
5/0	0.094
4/0	0.109
3/0	0.125
2/0	0.141
0	0.156
1	0.172
2	0.193

Table 11–3.

To prepare a hole for one of these pins, the parts to be pinned together are assembled. A hole approximately $\frac{1}{64}$ inch smaller than the small end of the pin is drilled through both parts. This hole is enlarged with a tapered reamer of a size that will allow the head of the pin to project from the hole when the pin is driven home.

The diameter of the small end of a taper pin is found by the following formula.

RULE 11-1   FINDING THE DIAMETER OF THE SMALL END OF
A TAPER PIN

Diameter of large end − (0.0208 × pin length) =
diameter of small end

Example 11-1   What drill size would leave reaming stock in a hole for a
$3/0$ taper pin? The pin length is $\frac{3}{4}$ in.

Solution   Diameter large end − (0.0208 × pin length)
= diameter small end

$$0.125 - (0.0208 \times .75) = d$$
$$0.125 - 0.0156 = d$$
$$0.109 = d$$

$0.109 - 0.015$ (for reaming stock) $= 0.094$ in.

Use the table of decimal equivalents (Figure 3-2) on page 61
to determine that the fraction closest to 0.094 is $\frac{3}{32}$
($\frac{3}{32} = 0.09375$). Thus, a $\frac{3}{32}$-in. diameter drill will be satisfactory.

Standard Milling Machine Taper

The **Standard Milling Machine taper** is a *fast,* or self-releasing, taper
found on milling machine arbors and adaptors for cutting tools to fit into the
machine spindle. The taper is $3\frac{1}{2}$ inches per foot and requires holding by a
drawbar. Table 11-4 gives identifying information for size.

STD. MILLING MACHINE TAPER	
No.	Diameter at Large End
30	1.250
40	1.750
50	2.750
60	4.250

Table 11–4.

Jarno Taper

The **Jarno taper** is a relative latecomer on the scene. It was the creation of a Brown & Sharpe employee. It is so simple that one number gives all dimensions necessary for its identification and manufacture. Many grinding machine and lathe headstock manufacturers have chosen the Jarno, which is 0.600 inch per foot for all sizes. The number of the taper is the key to all taper dimensions.

RULE 11-2   FINDING TAPER DIMENSIONS IN A NO. *N* JARNO TAPER

Large diameter $(D) = \dfrac{N}{8}$

Small diameter $(d) = \dfrac{N}{10}$

Length $(L) = \dfrac{N}{2}$

Example 11-2   Give the dimensions of a No. 10 Jarno taper.

$$D \quad {}^{\pi}/O \; JARNO \quad d$$

$$\longmapsto L \longmapsto$$

Solution $D = \dfrac{N}{8} = \dfrac{10}{8} = 1.250$ in.

$d = \dfrac{N}{10} = \dfrac{10}{10} = 1.000$ in.

$L = \dfrac{N}{2} = \dfrac{10}{2} = 5.000$ in.

Determining Amount of Taper

There are three parts to any taper: (1) the large diameter; (2) the small diameter; and (3) the length between the diameters. A machinist or toolmaker may find one of the following conditions on a blueprint of a tapered work-piece:

1. The amount of taper is given, but one of the diameters is not given. (See Example 11-3.)
2. The three parts of the taper are given, but not the taper. (See Example 11-4.)

RULE 11-3   FINDING A MISSING DIAMETER, GIVEN THE TAPER

Step 1   Multiply the taper per inch by the number of inches of tapered length.

Step 2   *To find the large diameter,* add the amount from step 1 to the small diameter.

To find the small diameter, subtract the amount from step 1 from the large diameter.

Example 11-3   Find the value of d.

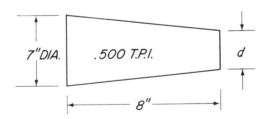

Solution

Step 1   Multiply the taper per inch (0.500) by the number of inches of tapered length.

0.500×8 in. $= 4.000$ in., the amount of taper

Step 2   Subtract the amount of taper from the large diameter.

$7.000 - 4.000 = 3.000$ in.

Thus, $d = 3.000$ in.

RULE 11-4   FINDING THE TAPER, GIVEN THE DIMENSIONS OF THE TAPER

Step 1   Subtract the smaller diameter from the larger.
Step 2   Divide the difference from Step 1 by the length of the taper.

Example 11-4   Here is a type of situation that is likely to be found on a blueprint. Find the taper per inch and the taper per foot.

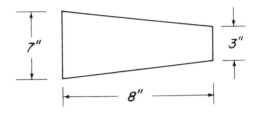

Solution

Step 1   7.000 − 3.000 = 4.000 total taper

Step 2 $\dfrac{4.000}{8 \text{ in.}}$ = 0.500 in./in. or 6.000 in./ft taper

---

Exercises 11-1

1. Taper pins have a taper of $\frac{1}{4}$ in. per foot. What is the taper per inch?
2. In Figure A, what is the *T.P.I.* of the tapered pin?

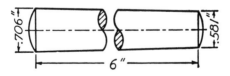

Figure A

3. A tapered reamer has a diameter of 0.162 in. at its small end and a diameter of 0.200 in. at its large end. The length of the tapered flutes is $1\frac{13}{16}$ in. What is the taper? The taper per inch?

4. If a Brown & Sharpe taper is turned on a piece $6\frac{1}{2}$ in. long and the diameter at the small end is 1.326 in., what is the diameter at the large end?

5. Figure B shows a shaped end of a pin that will be plunged by a form tool. What is diameter x?

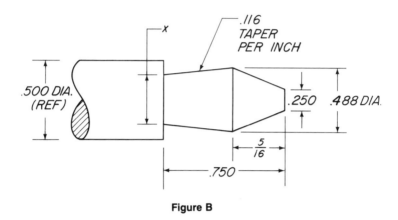

Figure B

6. What is the length x of the tapered shaft shown in Figure C?

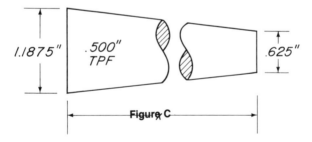

Figure C

7. Figure D represents a workpiece turned to a Jarno taper. Find the number of the taper and the small diameter.

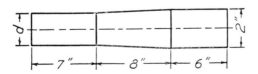

Figure D

8. In Figure E, the point of a conical workpiece has been faced off to a $4\frac{11}{16}$-in. length of tapered portion. What was the original length x of the tapered portion?

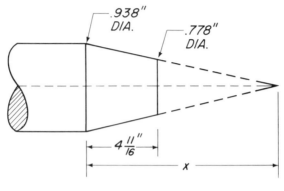

.938" DIA.

.778" DIA.

$4\frac{11}{16}$"

x

Figure E

9. In Figure F, what is the value of diameter x?

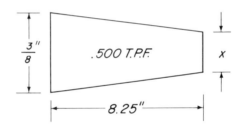

$\frac{3}{8}$"

.500 T.P.F.

x

8.25"

Figure F

10. Compute the large diameter of the inside taper (Brown & Sharpe) of the drill sleeve shown in Figure G.

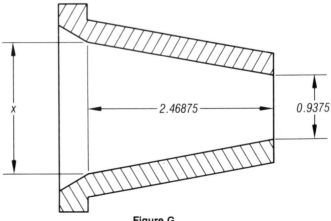

x

2.46875

0.9375

Figure G

11-2 Machining Tapers

OBJECTIVE
After completing this section, you should be able to:
1. Calculate the amount the tailstock should be offset.

Each of the primary machine tools (lathe, milling machine, grinder, drill press) has provisions for producing taper on workpieces. Sometimes it is performed by means of a rotary tool shaped to the exact taper needed. Other times, accessories or attachments for the machines make tapers possible.

The lathe is most often used for taper work. There are three common methods of turning tapers on a lathe:

Offsetting the tailstock

Taper attachment

Compound rest

The length of the taper and the number of workpieces determine which of the methods listed above is used. Numerous formed tools, both rotary and stationary, will produce a tapered hole. Among these are:

Tapered drills

Tapered reamers

Countersinks

Formed plunging tools

For boring in a lathe, taper can be produced by means of the compound rest and the taper attachment.

The **offset tailstock** method of turning tapers lends itself to *small* tapers and relatively *short* work. The tailstock of a metal-cutting lathe is set in line with the central axis of the headstock spindle. Diameters turned in this position (Figure 11-3) will be the same for the full length of the work.

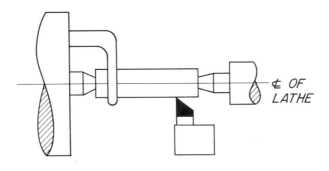

Figure 11-3

The tailstock can be adjusted sideways toward or away from the operator. This throws one end of the workpiece closer or farther away from the lathe cutting tool. (See Figure 11-4.)

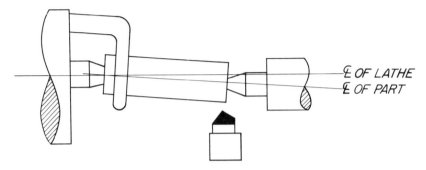

£ OF LATHE
£ OF PART

Figure 11-4

The amount the tailstock is offset depends on the *taper per inch* and the *total* length of the work in inches. Since the work revolves, each revolution of the piece doubles the amount that the tool cuts from one side. This means that the tool reduces the diameter by *twice* the amount of cut taken. The formula for finding the amount of offset to turn a taper is stated in Rule 11-5. (Often the workpiece is mounted on a mandrel that is longer than the piece being turned. This is why the formula states "total work length between centers" rather than "length of work.")

RULE 11-5   CALCULATING THE OFFSET OF THE TAILSTOCK

$$\text{Offset} = \frac{T.P.I. \times \text{total work length between centers}}{2}$$

The offset figured for a particular length of work is only good for that given length. A different length requires a new offset calculation.

Example 11-5   Determine the tailstock offset to produce the taper shown below.

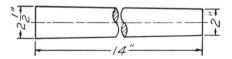

Solution   Offset $= \dfrac{\text{T.P.I.} \times \text{length between centers}}{2}$

First find *T.P.I.*:

$$\text{T.P.I.} = \dfrac{\text{difference in diameters}}{\text{length}}$$

$$= \dfrac{2.5 - 2}{14} = 0.0357 \text{ in.}$$

Then:

$$\text{Offset} = \dfrac{0.0357(14)}{2} = .250, \text{ or } \dfrac{1}{4} \text{ in.}$$

Example 11-6   Compute the tailstock offset for the workpiece below.

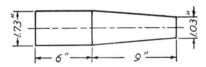

Solution $\text{T.P.I.} = \dfrac{1.73 - 1.03}{9} = 0.0778$

$$\text{Offset} = \dfrac{0.0778(15)}{2} = 0.5835 \text{ in.}$$

Using the decimal equivalent chart on page 61, find that 0.5835 is about $\frac{37}{64}$ in.

Note: The length is what is *between the centers*, not the tapered portion.

Exercises 11-2

---

1. Calculate the offset required to turn a taper of $\frac{3}{4}$ in./ft on a workpiece whose length is 18 in.

2. In Figure A, compute the tailstock offset required to turn the taper shown.

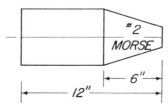

Figure A

3. If the tailstock is offset $\frac{1}{8}$ in., what taper per foot will be produced on a workpiece 10 in. long?

4. What amount of offset will be required to turn the taper shown in Figure B?

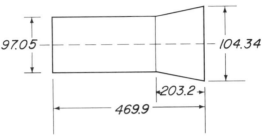

Figure B

5. The pulley shown in cross section in Figure C is turned to a diameter of 5 ins. It is then crowned (made higher in the middle) with a taper of $\frac{1}{2}$ in./ft from each end. If the pulley is mounted on a mandrel $5\frac{1}{4}$ in. long, figure the tailstock offset necessary to produce this crown.

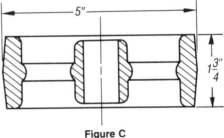

Figure C

6. Calculate the tailstock offset for turning for a taper of $\frac{7}{8}$ in./ft for a distance of 6 in. (The workpiece is 11 in. long.)

7. A 14-spline broach has a taper of $\frac{1}{16}$ in./ft. The broach is 16 in. long. Calculate the tailstock offset if the tapered part of the broach is 9 in. long.

8. Compute the tailstock offset for turning the tapered adjustable mandrel shown in Figure D.

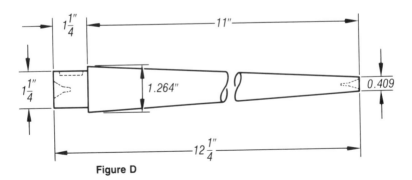

Figure D

11-3 Finding the Angle Corresponding to a Taper

OBJECTIVE

After completing this section, you should be able to:

1. Find the angle corresponding to a given taper.

Tapers that have a gradual slope are specified by dimensioning the work drawing in inches per foot. For tapers over 2 inches per foot, it is customary to dimension the included angle with the center line of the workpiece, although in some cases the angle that the chamfered surface forms with the center line of the work is so designated. Figure 11-5 shows both types of angular dimensioning.

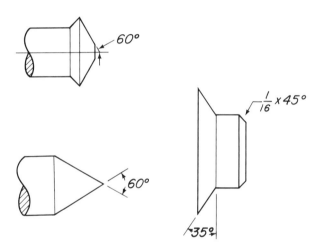

Figure 11-5

The angle corresponding to a given taper can be found by applying Rule 11-6.

RULE 11-6   FINDING THE ANGLE CORRESPONDING TO A TAPER

Step 1   Divide the taper per foot by 24.

Step 2   Find the angle whose tangent corresponds to the quotient in a table of natural functions.

Step 3   Double the angle.

Example 11-7   Find the angle corresponding to a taper of $\frac{1}{2}$ in. per foot.

Solution

Step 1 $\dfrac{0.500}{24} = 0.02083$

Step 2 $\tan 1° \, 12' = 0.02083$

Step 3 $2 \times (1° \, 12') = 2° \, 24'$

On blueprints originating in countries using the metric system of measurement, the taper is designated by **degree of slope** as an included angle. Rotary tools, such as tapered reamers, will give a percentage of taper per length.

Example 11-8   A tapered workpiece from a French blueprint is shown here. What is the large diameter D?

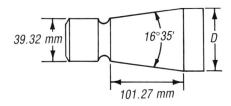

Solution   Solve the triangle derived from the drawing.

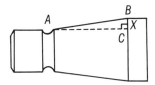

Triangle ABC is a right triangle. Angle A is half of the included angle of the taper. Also, $AC = 101.27$ mm.

$$\frac{1}{2} \times 16° \, 35' = 8° \, 17.5'$$

Then, in triangle ABC,

$$\tan 8° \ 17.5' = \frac{X}{101.27}$$

$$0.14574(101.27) = X = 14.759$$

And from this,

$$D = 39.32 + 2(14.759)$$
$$= 64.84 \text{ mm}$$

Exercises 11-3

1. If a workpiece 12 in. long is 2.005 in. and 1.03 in. in diameter at its ends, what is the included angle?

2. In Figure A, what is the value of the included angle?

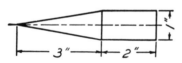

Figure A

3. Figure B shows a flat key. What is the value of angle A?

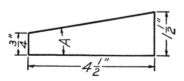

Figure B

4. A taper makes an angle of 12° 20' with the center line. What is the corresponding taper per foot?

5. If the taper per foot on a workpiece equals 1.75 in., what is the corresponding angle with the center line?

6. The workpiece shown in Figure C is a special taper for a French machine. At what degree of taper (half the included angle) would the taper attachment of an American lathe be set to cut this taper?

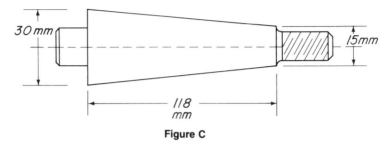

Figure C

7. Compute the setting of the taper attachment (half the included angle) for turning a taper on a work piece with a large diameter of 2.375 in., a small diameter of 2.125 in., and a length of 8 in.

11-4 Turning Tapers

OBJECTIVES

After completing this section, you should be able to:

1. Use the tailstock to exceed the limit of the taper attachment.
2. Find the angle needed to set the compound rest.
3. Find the amount of error in a desired taper.

The taper attachment will not cut tapers steeper than the range of swivel marked on the end of the bar, nor longer than the effective working length of the bar. There is, however, a way to exceed the limitations of the taper attachment by auxiliary use of the tailstock.

RULE 11-7 EXCEEDING THE LIMITATIONS OF THE TAPER ATTACHMENT

Step 1 Mount the work between centers.

Step 2 Set the taper attachment to its limit.

Step 3 Figure the tailstock offset for the *excess* of taper per *foot* beyond the normal range of the taper attach-ment.

Example 11-9   Suppose it is desired to turn a 3.500 *T.P.F.* on an 8-in. workpiece with a taper attachment calibrated to a limit of 3 in. *T.P.F.*

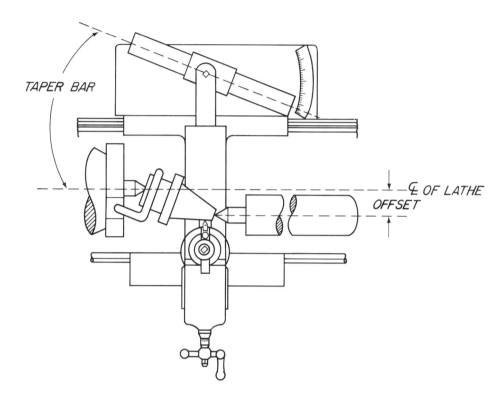

TAPER BAR

℄ OF LATHE

OFFSET

Solution

Step 1   Mount the workpiece between centers.

Step 2   Set the taper attachment to its 3-in. maximum.

Step 3   The remaining $\frac{1}{2}$-in. *T.P.F.* is figured.

$$T.P.I. = \frac{0.500}{12} = 0.042 \text{ in.}$$

$$\text{Offset} = \frac{0.042(8)}{2} = \frac{0.336}{2} = 0.168, \text{ or about } \frac{11}{64} \text{ in.}$$

Short and steep tapers can be plunged by an angled form tool or turned by the compound rest. The **compound rest** is a tool rest mounted on the cross slide of a lathe. It can be swiveled through 360° and its base is graduated in degrees. A witness mark serves as an indicator for the number of degrees turned.

Figure 11-6 shows a compound rest in normal position. The center line of the rest is at 90° to the lathe center line. There is a formula for determining the amount to set the graduated base for the correct angle. However, the graduations of the compound rest base are not standardized. On some lathes, the zero is placed on the horizontal center line of the lathe. On other lathes, it is not.

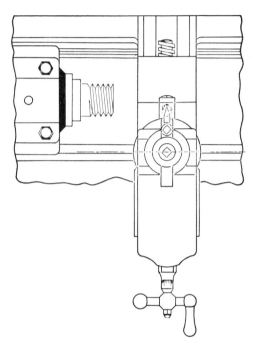

Figure 11-6

To avoid confusion on the part of the operator, the following commonsense approach is suggested.

RULE 11-8   SETTING THE COMPOUND REST

Loosen the compound rest and turn it until the tool travel is parallel with the lathe center line. (See Figure 11-7.) Next, look at the graduations on the base. Either zero or 90° will be lined up with the fixed witness mark. Swivel the compound rest *an amount equal to half the included angle called for on the blueprint.*

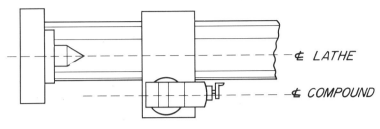

Figure 11-7

Example 11-10   The workpiece shown has a taper of 55°, dimensioned as shown. How many degrees off the center line of the lathe must the compound rest be set to produce the taper?

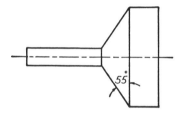

Solution   If the illustration is redimensioned, it becomes apparent that swiveling the compound rest 35° off the center line of the lathe gives the 55° desired.

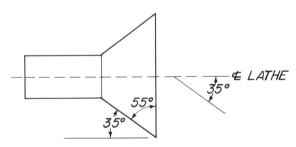

If either the tailstock offset or taper attachment is used to cut tapers, it is customary to make trial cuts, measure the diameters produced, and then adjust the tailstock or taper bar. For an outside taper, the machinist or toolmaker marks off a 1-inch distance on the surface of the workpiece and measures the two diameters. The difference between them should equal the amount of taper per inch required. (See Figure 11-8.) If it does not, further adjustment of the taper attachment or tailstock is necessary. When the taper per inch is correct, the work is turned down to 0.005 inch over the finished size of the small diameter. This leaves filing stock for fitting to a gauge.

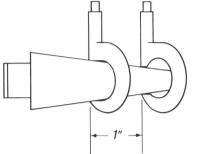

Figure 11-8

If a sample tapered piece is available, it may be possible to set it up in the lathe and run a dial indicator back and forth over the surface until adjustment of the tailstock or taper attachment causes the indicator to register zero change for the full tapered length of the sample piece.

Example 11-11   A trial cut is measured with a micrometer in a 3-in. interval and the diameters are $1\frac{9}{16}$ and $1\frac{1}{2}$ in., respectively. The desired taper per foot is $\frac{5}{16}$ in. Is the taper too steep or not steep enough?

Solution $T.P.1. = \dfrac{0.312}{12} = 0.026$ in.

In 3 in., the taper should be 0.078 in.

Measured diameter differences = 1.562 − 1.500, or 0.062 in.

The taper is not steep enough. The amount of error per inch is

$$\frac{0.078 - 0.062}{3} = 0.0053 \text{ in.}$$

If a taper can be corrected to register an error of no more or less than 0.002 inch, the error can be corrected by filing in the final fitting of the taper to a gauge, as shown in Figure 11-9. The usual practice is to draw a light chalk line along the length of the external taper, fit the ring gauge on the workpiece, and give the gauge a half-twist. If the chalk is disturbed *evenly* for its entire length, the taper is satisfactory. If the chalk is more heavily rubbed at the *small end,* the taper is *not steep enough.* If the taper is *too steep,* the chalk will rub heavily on the *large end* of the workpiece.

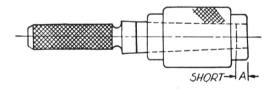

SHORT→A←

Figure 11-9

One of the more innovative developments in measuring instruments is the **taper micrometer.** Figure 11-10 shows an external taper micrometer. In Figure 11-11, an **internal taper micrometer** is illustrated.

Attached to the spindle on both micrometers is an extremely accurate 1-inch sine bar, which pivots on a post fixed to the frame. The total movement of the spindle is 0.300 inch. There are four standard micrometers to a set, which give a range from zero to 6-inch taper.

The obvious advantage of this type of tool is the elimination of any necessity for *removing the workpiece* from a machine. The work can be checked in the actual machine setup, direct readings in taper per inch made with the micrometer, and corrections made immediately.

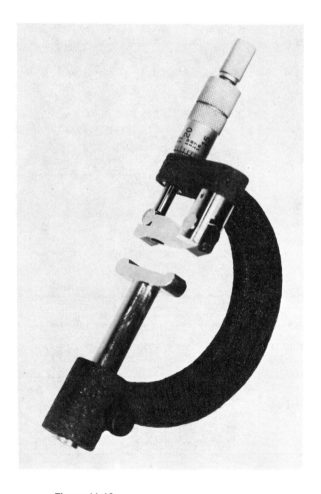

Figure 11-10

(Courtesy Taper Micrometer, TM Electronics Inc.)

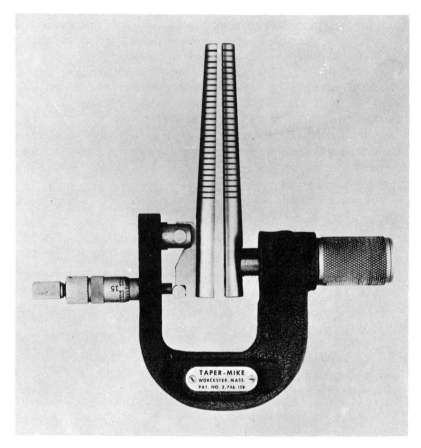

Figure 11-11

(Courtesy Taper Micrometer, TM Electronics Inc.)

Exercises 11-4

Determine the amount of error from the desired taper per inch, using the taper information given.

	Required Taper per Foot	Micrometer Reading	Distance Measured	Amount of Error
1.	$\frac{1}{4}$ in./ft	1.251 in. 1.232 in.	1 in.	
2.	Jarno taper	1.507 in. 1.359 in.	3 in.	
3.	Morse No. 4	2.398 in. 2.105 in.	6 in.	

	Required Taper per Foot	Micrometer Reading	Distance Measured	Amount of Error
4.	0.750 in./ft	1.804 in. 1.742 in.	1 in.	
5.	0.450 in./ft	2.669 in. 2.634 in.	1 in.	
6.	$\frac{1}{2}$ in./ft	1.252 in. 1.235 in.	1 in.	
7.	Morse No. 2	2.395 in. 2.110 in.	5 in.	
8.	0.850 in./ft	1.807 in. 1.745 in.	3 in.	
9.	0.950 in./ft	2.666 in. 2.631 in.	1 in.	

10. Using a taper attachment calibrated to a limit of 3 *T.P.F.*, calculate the tailstock offset for 3.250 *T.P.F.* on an 8-in workpiece.

11. Using a taper attachment whose limit is 3.5 *T.P.F.*, how much would you offset the tailstock on the lathe to cut a 3.625 *T.P.F.*? Length = 8 in.

12. In Figures A and B, how many degrees should the compound rest be swiveled from the center line?

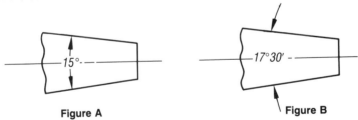

Figure A   **Figure B**

13. When machining the head of the flathead machine screw shown in Figure C, the included angle should be 82°. How many degrees off the center line of the lathe is the compound rest turned?

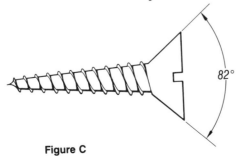

Figure C

Unit 11 Review Exercises

1. If the end diameters of a workpiece 1 ft long are $3\frac{1}{2}$ in. and 2 in., how much *T.P.F.* would the piece have?

2. The *T.P.F.* on a workpiece is 0.500 and the large diameter is 4 in. Find the small diameter if the length of the taper is 9 in.

3. Given the following dimensions from a blueprint, find the taper per inch.

    Large diameter = 1.25 in.
    Small diameter = 0.437 in.
    Length of taper = 3.125 in.

4. Determine the tailstock offset to produce a taper from the following dimensions.

    Large diameter = 2.375 in.
    Small diameter = 2.0625 in.
    Length of taper = 4.000 in.
    Total length of work = 9.375 in.

5. Calculate the offset required to turn a taper of $\frac{3}{4}$ in./ft if the workpiece is 19 in.

6. Find the angle corresponding to a taper of $\frac{5}{8}$ in./ft. (See Figure A.)

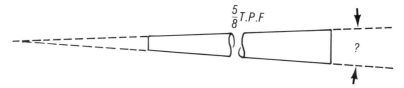

7. If a workpiece 9 in. long is 1.505 in. and 2.675 in. in diameter at its ends, what is the included angle?

8. Suppose it is necessary to turn a 3.875 *T.P.F.* on a 12 in. workpiece with a taper attachment whose limit is 3 in. How much must the tailstock be offset?

9. How many degrees off the center line of the lathe must the compound rest be swiveled for the following included angles on a workpiece: 60°, 15°, and 25° 35′?

10. A trial cut is measured with a micrometer every 3 in. with these results: 1.418 in., 1.290 in., and 1.162 in. The desired *T.P.F.* is $\frac{7}{8}$ in. Is the actual taper too steep or not steep enough?

Gear Speeds

Well into this century, manufacturing plants supplied motive power to their machines through a complicated system of belts connected to overhead shafts. The prime mover was steam. In fact, the engine lathe received its name from the steam engine that drove it. Apprentices in those early days received instruction in belt lacing and calculation of pulley speeds.

Many present-day, smaller machines have belts incorporated in their design. For example, a bench lathe that is belt-driven has certain advantages. If the belt can slip, damage to the machine or broken tools are less likely to occur. Nevertheless, the use of belts and pulleys in the machine shop has decreased since overhead pulleys and shafts have been replaced by individual motor drives. Gears are a better device for providing driving power or traction—especially where heavy loads must be overcome.

12-1 Gear Speeds and Simple Gears

OBJECTIVES

After completing this section, you should be able to:

1. Use the general speed formula.
2. Find gear speeds in a simple gear train.

Knowledge of the speed of gears in mesh is knowledge that most machinists, auto mechanics, engineers, and inventors require at one time or another. Whenever a machinist sets a certain number of revolutions per minute or a feed rate on a machine, he or she is choosing a combination of gears that will give the desired results.

A belt running around a pulley can slip. If the belt is replaced by teeth and the pulleys are brought together in mesh, the slippage is eliminated. Thus, when you think of a gear as a pulley with teeth on its rim, the calculation of gear speeds becomes simple.

The speed of a gear is inversely proportional to the number of teeth. Thus, the more teeth on a gear, the fewer revolutions per minute (rpm's) the gear makes. If T and S are the number of teeth and the speed of the driver, and t and s are the number of teeth and speed of the driven gear, then

$$\frac{T}{t} = \frac{s}{S}$$

Cross-multiplying gives the general speed formula stated in Rule 12-1.

RULE 12-1   GENERAL SPEED FORMULA

The general speed formula for two gears in mesh is $T \times S = t \times s$, where T is the number of teeth on the driver, S is the speed of the driver, t is the number of teeth on the driven gear, and s is the speed of the driven gear.

Example 12-1   Two gears in mesh have 20 and 30 teeth. If the 20-tooth gear rotates at 150 rpm, what is the speed of the larger gear?

Solution $T \times S = t \times s$

Transposing,

$$S = \frac{t \times s}{T} = \frac{20(150)}{30} = 100 \text{ rpm.}$$

In a **simple gear train,** two or more gears are in mesh, but each gear in the train is on a *different* shaft. In such a train, the first gear in the train is the driver and the last is the driven gear. *The speeds of the driver and the driven gears in a simple gear train are inversely proportional to the number of teeth.* Hence, the formula $T \times S = t \times s$ applies to a simple gear train.

If a simple gear train is composed of two meshing gears, the relationship of the number of teeth and speed of the gears is given by the basic speed formula $T \times S = t \times s$. The inverse proportion between the speeds and the number of teeth is clearly illustrated by the following example.

Example 12-2   Suppose two gears with 36 and 48 teeth are meshed and the speed of the 36-tooth gear is 300 rpm. Find the speed of the 48-tooth gear.

Solution $s = \dfrac{T \times S}{t} = \dfrac{36 \times 300}{48} = 225 \text{ rpm}$

As a check, note that

$$\frac{T}{t} = \frac{36}{48} = \frac{225}{300} = \frac{s}{S}$$

It frequently happens that motion must be transmitted by gears from one shaft to another when the shafts are too far apart for the gears to mesh. Large gears might span the distance between shafts, but large gears take up space. The use of large gears can be avoided by the use of one or more idlers. An **idler gear** does not affect the speed of the driver (the first gear in a simple train) or the speed of the driven gear (the last gear in a simple train).

When two gears are in mesh, they rotate in *opposite directions*. In Figure 12-1, two idlers are placed between the driver A and the driven gear D. These two idlers will cause the driver and driven gears to rotate in opposite directions, but will not affect their relative speeds.

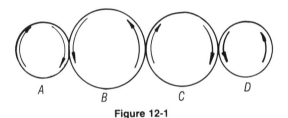

Figure 12-1

Example 12-3   The simple gear train below is composed of a driver A, an idler B, and a driven gear D. The number of teeth on each gear is shown, and the direction of rotation of gear A is clockwise. If the driven gear D is rotating at 600 rpm, what are the speed of A and directions of rotation of B and D?

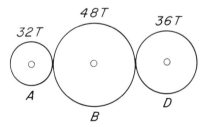

Solution $T \times S = t \times s$,   so $S = \dfrac{t \times s}{T}$

$$S = \frac{36(600)}{32} = 675 \text{ rpm}$$

Since the rotation of A is clockwise, B rotates counterclockwise and D rotates clockwise.

In computing the number of teeth or speed of the driver or driven gear in a simple gear train, disregard all idler gears. In the determination of the direction of rotation of driver or driven gear, *the idler gears must be considered*. Idler gears are both driver and driven gears.

Too much emphasis is traditionally placed on the direction of rotation of gears in a gear train. What is often overlooked is the fact that if the driven gear does not rotate in the desired direction, a reversing switch on the electric driving motor will solve the problem.

Exercises 12-1

1. A 20-tooth gear turning at 200 rpm drives a gear with 40 teeth. What is the speed of the driven gear?

2. In Figure A, gear A drives gear B. If B makes 150 rpm, what is the speed of gear A?

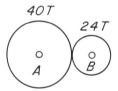

Figure A

3. A driver gear with 24 teeth rotates at 400 rpm, and the gear with which it meshes is to rotate at 300 rpm. How many teeth must the driven gear have?

4. When two gears are geared "even," they have the same number of teeth. What is the speed ratio?

5. If a 24-tooth gear and a 64-tooth gear are meshed, what is the resulting speed ratio?

6. Driver and driven gears are required to rotate at 400 and 500 rpm, respectively. The gears available have the following number of teeth: 20, 24, 30, 36, 40, 48, 50, 56, 60, 64, 70, 72, 80, and 84. Find pairs of gears that give the required speed ratio.

7. Two gears in mesh are rotating at 400 and 700 rpm. How many teeth must the gear rotating at 400 rpm have if the other gear has 64 teeth?

8. In Figure B, gear A drives the 44-tooth gear through an idler with 20 teeth. If A rotates at 275 rpm, calculate the speed of the driven gear.

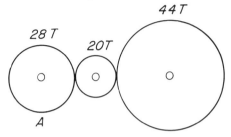

Figure B

9. Two gears with 72 and 84 teeth, respectively, are connected by an idler. If the 72-tooth gear rotates at 126 rpm, what is the speed of the 84-tooth gear?

10. Two gears with speeds of 200 and 600 rpm are connected by an idler. If the driver with 48 teeth is rotating at 200 rpm, how many teeth does the driven gear have?

11. In the simple gear train shown in Figure C, the driven gear is rotating at 800 rpm. What is the speed of the driver A?

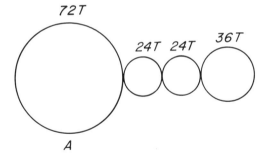

Figure C

12. A simple gear train contains two idlers. If the ratio of the teeth on the driver and driven gears is 2 to 1, what is the ratio of their speeds?

13. In Figure D, if S makes one complete revolution, how many turns will B make?

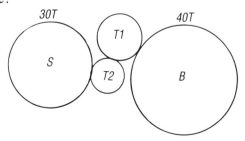

Figure D

14. In Figure D, if *B* is required to make one-half turn for each turn of *S*, how many teeth should *B* have if *S* has 36 teeth?

15. In a simple gear train, gear *A* has 24 teeth and rotates at 72 rpm, gear *B* has 36 teeth, and gear *C* has 48 teeth. Gear *A* drives gear *B*, and gear *B* drives gear *C*. What is the rpm of gear *C*?

12-2 Compound and Worm Gearing

OBJECTIVES

After completing this section, you should be able to:

1. Find gear speed in compound gearing.
2. Find gear speed in worm gearing.

Compound Gears

In a **compound gear train,** *two gears of the train are keyed to the same shaft,* and as a result, one is driven and the other is a driver. Compound gearing is used to reduce space and obtain speeds that would be impossible with simple gearing.

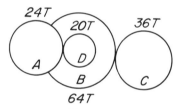

Figure 12-2

Figure 12-2 represents a compound gear train with gears *B* and *D* keyed on the same shaft. If gear *A* is the *driver* in the compound gear train, *D* is a *driven* gear; since *B* and *D* are keyed to the same shaft, *B* is a *driver* and *C* is a *driven* gear.

RULE 12-2 FINDING COMPOUND GEAR SPEED

In a compound gear train, the product of the number of teeth of the driving gears and the number of revolutions per minute of the first driver equals the product of the number of teeth of the driven gears and the number of revolutions per minute of the last driven gear. This can be written as

rpm of final driven gear

$= \dfrac{\text{product of all driving teeth} \times \text{rpm of first driving gear}}{\text{product of all driven teeth}}$

Example 12-4 In the compound gear train shown in Figure 12-2, compute the speed of C if the driver A rotates at 300 rpm. The first driver is A and the second B; the first driven gear is D and the second, or last, is C.

Solution $\text{rpm} = \dfrac{300 \times 24 \times 64}{20 \times 36}$

$\qquad\qquad = 640$

Worm Gears

It is sometimes necessary to compute the parts of a gear train when the train consists of a worm and gear. The computation is performed in the same way as with spur gears, except that the worm thread is regarded as a gear. If the worm is single-threaded, it is considered a gear with one tooth; if it is double-threaded, it is considered a gear with two teeth; and so on.

Figure 12-3 represents a worm and gear. The worm has a thread cut on its surface that meshes with the teeth of the worm gear. The worm gear makes possible a large reduction in speed with an increase in the power transmitted.

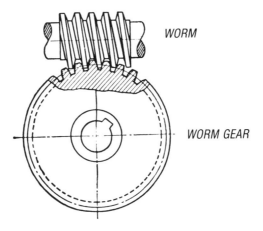

WORM

WORM GEAR

Figure 12-3

RULE 12-3 FINDING WORM GEAR SPEED

The speed of the worm times the thread number equals the speed of the worm gear times the number of teeth on the worm gear.

Example 12-5 If the worm below is single-threaded and rotates at 80 rpm, find the speed of the worm gear if it has 40 teeth.

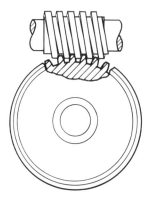

Solution rpm of gear × 40 = 80 × 1

$$\text{rpm of gear} = \frac{80 \times 1}{40} = 2$$

In this case, the speed was reduced from 80 to 2 rpm, or by a ratio of 40 to 1.

Exercises 12-2

1. In the compound gear train shown in Figure A, the gear A is the first driver and rotates at 200 rpm. What is the speed of the 40-tooth gear?

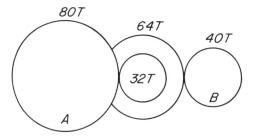

Figure A

2. In Figure B, if A is the first driver in the compound gear train, find the number of teeth required on gear A so that A and B will rotate at 100 and 300 rpm, respectively.

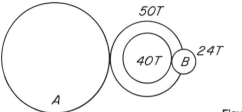

Figure B

3. In Figure C, how many turns of A will be made for each turn of B?

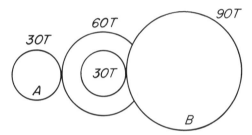

Figure C

4. In Figure D, if A is required to make four turns for each revolution of gear B, gear B will have to be replaced by a gear with how many teeth?

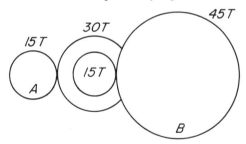

Figure D

5. In Figure E, gear A is the first driver in the compound gear train. If gear A makes one revolution, how many revolutions will gear B make?

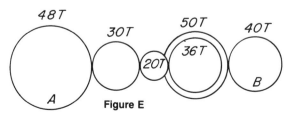

Figure E

6. In Figure F, if gear *A* rotates at 300 rpm, at what speed will gear *B* rotate?

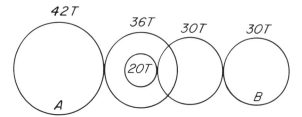

Figure F

7. In Figure G, a gear train for thread cutting in a lathe is represented. Gears *C* and *B* can be replaced by gears with different numbers of teeth. If gear *B* makes one revolution, how many revolutions will gear *A* make?

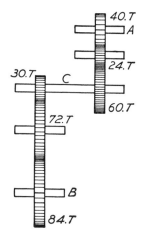

Figure G

8. In Figure H, how many teeth should there be on gear *C* so that *A* will make one turn for one-third of a turn of *B*?

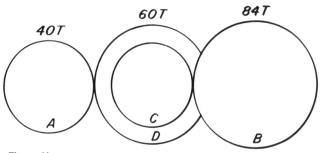

Figure H

9. In Figure I, if gear *A* rotates at 750 rpm and gear *B* rotates at 375 rpm, find the number of teeth required on the stud-shaft gear *C*.

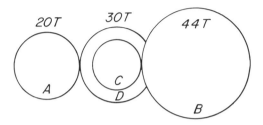

Figure I

10. A worm rotates at 120 rpm and it is single-threaded. If it meshes with a 60-tooth worm gear, what is the speed of the gear?

11. An 80-tooth worm gear meshes with a double-threaded worm. If the worm gear rotates at 5 rpm, what is the speed of the worm?

12. A triple-threaded worm revolves at 200 rpm. What will be the speed of the worm gear if the gear has 60 teeth?

Unit 12 Review Exercises

1. Gears *A* and *B* have 38 and 57 teeth, respectively. If gear *A* rotates at 120 rpm, what will be the speed of gear *B?*

2. A simple gear train is composed of a driver, an idler, and a driven gear. The driver has 24 teeth, the idler has 52 teeth, and the driven gear has 40 teeth and is rotating at 600 rpm. What is the speed of the driver?

3. In exercise 2, the driver is rotating in a clockwise direction. In what direction is the idler rotating? The driven gear?

4. Two gears in mesh are rotating at 300 rpm and 600 rpm. How many teeth must the gear rotating at 300 rpm have if the other gear has 54 teeth?

5. The speeds of two gears are in the ratio of 1 to 3. If the faster one rotates at 180 rpm, find the speed of the slower one.

6. Two gears are to have a speed ratio of 2.5 to 3. If the larger gear has 72 teeth, how many teeth must the smaller gear have?

7. In Figure A, the speed of gear *A* is 72 rpm. How many teeth must gear *D* have if it is to rotate at 20 rpm?

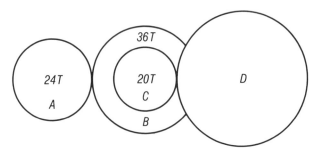

Figure A

8. A worm is in mesh with a worm gear that has 27 teeth. If the speed of the worm is 135 rpm and the speed of the worm gear is 15 rpm, what is the thread number of the worm?

9. In Figure B, how many rpm will gear *D* make, if gear *A* rotates at 180 rpm?

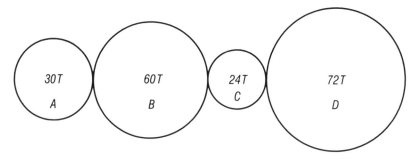

Figure B

10. In Figure B, under the conditions stated in exercise 9, what is the speed of gear *C*?

Speeds and Feeds

If an unbroken chip is peeled from a workpiece by a cutting tool for 1 minute, the length of that chip can be measured in feet or meters. To give a fair test of the amount of metal that can be removed in a minute's time, the workpiece must be moving at a rate sufficient to remove a maximum length of chip without burring or destroying the tool. Such a rate is known as the **cutting speed**.

The meaning of cutting speed depends on the machine considered. On a *lathe,* a *boring mill,* or similar machines, it means the number of feet or meters on the surface of the workpiece that pass the tool in 1 minute. On a *drill press* or *milling machine,* the tool is rotating. Therefore, cutting speed is the speed of a point on the circumference of the drill or milling cutter in feet or meters per minute.

13-1 Cutting Speeds

OBJECTIVES

After completing this section, you should be able to:

1. Calculate rpm for a given cutting speed.
2. Calculate drill speeds.
3. Convert English units of cutting speed to metric units.

The types of material that are cut in a modern machine shop are numerous. These materials range from hard to soft in their resistance to cutting. Obviously, wood and metal cannot be cut at the same cutting speed. There are several other factors that may affect the cutting speed of any given material:

The kind of cutting tool material

Whether the cut is a rough or finish cut

The feed rate per minute

The condition of the machine

The rigidity of the setup

This should make clear that, at best, a *range* of acceptable cutting speeds can be offered for a given job. The proper selection of a cutting speed within that range will depend on the machinist's good judgment and experience. Such judgment will be based on all the factors listed above.

The kinds of metals most frequently cut in the average machine shop are listed in Table 13-1, together with reasonable *starting speeds* for high-speed and carbide tools. (A cemented carbide tool has the property of *real hardness,* which means that it can be used at a cutting speed that will heat the cutting edge to a dull red without burning the tool. This property makes possible higher revolutions per minute.) However, the manufacturing engineer or machinist should consult technical literature provided by tooling suppliers for the latest available information.

	Starting Speed in Feet per Minute	
	---	---
Material	High-Speed Steel	Carbide
Stainless steel	50	150
Cast steel	50	150
Cast iron	75	225
High-carbon steel	50	150
Medium-carbon steel	75	200
Low-carbon steel	125	375
Brass	300	800
Magnesium	600	1000
Aluminum	600	1000
Copper	300	1000

Table 13-1

The cutting speeds listed in Table 13-1 are not arbitrary, but have been selected on the basis of common sense and experience. The cutting speed range for cast iron, for example, can vary from 40 to 100 feet per minute, depending on the pig iron content of the cast iron. It is better to start at a low cutting speed to avoid burned tools and then adjust the cutting speed upward if needed. The color of the chip produced will indicate when the cutting speed is approaching the upper limit of its range.

Methods people, process engineers, estimators, and expert shop workers all use cutting speeds to determine the proper number of revolutions per minute for work or tools. An experienced machine operator who has worked with the same materials year in and year out can simply look at a turning workpiece or tool and judge roughly whether it is turning too fast. But the first time the operator must machine an unfamiliar material, guesswork can cause trouble.

The number of revolutions per minute for a given material and diameter is found from the formula below.

$$\text{rpm} = \frac{12 \times \text{cutting speed in surface feet per minute}}{\text{diameter of what is turning} \times \pi}$$

Since 12 divided by π is approximately 4, the above formula can be simplified to one that is easily remembered and used.

RULE 13-1 CALCULATING REVOLUTIONS PER MINUTE

$$\text{rpm} = \frac{4C.S.}{D}$$

where *C.S.* is cutting speed in feet per minute and *D* is diameter in inches.

Example 13-1 What is the starting rpm of a 3-in.-diameter brass bar to be turned with a high-speed tool? (Use Table 13-1.)

Solution $\text{rpm} = \dfrac{4C.S.}{D} = \dfrac{4(300)}{3} = \dfrac{1200}{3} = 400$

Example 13-2 What is a safe starting speed for a 5-in.-diameter milling cutter with carbide-inserted teeth to mill low-carbon steel?

Solution $\text{rpm} = \dfrac{4C.S.}{D} = \dfrac{4(375)}{5} = \dfrac{1500}{5} = 300$

The machinist sets the number of revolutions per minute to the nearest speed-dial reading on the machine corresponding to the value derived from the formula. Most machinists know the old rule, "The larger the diameter, the slower the speed." This is based on the effect that the size of a diameter has on revolutions per minute. A larger diameter has more feet of material on its surface, or circumference, to move past the tool in a given minute, so it has a slower speed.

RULE 13-2 FINDING THE METRIC EQUIVALENT OF A CUTTING SPEED

The conversion formula is m/min = (ft/min)(0.3048).

Example 13-3 Convert a cutting speed of 50 ft/min to its metric equivalent.

Solution m/min = (ft/min)(0.3048)
    = 50(0.3048)
    = 15.24, or about 15 m/min

The geometry of an ordinary two-lipped **twist drill** requires that special care be given to the choice of the correct number of revolutions per minute. The cutting end of a drill consists of two cutting edges separated by a "dead," or noncutting, chisel point. This chisel point is the exposed tip of the drill's backbone, or the portion left after flutes are milled into its diameter. (See Figure 13-1.)

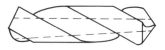

Figure 13-1

Since the chisel point does not cut, its resistance must be overcome by increased pressure on the drill. For this reason, both the proper speed and feed must be chosen with care for a twist drill or any other type of fluted rotary tool.

RULE 13-3 CALCULATING DRILL SPEEDS

Good shop practice requires that drilling be done at 75% of normal cutting speed for a given material and diameter.

Example 13-4 Calculate the proper rpm for a high-speed twist drill cutting a 2-in. diameter in low-carbon steel.

Solution $\text{rpm} = \dfrac{4C.S.}{D} = \dfrac{4(125)}{2} = 250$

$$250 \times 0.75 = 188 \text{ rpm}$$

Exercises 13-1

Find the missing number of revolutions per minute for each of the work-pieces. Round off fractional revolutions per minute to the nearest whole number.

Conversion factor: $\dfrac{m/min}{0.3048} = ft/min$

	Materials	Diameter in Inches	Cutting Speed		rpm
			Feet	Meters	
1.	Brass	$2\frac{1}{2}$	350		
2.	Low-carbon steel	3		38	
3.	Cast iron	4.5		69	
4.	Magnesium	2	600		
5.	Med.-carbon steel	8.75	200		
6.	Stainless steel	$3\frac{5}{8}$		46	
7.	Copper	1	300		
8.	High-carbon steel	$1\frac{1}{2}$	150		
9.	Cast steel	6	150		
10.	Aluminum	8		305	

Fill in the correct number of revolutions per minute for high-speed drills.

	Drill Diameter in Inches	Material	rpm
11.	$\frac{1}{2}$	Low-carbon steel	
12.	$1\frac{1}{2}$	Cast iron	
13.	0.875	Brass	
14.	$\frac{3}{8}$	High-carbon steel	
15.	0.750	Cast iron	
16.	$\frac{1}{4}$	Copper	
17.	1	Med.-carbon steel	
18.	$1\frac{5}{32}$	Low-carbon steel	

13-2 Feed Rates and Cutting Time

OBJECTIVES

After completing this section, you should be able to:

1. Find cutting time.
2. Find rpm and feed rate for a cutting problem.
3. Convert English units of feed rate to metric units.
4. Find gang-milling feeds and speeds.

The **feed** of a lathe tool is the distance it travels along the work for each revolution of the work. The amount of feed depends on the depth and nature of the cut. A roughing cut is deeper than a light-finish cut. More metal being removed per revolution of the work requires a low number of revolutions per minute and a *fast* rate of feed. A finish cut can be made at much higher speed, but the feed should be *slower* so that tool marks are not left in the work.

The rate of feed for lathe tools is given as some number of thousandths of an inch per revolution of the work. A large variety of feeds are available by setting feed tumbler levers on the feed box of a lathe headstock. No attempt can be made to give a hard and fast rule for lathe-tool feed, as there are too many variables. Each case must be decided based on its own facts.

Once a definite speed and feed rate has been established for a given cut, this information can be used for both lathe and milling operations to set a cutting time or standard. Estimators and time-standards people use this information in their daily work.

If a tool feeds past work at X thousandths of an inch for each revolution, the product of the feed and rpm will give the distance covered in 1 minute's time, or the **feed rate per minute.** The length of the cut divided by the feed rate per minute gives the **cutting time.**

RULE 13-4   FINDING CUTTING TIME

$$\text{Cutting time} = \frac{\text{length of cut}}{\text{feed rate per minute}}$$

Example 13-5   A lathe tool feeds 0.008 in. per revolution. The rpm is 275. The longest cut is $5\frac{1}{2}$ in. How long will it take for the tool to travel this length?

Solution $0.008 \times 275 = 2.200$ in./min

$$\frac{\text{Length of cut}}{\text{Feed rate per min}} = \frac{5.500}{2.200} = 2\frac{1}{2} \text{ min}$$

The feed rate on a milling machine is in *inches per minute of table travel.* This means that a feed rate of 1 inch per minute will move a piece 1 inch long under the cutter in exactly 1 minute.

To find the proper feed rate for a milling cutter, three things must be known:

rpm of the cutter, number of teeth on the cutter, and **chip load per tooth.**

The chip load per tooth for a given cutter will depend on which kind of cutter is used. A large, sturdily built cutter can "bite off" more metal per tooth than can a thin, fragile, saw-type cutter. There are six main types of cutters, classified according to their uses and construction:

1. *Metal-cutting saws*
   Slitting saws
   Screw-slotting saws
2. *End Mills*
3. *Form cutters*
   Angle cutters
   Gear cutters
   Convex-concave cutters
   Corner-rounding cutters
   Special-form cutters
   Tee-slot cutters
   Dovetail cutters

4. *Side cutters*
   Staggered-tooth cutters
   Half-side cutters
   Regular (straight) cutters
5. *Plain cutters*
   Helical-tooth cutters
   Woodruff-key cutters
   Slab cutters
6. *Face, or "butt," mills*

Your shop instructor will have examples of each type to show you. A glance at a cutter is usually sufficient to place it in one of the six categories listed above. Each type of cutter carries a recommended chip load or feed per tooth. Good *starting* feeds for the six types of cutters are listed in Table 13-2.

CHIP LOADS	
Type of Cutter	**Load per Tooth**
Metal-cutting saws	0.001 in.
Form cutters	0.002 in.
End mills (over $\frac{1}{2}$ inch diameter)	0.003 in.
Plain cutters	0.005 in.
Side cutters	0.007 in.
Face mills	0.010 in.

Table 13-2

The feed rates given in Table 13-2 are based on an average cut of 0.100 inch in depth and will vary with the setup of machine and work, as well as the type of material being machined. Therefore, these values are only approximate for the purpose of establishing relative chip load for the following examples.

The formula for finding the starting rate of feed in inches per minute for a milling cutter is stated in Rule 13-5.

RULE 13-5 FINDING THE STARTING RATE OF FEED (INCHES PER MINUTE)

Feed rate = chip load per tooth
× number of teeth × rpm

Example 13-6 A high-speed-steel side cutter 6 in. in diameter with 30 teeth must cut a $\frac{1}{8}$-in. deep slot in low-carbon steel. What rpm and feed rate are needed?

Solution A side cutter calls for a 0.007-in. chip load per tooth.

$$\text{rpm} = \frac{4C.S.}{D} = \frac{4(125)}{6} = 83$$

Feed rate = chip load × teeth × rpm
= 0.007 × 30 × 83
= 17.43, or about $17\frac{7}{16}$ in./min

Example 13-7 Find the rpm and feed rate for a $3\frac{1}{2}$-in.-diameter gear cutter with 12 teeth to cut cast iron.

Solution $\text{rpm} = \frac{4C.S.}{D} = \frac{4(75)}{3.5} = 86$

feed rate = 0.002 × 12 × 86 = 2 in./min

Inches per minute can be converted to *centimeters per minute*. The conversion factor is 1 inch per minute = 2.54 centimeters per minute.

Example 13-8 Convert 5.5 in./min to its approximate metric equivalent.

Solution 5.5 × 2.54 = 14 cm/min

When two or more milling cutters are mounted on the same arbor to mill several surfaces at the same time, the setup is called *gang milling*. Figure 13-2 illustrates a typical gang-milling operation.

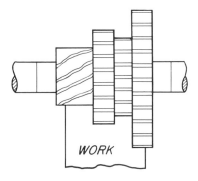

Figure 13-2 Gang milling.

In gang milling, there must not be too much variation between cutter diameters or number of teeth. The problem is that the greatest rpm allowable is the one that the *largest* cutter in the setup can safely take. On the other hand, the feed rate is determined by the cutter with the *least* number of teeth, since this cutter must keep up the same pace as the others in the gang setup. The guidelines for gang-milling speeds and feeds are stated in Rule 13-6.

RULE 13-6   FINDING GANG-MILLING FEEDS AND SPEEDS

The rpm is determined by the diameter of the largest cutter.
Feed rate is determined by the cutter with the fewest teeth.

Example 13-9   A set of three high-speed-steel side cutters are to gang mill three separate surfaces simultaneously on a low-carbon-steel block. Determine the feed rate and rpm for the gang.
Cutter 1 has diameter of 6 in. and 14 teeth.
Cutter 2 has diameter of 8 in. and 18 teeth.
Cutter 3 has diameter of $4\frac{1}{2}$ in. and 12 teeth.

Solution   Gang rpm $= \dfrac{4C.S.}{D} = \dfrac{4(125)}{8} = 63$

Gang feed rate $=$ chip load $\times$ number of teeth $\times$ rpm
$= 0.007 \times 12 \times 63$
$= 5.292$, or about $5\dfrac{19}{64}$ in./min

Exercises 13-2

Fill in the cutting times.

	rpm of Work	Feed Rate of Tool	Length of Cut	Cutting Time in Minutes
1.	300	0.0025	2 in.	
2.	55	0.005	$1\frac{3}{4}$ in.	
3.	275	0.003	5 in.	
4.	150	0.004	$3\frac{1}{4}$ in.	
5.	800	0.002	$\frac{3}{4}$ in.	
6.	74	0.010	$2\frac{7}{8}$ in.	
7.	325	0.006	$6\frac{1}{8}$ in.	
8.	450	0.0045	12 in.	
9.	760	0.0015	3.625 in.	
10.	130	0.008	$1\frac{1}{2}$ in.	

You are a machine shop supervisor. Supply the rpm and feed rates for the following milling jobs.

	Surface Feet per Minute	Dia. of Cutter	Type of Material	Chip Load	No. of Teeth	rpm	Feed Rate
11.	50	2 in.	Cast steel	0.003	8		
12.	150	0.500 in.	High-carbon steel	0.002	4		
13.	75	6 in.	Med.-carbon steel	0.007	14		
14.	1000	6 in.	Aluminum	0.001	42		
15.	125	$4\frac{1}{2}$ in.	Low-carbon steel	0.010	10		
16.	600	$2\frac{1}{2}$ in.	Magnesium	0.005	16		
17.	300	1 in.	Brass	0.002	3		
18.	200	8 in.	Med.-carbon steel	0.010	16		

The following milling jobs must be performed on a machine with feed rate calibrated in centimeters. Calculate the rpm and feed rates in centimeters per minute.

Conversion factors:

$$\frac{m/min}{0.3048} = ft/min \qquad \frac{cm/min}{2.54} = in./min$$

	Surface Meters per Minute	Cutter Teeth	Dia. of Cutter (mm)	Chip Load (mm)	rpm	Feed in cm/min
21.	46	4	76.20	.05		
22.	305	42	152.4	.03		
23.	183	16	63.5	.13		
24.	91	3	25.4	.05		
25.	38	10	114.3	.25		
26.	23	14	152.4	.18		

27. A carbide insert can be fed at 0.020 in. per revolution, using an rpm of 3750. Assuming the length of cut is 4.250 in., how long will this cut take?

28. Given a rapid feed rate of 475 in./min for a CNC mill, how long will it take to complete a total of 273.437 in. of rapid travel movement?

29. Using solid carbide end mills to side cut at 0.005-in. load per tooth, what will be the feed rate (in inches per minute) for a 1-in.-diameter cutter with two teeth rotating at 4000 rpm?

30. Under conditions noted in exercise 29, how long would it take to mill the 7-by-5-in. rectangular part shown in Figure A? (*Note:* There will be cutter overtravel of $\frac{1}{2}$ in. on each side.)

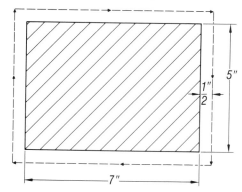

Figure A

Unit 13 Review Exercises

1. What would be the starting rpm of a 1.500-in.-diameter aluminum bar to be turned with a carbide insert at 1500 ft/min?

2. A 1.500-in.-diameter aluminum bar will be drilled using a 0.6875-in.-diameter high-speed-steel drill. Use Table 13-1 and Rule 13-3 to calculate a safe rpm. (Round to nearest 100 rpm.)

3. Convert a cutting speed of 225 ft/min to its metric equivalent.

4. A lathe tool feeds 0.009 in. per revolution. How long will a cut measuring 2.300 in. take at 2800 rpm?

5. Calculate the starting rate of feed for a 0.750-in.-diameter carbide end mill with two teeth using a chip load of 0.005 in. per tooth at 3400 rpm.

6. Convert the starting feed for the 0.750-in.-diameter end mill in exercise 5 to centimeters per minute.

7. In the ganged set of high-speed-steel side cutters shown in Figure A, the two outside cutters are 6 in. in diameter, and both have 24 teeth. The cutter between them is 4.750 in. in diameter and has 18 teeth. What would be the gang rpm to mill an aluminum workpiece?

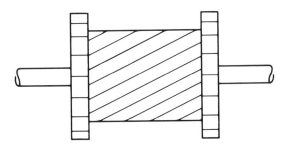

Figure A

8. In exercise 7, what would be the gang feed rate?

9. What would be the starting rpm to machine an 8-in.-diameter low-carbon-steel bar using a carbide insert at 625 ft/min?

10. Using a chip load of 0.005 in. per tooth on a high-speed-steel end mill with four teeth, what rpm should be used for a starting feed rate of 6 in./min?

Screw Threads

There are six fundamental machines: the wedge, the wheel, the pulley, the lever, the inclined plane, and the screw. A screw thread is a helical groove wound around a cylinder. It can be used for locking, fastening, transmitting motion, and measuring. The first screws appear to have been made by coiling a wire of the correct size around a rod, hammering soft material around the wire, and unscrewing the wire.

When the beginner machinist "chases" a thread for the first time with a single-point tool in an engine lathe, he or she reaches a plateau of accomplishment that reveals how rewarding the art of metal cutting can be. While the actual mechanics of thread cutting can only be learned by practice at a machine, this chapter will furnish the basic mathematics that the machinist needs to cut a thread.

14-1 Parts of a Thread

OBJECTIVE

After completing this section, you should be able to:

1. Identify parts of a thread.

Each part of a thread has a name. Some parts, such as the helix angle, are more important to the designer than to the machinist. Only those parts referred to in formulas used by machinists will be presented. Figure 14-1 shows those parts.

MAJOR DIAMETER (D)

The **major diameter** is the most important part of the bolt or screw. It is the diameter on which the thread is placed. The major diameter will have a

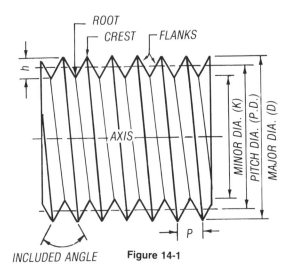

INCLUDED ANGLE   **Figure 14-1**

nominal size with a *minus* tolerance. The mating nut will have the same nominal size with a *plus* tolerance so that a satisfactory assembly will result. Major diameter is sometimes also called *thread diameter* or *outside diameter* (*O.D.*).

MINOR DIAMETER (*K*)

The **minor diameter** is the diameter formed by the bottom of an external thread and was formerly called the *root diameter*. The hole in a nut is always slightly larger than the minor diameter of the screw. This prevents interference between mating parts.

THREAD DEPTH (*h*)

The *vertical* height of thread from major to minor diameter is called the **thread depth.** Its actual value will be determined by the type of thread and fit desired.

FLANK, CREST, AND ROOT

The **flank** is the side of the thread. The **crest** is the top surface connecting two flanks. The **root** is the bottom surface connecting two flanks.

PITCH

The **pitch** of a thread is the actual linear distance from a point on one thread to the *corresponding* point on the next thread. It is found, in the English system, by dividing the number of threads per inch into 1 inch. The metric system identifies the thread pitch by a fixed number of millimeters. (See the discussion of metric thread, page 342.)

LEAD

The distance that a screw thread advances in one turn of the screw is called the **lead.** On a single-thread screw, lead and pitch are identical. On a double-thread screw, the lead is twice the pitch. On a triple-thread screw, the lead is three times the pitch, and so forth.

PITCH DIAMETER (*P.D.*)

A thread is cut to a theoretical depth from the major diameter of the screw. But because the major diameter will vary from tolerance change, another method of measuring the thread depth must be found. The surface of an imaginary cylinder of *one size only* will cut the threads at a point where the width of the threads and grooves are *equal*. The diameter of such a cylinder is called the **pitch diameter.**

Figure 14-2 shows a correct pitch diameter superimposed on a thread. It should be emphasized that there is only *one* correct pitch diameter. Other pitch diameters can be drawn, but they do not cut the threads and grooves equally. Figure 14-3 illustrates two incorrect diameters.

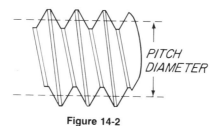

Figure 14-2

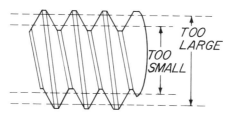

Figure 14-3

The cylinder that forms the pitch diameter is imaginary, but the points on the flanks of the thread where it would touch can be measured if the thread is of the correct shape.

INCLUDED ANGLE

The **included angle** is the angle formed by the flanks.

The *shape* of threads has varied through the centuries. There was no significant agreement on a common shape or size until the Industrial Revolution made it imperative that some standards be established. Since those early days, several practical thread forms have been developed and refined. They all required standardization of included angles, diameters, and pitches. Modern practice requires that at least three things be listed to identify a screw: the major diameter, the pitch (or number of threads per inch, from which pitch can be calculated), and the name (which identifies the included angle). Thus, an Acme bolt of a certain size would be identified on a print as:

$$1\tfrac{1}{4}''\text{--}10 \text{ Acme}$$

The major diameter is $1\frac{1}{4}$ inches, the number of threads is 10, and the form of the thread is Acme, which has an included angle of 29°.

In the sections that follow, the most common thread forms in use will be discussed, together with the necessary mathematics. The machinist or toolmaker should become thoroughly familiar with this information so as to understand what makes these thread forms unique.

Exercises 14-1

1. What would be the major diameter for cutting the following size threads: $1''$–8 NC, $\frac{1}{2}''$–13 NC, and $\frac{1}{4}''$–20 NC?

2. The minor diameter of a screw thread is equal to the major diameter minus twice the depth of thread. Find the minor diameter of a $\frac{1}{2}''$–13 NC thread if the depth of thread is $\frac{3}{64}$ in.

3. The thread depth of Acme thread is found by multiplying the pitch by 0.5. Find the depth of Acme threads having the following pitch: $\frac{1}{2}$, $\frac{1}{5}$, $\frac{1}{4}$, and $\frac{1}{8}$.

4. The formula for finding the crest width for a modified 60° vee thread is $\frac{1}{8}$ the pitch, or $\frac{1}{8} P$. Find the crest width for threads with the following pitch: $\frac{1}{20}$, $\frac{1}{13}$, $\frac{1}{24}$, and $\frac{1}{10}$.

5. A single screw thread has 12 threads per inch. What is its pitch? What is its lead?

6. What is the lead of a triple $\frac{1}{8}$-pitch thread?

14-2 American National Standard Thread and Unified Thread

OBJECTIVES

After completing this section, you should be able to:

1. Identify the American National Standard thread.
2. Find specified dimensions of the American National Standard thread.
3. Find the measurement over the wires for the American National Standard thread.
4. Describe the Unified thread system.

Figure 14-4 illustrates the thread form favored by more than 95% of American manufacturers. Its included angle is 60°. This thread form evolved from the sharp "vee" (see Figure 14-1), which is rarely used now except for permanent assemblies. The points of the sharp vee become flattened under pressure and are difficult to cut with tools that can keep their cutting edges pointed for any length of time.

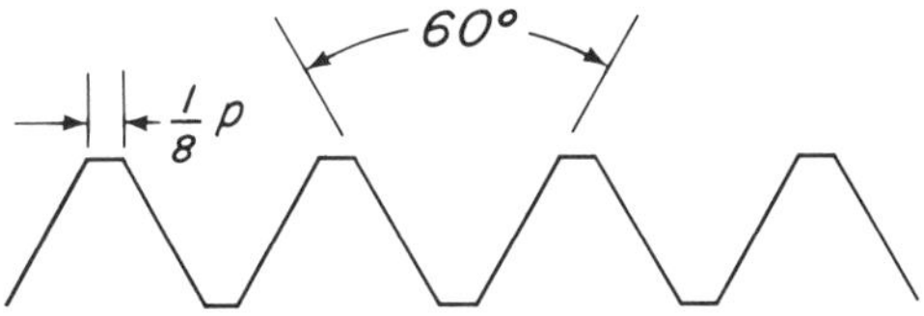

Figure 14-4

The American National Standard (ANS) thread is the old sharp vee with flattened crests and roots. It is referred to as a *modified* thread because of its flattened crests and roots. Its included angle of 60° is a wide one, capable of being held to very accurate dimensions. A 60° angle is strong, with good locking characteristics. The flat is $\frac{1}{8}$ of the pitch. A tool cutting this angle can hold its point in spite of cutting forces without breaking prematurely.

The American National Standard is divided into two main series—a coarse and a fine series. This simply means that for a given diameter, the screw or bolt can have one of two pitches. The choice of series depends on the purpose for which the screw will be used. The electrical and watchmaking industries use an extrafine series. Table 14.1 lists the National Coarse (NC) and National Fine (NF) series by size and number of threads per inch (N).

NC (National Coarse)		NF (National Fine)	
Size	N	Size	N
0	—	0	80
1	64	1	72
2	56	2	64
3	48	3	56
4	40	4	48
5	40	5	44
6	32	6	40
8	32	8	36
10	24	10	32
12	24	12	28
$\frac{1}{4}$	20	$\frac{1}{4}$	28
$\frac{5}{16}$	18	$\frac{5}{16}$	24
$\frac{3}{8}$	16	$\frac{3}{8}$	24
$\frac{7}{16}$	14	$\frac{7}{16}$	20
$\frac{1}{2}$	13	$\frac{1}{2}$	20
$\frac{9}{16}$	12	$\frac{9}{16}$	18
$\frac{5}{8}$	11	$\frac{5}{8}$	18
$\frac{3}{4}$	10	$\frac{3}{4}$	16
$\frac{7}{8}$	9	$\frac{7}{8}$	14
1	8	1	14
$1\frac{1}{8}$	7	$1\frac{1}{8}$	12
$1\frac{1}{4}$	7	$1\frac{1}{4}$	12
$1\frac{3}{8}$	6	$1\frac{3}{8}$	12
$1\frac{1}{2}$	6	$1\frac{1}{2}$	12
$1\frac{3}{4}$	5		
2	$4\frac{1}{2}$		

Table 14–1

The following rules give several formulas for working with modified 60° thread. You should become thoroughly familiar with them. The accompanying tables illustrate use of the rules.

RULE 14-1   FINDING PITCH (*P*) OF THREAD

$$P = \frac{1}{N}$$

(See Table 14-2.)

RULE 14-2   FINDING SINGLE DEPTH (*h*) OF THREAD

$h = 0.6495P$

(See Table 14-3.)

No. of Threads per Inch (*N*)	Pitch $\frac{1}{N}$
32	0.0312
16	0.0625
10	0.1000
13	0.0769

Table 14–2

N	*P*	*h*
32	0.0312	0.0203
16	0.0625	0.0406
10	0.1000	0.0649
13	0.0769	0.0499

Table 14–3

RULE 14-3   FINDING PITCH DIAMETER (*P.D.*)

P.D. = *O.D.* − single depth of thread (*h*)
(See Table 14-4.)

O.D.	*N*	*h*	*P.D.*
$\frac{5}{16}$	24	0.0271	0.285
$\frac{3}{8}$	16	0.0406	0.334
$\frac{1}{4}$	20	0.0325	0.217
$\frac{1}{2}$	13	0.0499	0.450

Table 14–4

Tap drill size is such that enough metal is left in the resulting hole to provide for approximately 75% engagement of thread. More than this is not desirable because of possible tap breakage in tough materials. Even if the drill is supported by a guide bushing, as in a drill jig, it can be expected to cut slightly oversize. In selecting a tap drill, common practice is to choose the commercial drill *closest* to the recommended size or a few thousandths of an inch larger. Commercial drill charts that include metric drill sizes give the best range for selection.

RULE 14-4   FINDING TAP DRILL SIZE (*T.D.S.*)

T.D.S. = *O.D.* − *P*

(See Table 14-5.)

O.D.	*N*	*P*	*T.D.S.*
#10(0.190)	32	0.0312	0.159
$\frac{3}{8}$	16	0.0625	0.312
$\frac{3}{4}$	10	0.1000	0.650
$\frac{1}{2}$	13	0.0769	0.423

Table 14–5

Example 14-1   Find the commercial tap drill size for $\frac{1}{2}''$–20 NF bolt.

Solution $T.D.S. = O.D. - P = 0.500 - 0.050 = 0.450$ in.

A commerical drill chart such as the one in the back of the book (Table I) would indicate that a $\frac{29}{64}$-diameter drill equals 0.4531. For general work, this is acceptable.

The tap drill formula can be applied to any diameter. The depth of a 60° modified thread remains the same, no matter how large a diameter it is located on. Thus, if a $5\frac{7}{8}$-diameter hole must be bored to a size sufficient for 12 threads per inch, the same tap drill size formula applies for boring as well as drilling.

$T.D.S. = O.D. - P = 5.875 - 0.0833 = 5.7917$ bore diameter

The common practice in single-point threading in a lathe is to swing the compound rest $29\frac{1}{2}°$ off 90° to the axis of the workpiece. The point of the tool is ground to have a 60° included angle. Moving it along a path of $29\frac{1}{2}°$ gives a slight clearance between the right-hand side of the tool and the thread groove, as shown in Figure 14-5.

The tool will obviously move along a path greater in length than the vertical depth of thread. The formula to find the approximate depth to feed the tool is stated in Rule 14–5.

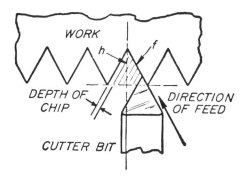

Figure 14-5

RULE 14-5   FINDING APPROXIMATE DEPTH TO FEED A TOOL MOVING ALONG A $29\frac{1}{2}°$ PATH

Compound rest travel = 0.75 × pitch (*P*)

Example 14-2   With the compound rest on $29\frac{1}{2}°$, how far must the tool be moved to arrive at the vertical depth of a 16-pitch thread?

Solution   Compound rest travel = $0.75P = 0.75(0.0625) = 0.0469$

The geometric justification for moving $\frac{3}{4}$ pitch is shown in Figure 14-6.

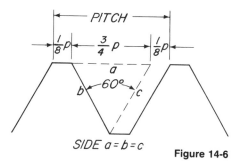

Figure 14-6

There are several ways to measure the size of a completed thread. Depending on the accuracy desired, the following methods are used:

Cutting thread to fit a mating part   Using the three-wire system

Checking with a ring or plug gauge   Using an optical comparator

Using a thread-pitch micrometer

For the occasional thread produced to order by a machinist or toolmaker, thread-pitch micrometers and the three-wire system are adequate. The **thread micrometer** has an anvil that is vee-shaped, and the end of the spindle is conical. The included angle of the cone is 60°, as is the angle of the vee on the anvil. The point of the cone is removed and the bottom of the vee is carried low, so that the micrometer will clear the top of the thread on one side and the bottom of the thread on the opposite side. The fit is then on the sides of the thread. The cone and anvil contact the flanks of the thread. The diameter reading indicated on the micrometer is the pitch diameter of the thread.

The thread-pitch micrometer does not give an exact reading, since the vee on the anvil is a fixed size and cannot adjust to the thread slant (helix angle) of all pitches. Thread micrometers are available for measuring pitches up to $4\frac{1}{2}$ threads per inch.

When the **three-wire system** is used to measure a screw thread, the pitch diameter is not measured but is *tested* by measuring the distance M across three wires placed in the thread groove. Figure 14-7 illustrates this.

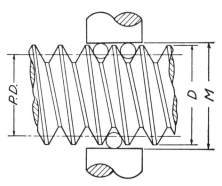

Figure 14-7

RULE 14-6   FINDING THE MEASUREMENT OVER THE WIRES
M FOR THE AMERICAN NATIONAL STANDARD
THREAD

$M = D - 1.515P + 3G,$

where M is the measurement over wires, D is the
thread diameter, P is the pitch of the thread, and G is
the wire diameter.

Best wire diameter is one that has *pitch-line contact*. The three-wire measurement can be made with a micrometer where an accuracy in thousandths of an inch will suffice. Rule 14–7 gives the formula to find the best wire diameter. This formula takes into account the amount that is missing from a sharp vee thread. Therefore, the micrometer reading is equal to the outside diameter of a sharp vee thread.

RULE 14-7   FINDING THE BEST WIRE DIAMETER G

$G = 0.57735P$

Example 14-3   Find the measurement over the wires for a $\frac{5}{16}''$–18 NC thread.

Solution $M = D - 1.5155P + 3G$, where $G = 0.57735P$
$= 0.03208$ in.
$M = 0.3125 - 1.5155(0.0556) + 3(0.03208)$
$= 0.3244$ in.

Unified and American Threads

The American National Standard thread has an included angle of 60° and flattened roots and crests. The British Standard Whitworth thread has an included angle of 55° and rounded roots and crests. Interchangeability problems during World War II led to the establishment of the Unified thread system, which allows for interchangeability of parts among the United States, Great Britain, Canada, and several other nations.

The Unified thread has an included angle of 60°. The crest of the internal thread (nut) must be flattened. However, the other roots and crests may be either rounded or flattened.

Tolerances have been relaxed for greater ease of assembly. With sufficient built-in allowance, a rounded crest of one thread will not interfere with the flattened root of its mating thread. This is exactly the situation in Figure 14-8, which shows an American nut and a British screw. Figure 14-9 shows a British nut and an American screw.

AMERICAN NUT	BRITISH NUT
BRITISH SCREW	AMERICAN SCREW
Figure 14-8	**Figure 14-9**

Unified threads are interchangeable with American National Standard threads of the same diameter and pitch. They have the same numbering systems for their coarse and fine series. The letters *UN* substitute for the letter *N*. UNF means "Unified Fine." A 12 UNF fastener has 28 threads per inch.

Some of the formulas for thread dimensions are different in the Unified system. For example, in an American National Standard thread, depth = .6495*P*. However, Unified threads have two different thread depths, one for internal threads and one for external threads:

Depth of internal thread $= 0.5413P$

Depth of external thread $= 0.6134P$

Exercises 14-2

Fill in the missing values for American National Standard threads.

O.D. = thread diameter
P.D. = pitch diameter
P = pitch of thread
h = depth of single thread
N = threads per inch
T.D.S. = tap drill size

	O.D.	*N*	*h*	*P*	*P.D.*	*T.D.S.*
1.	$\frac{1}{4}$	28				
2.	$\frac{5}{16}$	18				
3.	$\frac{3}{8}$	24				
4.	$\frac{7}{16}$	14				
5.	$\frac{5}{8}$	11				
6.	1	8				
7.	$\frac{1}{4}$	20				

Find each value. This will provide a handy reference for measurements using the three-wire system.

	Threads per Inch (*N*)	Best Wire Size (*0.57735P*)	Threads per Inch (*N*)	Best Wire Size (*0.57735P*)
8.	80	0.0072	16	
9.	72	0.0080	14	
10.	64		13	
11.	56		12	
12.	48		11	
13.	44		10	
14.	40		9	
15.	36		8	
16.	32		7	
17.	28		6	
18.	24		5	
19.	20		$4\frac{1}{2}$	
20.	18		4	

21. Find the pitch of a 12 UNF thread.

22. Find the depth of internal thread of a 12 UNF thread.

14-3 Machine-Screw Diameters

OBJECTIVES

After completing this section, you should be able to:

1. Find the outside diameter of a machine screw.
2. Find the tap drill size for a machine screw.

Screws with a diameter less than $\frac{1}{4}$ inch are called **machine screws.** A series of numbers from 0 to 12 is used to specify the sizes of these screws. These numbers give no indication of the outside diameter of a particular screw. A simple rule will help determine the outside diameter of any given machine screw.

> **RULE 14-8 FINDING THE OUTSIDE DIAMETER OF A MACHINE SCREW**
>
> Multiply the screw number by 0.013 and add 0.060.

Example 14-4   What is the outside diameter of a No. 10 machine screw?

Solution $10 \times 0.013 + 0.060 = 0.130 + 0.060 = 0.190$ in.

Note that the outside diameter of a No. 0 screw is 0.060 inch and the change in outside diameter between numbers is approximately 0.013 inch.

To find the tap drill size, use the formula

$$T.D.S. = O.D. - P$$

For a No. 10 screw with 24 threads per inch, the pitch is $\frac{1}{24}$, or about 0.042. The tap drill size is therefore $0.190 - 0.042 = 0.148$. From Table I in the back of the book, the tap drill size corresponding to the next greater value is No. 25.

Exercises 14-3

For future reference, calculate the outside diameters and tap drill sizes of these machine screws. (Use Table I to find the tap drill.)

	MACHINE SCREW SIZES			
	No. of Screw	**Threads per Inch**	**Outside Diameter**	**Tap Drill Size (T.D.S.)**
1.	0	80	0.060	
2.	1	64		
3.	1	72		
4.	2	56		
5.	2	64		
6.	3	48		
7.	3	56		
8.	4	40		

MACHINE SCREW SIZES				
	No. of Screw	Threads per Inch	Outside Diameter	Tap Drill Size (*T.D.S.*)
9.	4	48		
10.	5	40		
11.	5	44		
12.	6	32		
13.	6	40		
14.	8	32		
15.	8	36		
16.	10	24		
17.	10	32		
18.	12	24		
19.	12	28		

14-4 Square and Acme Threads

OBJECTIVES

After completing this section, you should be able to:

1. Identify square and Acme threads.
2. Use working formulas for the Acme thread.

Square Thread

The **square thread** in its original form is shown in Figure 14-10. It was used to transmit motion and was a very good thread for the reduction of friction. Its straight sides have no wedging action. However, its sharp corners were potential stress points for cracks and were very difficult to produce with tools that would retain their sharp corners. Taps to cut square threads are available, but are always classified as "special," with a corresponding increase in price.

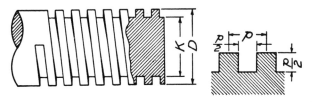

Figure 14-10

For these reasons, the original square thread has been modified into one with a 10° included angle or 5° flank angle. As such, it is still used in jacks and other mechanisms to counteract loads.

Acme Thread

The sloping sides of the **Acme thread** (Figure 14-11) reduce friction and allow for easy engagement. For this reason, the Acme thread can be found in the half-nut mechanisms on engine lathes where thread chasing requires smooth engagement of the lead screw. The truss design of the Acme makes it strong for transmitting heavy loads. Very accurately ground Acme screws are used in numerically controlled equipment and are known as **translating screws.**

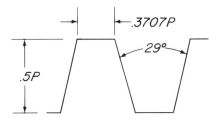

Figure 14-11

Table 14-6 lists sizes and pitches for the **general purpose Acme.**

O.D.	*N*	*O.D.*	*N*
$\frac{1}{4}$	16	$\frac{5}{8}$	8
$\frac{5}{16}$	14	$\frac{3}{4}$	6
$\frac{3}{8}$	12	$\frac{7}{8}$	6
$\frac{7}{16}$	12	1	5
$\frac{1}{2}$	10	2	4

Table 14—6

The Acme thread has its own set of working formulas.

RULE 14-9 FINDING SINGLE DEPTH OF THREAD (*h*)

$h = 0.5P$

(See Table 14–7.)

N	*P*	*h*
16	0.063	0.032
10	0.100	0.050

Table 14–7

RULE 14-10 FINDING TAP DRILL SIZE

$T.D.S. = (O.D. - P) + 0.020$

(See Table 14–8.)

O.D.	*N*	*T.D.S.*
$\frac{1}{4}$	16	0.208 in.
$\frac{1}{2}$	10	0.420 in.

Table 14–8

In the formula for tap drill size, 0.020 is added to provide clearance for the *minor* diameter of the mating screw. The clearance for the *major* diameter is provided by an oversize condition of the tap diameter. Ordinarily, a set of three progressively larger taps is used to thread an Acme hole.

One of two methods is practiced by the machinist in the ordinary machine shop to check the pitch diameter of an Acme thread.

1. A pair of gear-tooth calipers can be set to the correct width of the tooth at the pitch diameter and used as a check on the depth of these points from the tooth crest.

2. A single wire of a calculated diameter can be inserted into the thread groove. When this wire is flush with the top of the thread, the thread depth is correct. (See Figure 14-12.)

Figure 14-12

RULE 14-11   ONE-WIRE MEASURING OF PITCH DIAMETER

$$G = \frac{0.4872}{N}$$

where N is the number of threads per inch and G is the diameter of wire to be used.

Exercises 14-4

Fill in the missing values.

	O.D.	N	h	T.D.S.	Single Measuring Wire for P.D.
1.	$\frac{5}{8}$	8			
2.	$\frac{3}{4}$	6			
3.	1	5			
4.	2	4			
5.	$\frac{7}{16}$	14			
6.	1	8			
7.	$1\frac{1}{2}$	4			
8.	$1\frac{1}{4}$	16			
9.	$\frac{3}{8}$	12			
10.	$\frac{1}{2}$	10			

14-5 Worm and Metric Threads

OBJECTIVES

After completing this section, you should be able to:

1. Identify worm and metric threads.
2. Use working formulas for the worm thread.
3. Convert to American tap drill sizes from metric.

Worm Thread

A **worm thread** is shown in Figure 14-13. It has the same included angle and appearance as the Acme thread, but is deeper. The pitch of the worm thread is the same as that of the worm wheel, with which it will mate. The diameter of worms is not standardized.

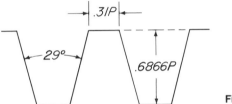

Figure 14-13

RULE 14-12 FINDING THE DEPTH OF A THREAD (h)

$$h = \frac{0.6866}{N} = 0.6866P$$

(See Table 14-9.)

RULE 14-13 FINDING THE WIDTH OF A TOOL (w)

$$w = 0.310P$$

(See Table 14-10.)

N	h
4	0.1717
6	0.1144
8	0.0858
10	0.0687

Table 14-9

N	P	w
4	0.250	0.078
6	0.169	0.052
8	0.125	0.039
10	0.100	0.031

Table 14-10

Metric Thread

Figure 14-14 illustrates the International Metric thread form. It differs from the American National thread in the shape of the thread root. The rounded metric root gives a stronger thread. Mechanics have been using this thread for years on spark plugs.

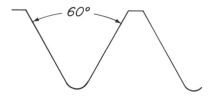

Figure 14-14

The pitch of American threads is specified as a number of threads per inch, but metric threads are given as a number of *millimeters from one thread to the next*. In the Metric Coarse series of thread, only the diameter is given; the number of threads is not specified.

Rule 14-13 gives a formula to convert from metric sizes for American tap drills.

RULE 14-13   FINDING AMERICAN TAP DRILL SIZE FOR
METRIC BOLTS

American $T.D.S. = (O.D._{mm} - P_{mm}) \times 0.0394$

Example 14-5   What is the correct tap drill for a 3-mm bolt with a 0.60-mm pitch?

Solution $T.D.S. = (3.00 - 0.60) \times 0.0394$
$= 2.40 \times 0.0394$
$= 0.0946$ in., or a $\frac{3}{32}$-in.-diameter drill

In the tables in the back of this book, you will find the standard sizes and pitches for metric bolts and screws. These listings are according to the recommendations of the *ISO*, or *International Standards Organization*, which was given the task of determining what metric measurements will be used in the future. This organization took as its model the metric bolt sizes adopted by *DIN*, the German engineering association, since these were the best European standards available.

Exercises 14-5

Fill in the depth of thread and the width of the threading tool required to cut a worm thread.

	N	Width	Depth of Thread	N	Width	Depth of Thread
1.	13			8		
2.	4			20		
3.	6			10		
4.	15			12		
5.	14			16		
6.	7			9		

7. What American tap drill size should be used for a metric thread 5 mm in diameter with a .8 mm pitch?

8. What American tap drill size should be used for a metric thread 26 mm in diameter with a 3-mm pitch?

Unit 14 Review Exercises

1. What is the pitch of a double-threaded screw having a lead of $\frac{5}{32}$ in.?

2. With the compound rest set at $29\frac{1}{2}°$, how much must the tool be moved to arrive at the vertical depth for threads with the following pitch: $\frac{1}{16}$, $\frac{1}{10}$, $\frac{1}{13}$, $\frac{1}{20}$, and $\frac{1}{24}$?

3. Find the single-wire diameter for a $1''$–18 Acme thread and the best wire diameter for a $\frac{3}{4}''$–10 NC thread.

4. In the unified thread system, what is the shape of the crest and root of a British nut? An American nut?

Complete the table for NC thread.

	Thread Size	P	h	P.D.	T.D.S.
5.	5/16″–18				
6.	7/16″–14				
7.	3/8″–16				

Complete the table for Acme thread. (Hint: P.D. = O.D. − h.)

	Thread Size	P	h	P.D.	T.D.S.
8.	1″–8				
9.	7/8″–9				
10.	5/8″–11				

Gear Trains
and Indexing

Years ago, quite a few machine shops cut their own spur, bevel, and worm gears. Today, however, most nonmolded gears are produced on special gear-generating machines capable of high production. Nonstandard gears in small quantities can be very expensive to make. Nevertheless, a small machine shop will still need an occasional spur gear that cannot be purchased as a stock item from a mill supply house. A simple spur gear can be made by an experienced machinist or toolmaker who understands basic gearing and indexing. Such a person is a valuable asset to any shop.

This unit will emphasize the mathematics necessary to compute a gear train of two spur gears and cut those gears with the help of a lathe, milling machine, formed cutter, and dividing head. Bevel and worm gears are not covered, since their production is much more complex and requires greater skill than the average machine shop trainee is expected to demonstrate. This unit will also discuss indexing, which is an operation that makes a specified number of cuts on a workpiece.

15-1 Gear Systems

OBJECTIVES

After completing this section, you should be able to:

1. Identify parts of a gear.
2. Find the outside diameter of a gear.
3. Find the center-to-center distance in mating gears.

Spur gears have straight teeth parallel to the gear axis. They are cylindrical in shape and operate on parallel shafts. Figure 15-1 is a sectional view of a gear blank and a partial diagram of spur gear teeth.

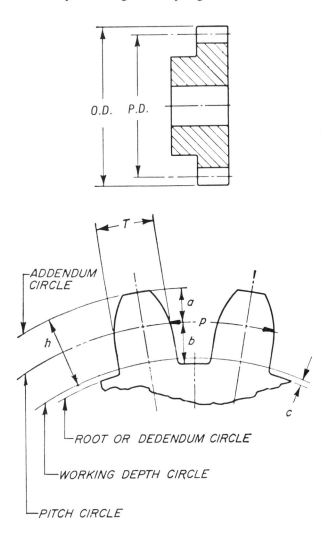

Figure 15-1

The following definitions and symbols apply to Figure 15-1.

ADDENDUM (*a*)

The **addendum** is the length of the tooth from the pitch circle to the outside.

DEDENDUM (*b*)

The **dedendum** is the length of the tooth from the pitch circle to the base, or root. It is equal to the addendum plus clearance.

CLEARANCE (*c*)

The **clearance** is the distance between the end of a tooth and the bottom of the mating gear.

OUTSIDE DIAMETER (*O.D.*)

The **outside diameter** is the overall diameter of the gear.

PITCH DIAMETER (*P.D.*)

The **pitch diameter** is the diameter of the pitch circle.

PITCH CIRCLE

The **pitch circle** separates each tooth into addendum and dedendum. The pitch circles of two mating gears are the diameters that would be tangent to each other before the teeth are added.

CIRCULAR PITCH (*P*)

The **circular pitch** is the distance from a point on one tooth to the corresponding point on the next tooth measured along the pitch circle.

WHOLE DEPTH (*h*)

The **whole depth** is the depth of engagement of two mating gears. It is equal to the sum of two addendums.

CHORDAL THICKNESS (*T*)

Chordal thickness is the tooth thickness measured at the pitch circle.

There are two systems for establishing the spacing and size of teeth on a gear relative to the size of the gear. The older of these is called the **circular-pitch** system. In this system, the distance from a point on one tooth to a corresponding point on the next tooth, *measured along the pitch circle,* is given a number in inches or fractions of an inch.

This system works as though a pair of dividers were set to a certain spacing and teeth were stepped off along the pitch circle using this spacing. In fact, this is exactly how gear teeth were laid off on large cast-iron gears in former times. The circular-pitch system may be used when gear teeth are fine or when a given center distance between mating gears cannot be met by standard gears. The more frequently used system is that of diametral pitch.

Diametral pitch is a code for designating tooth size, number of teeth, and the relative size of a gear. It is the number of teeth a gear has for each inch of pitch diameter. A diametral pitch gear of 10 is a gear that has 10 teeth for each inch of pitch diameter. Bearing in mind that a gear with pitch diameter of 1 inch would be approximately the size of a 25-cent piece, it may be seen that 10 teeth on a circle of this size would be of medium size.

If a 10-diametral-pitch gear has 30 teeth, the gear they fit will be about three times the size of a quarter. Figure 15-2 shows 2-, 4-, 8-, and 12-diametral-pitch teeth drawn full size. Note that the size *decreases* considerably as the diametral pitch *increases*.

You can readily see that 24-pitch or 36-pitch teeth would be small. You would not expect to cut 2-diametral-pitch teeth on a gear with a small diameter or 30-diametral-pitch teeth on a gear with a large diameter.

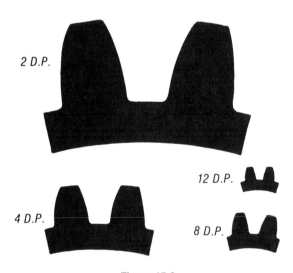

2 D.P.

12 D.P.

4 D.P.

8 D.P.

Figure 15-2

The **outside diameter** of a gear is important because the gear blank is first turned to this dimension before the teeth are cut. The following formula is used to find the outside diameter.

RULE 15-1   FINDING THE OUTSIDE DIAMETER OF A GEAR

$$O.D. = \frac{N + 2}{D.P.}$$

where N is the number of teeth and $D.P.$ is the diametral pitch.

Example 15-1   If a 12-pitch gear has 58 teeth, find its outside diameter.

Solution $O.D. = \dfrac{N + 2}{D.P.} = \dfrac{58 + 2}{12} = 5$ in.

This formula can be transposed to find the number of teeth or diametral pitch, provided the other two terms are known.

RULE 15-2   FINDING THE NUMBER OF TEETH OR DIAMETRAL PITCH WHEN TWO TERMS ARE KNOWN

$N = (O.D.)(D.P.) - 2$ $D.P. = \dfrac{N + 2}{O.D.}$

Example 15-2   A 54-tooth gear has an outside diameter of 4.666 in. What diametral pitch is this gear?

Solution $D.P. = \dfrac{N + 2}{O.D.} = \dfrac{56}{4.666} = 12$

Example 15-3   A 14-diametral-pitch gear has an outside diameter of 1.571 in. How many teeth will this gear have?

Solution $N = (O.D.)(D.P.) - 2 = (1.571)(14) - 2 = 20$ teeth

To help picture what is meant by the **center-to-center** distance, think of two gears in mesh. The smaller of the two mating gears is called the **pinion.** The larger of the two gears is simply called the **gear.** The pinion is the **driver** and the gear is the **driven** gear. Since spur gears operate on parallel shafts, you are really trying to find the distance between the centers of two parallel shafts.

RULE 15-3   FINDING CENTER-TO-CENTER DISTANCE C OF TWO MESHING SPUR GEARS USING THE DIAMETRAL-PITCH SYSTEM

$C = \dfrac{N_g + N_p}{2D.P.}$

where N_g is the number of teeth on the gear, N_p is the number of teeth on the pinion, and $D.P.$ is the diametral pitch.

Example 15-4   What is the center-to-center distance when a 12-pitch, 24-tooth pinion meshes with a 12-pitch, 48-tooth gear?

Solution $C = \dfrac{N_g + N_p}{2D.P.}$

$$= \frac{48 + 24}{24}$$

$$= \frac{72}{24} = 3 \text{ in.}$$

Exercises 15-1

Fill in the missing values for the gears listed.

	O.D.	*N*	*D.P.*
1.	2.800	12	
2.	6.500		10
3.		44	18
4.	3.125	48	
5.		36	24
6.	9.167		12
7.	5.429		14
8.	3.091	66	
9.		92	8

Find the center-to-center distance of the following meshed gears.

10. 6 *D.P.*, 18 and 48 teeth   **11.** 8 *D.P.*, 36 and 64 teeth

12. 5 *D.P.*, 30 and 42 teeth   **13.** 12 *D.P.*, 40 and 70 teeth

14. 10 *D.P.*, 24 and 60 teeth   **15.** 4 *D.P.*, 24 and 24 teeth

16. 16 *D.P.*, 36 and 54 teeth   **17.** 9 *D.P.*, 69 and 36 teeth

18. In Figure A, the diametral pitch is 12. What is the center-to-center distance between the two gears shown?

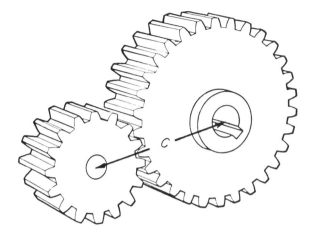

Figure A

19. In Figure B, if the diametral pitch is 10, what is the center distance C?

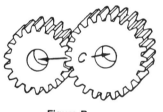

Figure B

15-2   Designing a Gear Train

OBJECTIVE
After completing this section, you should be able to:

1. Find the circular pitch needed for a specific gear train.

In previous problems, each set of meshing gears had a combination of teeth that resulted in a certain ratio or speed. For example, two gears with 42 and 21 teeth, respectively, would result in a ratio of 2 to 1, or—as it is usually expressed—2 : 1. This simply means that one gear turns twice as fast as the other.

When designing gear trains, some of the following factors must be considered:

Space available for the gears (center to center)
Speed ratio to be developed (size of gears)
Size of tooth for strength (diametral pitch)

Assume that a designer has a gear box of a size that restricts the system to a center distance between shafts of 8 inches. The designer wishes to develop a speed ratio of 3 to 1 between these shafts and, for strength in overcoming load, chooses a 24-diametral-pitch gear tooth. The design procedure is as follows:

Center-to-center distance = 8 inches
Pitch diameter of pinion = $(P.D.)_1$
Pitch diameter of gear = $(P.D.)_2$
Twice the center distance = $(P.D.)_1 + (P.D.)_2$

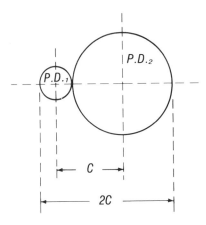

Figure 15-3

In Figure 15-3, it is obvious that the pitch diameter of each gear is a proportional part of twice the center distance. Therefore, if the ratio is 3 : 1,

$$(P.D.)_1 = \frac{1}{4} \text{ of } 2C \text{ and } (P.D.)_2 = \frac{3}{4} \text{ of } 2C$$

$2C = 16$ in.

$$\frac{1}{4} \times 16 = \frac{16}{4} = 4 \text{ in.} = \text{pitch diameter of pinion} = (P.D.)_1$$

$$\frac{3}{4} \times 16 = 12 \text{ in.} = \text{pitch diameter of gear} = (P.D.)_2$$

RULE 15-4   FINDING THE NUMBER OF TEETH IN EACH GEAR OF A GEAR TRAIN

N = pitch diameter × diametral pitch

Number of teeth in pinion (N_p) = 4 × 24 = 96

Number of teeth in gear (N_g) = 12 × 24 = 288

$$O.D.\text{ of pinion} = \frac{N_p + 2}{D.P.} = \frac{98}{24} = 4.083 \text{ in.}$$

$$O.D.\text{ of gear} = \frac{N_g + 2}{D.P.} = \frac{290}{24} = 12.083 \text{ in.}$$

If the shafts are a specified distance apart and it is desired to operate the gears at a specified ratio, it may not be possible to use two gears of standard diametral pitch. In this instance, use the following formula for converting center-to-center distance to circular pitch.

RULE 15-5   CONVERTING CENTER-TO-CENTER DISTANCE TO CIRCULAR PITCH

$$P = \frac{2\pi C}{(N_p + N_g)}$$

where P is the circular pitch and C is the center-to-center distance.

Example 15-5   What must be the circular pitch of two gears that are to be operated on shafts 5.963 in. apart, if the pinion has 18 teeth and the gear has 42 teeth?

Solution $P = \dfrac{2\pi C}{(N_p + N_g)} = \dfrac{2(3.1416)(5.968)}{18 + 42}$

$$= \frac{37.5}{60} = 0.625, \text{ or } \frac{5}{8} \text{ pitch}$$

Exercises 15-2

1. What is the circular pitch of two gears that are to be operated on shafts 7.957 in. apart if the gears have 64 and 36 teeth?

2. If two parallel shafts are 10.23 in. apart, what must be the circular pitch of a gear and pinion with 50 and 25 teeth mounted on these shafts?

3. Two spur gears with 44 and 36 teeth, respectively, are to operate on parallel shafts 4.774 in. apart. What circular pitch should be chosen for these gears?

Find the number of teeth on the pinion (N$_p$) and the number of teeth on the gear (N$_g$).

	P.D. of Pinion	P.D. of Gear	Diametral Pitch	N_p	N_g
4.	8 in.	$3\frac{1}{4}$ in.	12		
5.	$5\frac{1}{4}$ in.	$9\frac{3}{8}$ in.	8		
6.	3.3 in.	7 in.	10		
7.	$2\frac{2}{3}$ in.	$7\frac{2}{3}$ in.	15		
8.	$5\frac{1}{4}$ in.	$7\frac{1}{2}$ in.	8		

Change center-to-center distance to circular pitch.

	N_p	N_g	Center-to-Center Distance	Circular Pitch
9.	25	35	5.625 in.	
10.	42	75	7.312 in.	
11.	45	62	10.7 in.	
12.	32	100	$5\frac{1}{2}$ in.	
13.	30	45	15 in.	

15-3 Converting Speed Ratios to Gear Combinations

OBJECTIVE

After completing this section, you should be able to:

1. Use continued fractions to convert speed ratios to gear combinations.

If it becomes necessary to translate a given speed ratio into an equivalent pair of gears, the appropriate pitch diameters can be found by using **continued fractions.** The technique of continued fractions gives several fractions that are approximations of the original fraction.

Consider the fraction $\frac{5}{8}$. It has a decimal equivalent of 0.625. The following example shows how to do continued fractions by using a known fraction and its decimal equivalent.

Example 15-6   Find approximate fractions for $\frac{5}{8}$.

Solution

Step 1   Divide the numerator into the denominator.

$$5\overline{)8}(1$$
$$\underline{5}$$
$$3$$

Step 2   Divide the remainder into the preceding divisor.

$$5\overline{)8}(1$$
$$\underline{5}$$
$$3\overline{)5}(1$$
$$\underline{3}$$
$$2$$

Continue dividing remainders into preceding divisors, until the remainder is zero.

$$5\overline{)8}(1 \text{ } partial \text{ } quotient$$
$$\underline{5}$$
$$3\overline{)5}(1 \text{ } partial \text{ } quotient$$
$$\underline{3}$$
$$2\overline{)3}(1 \text{ } partial \text{ } quotient$$
$$\underline{2}$$
$$1\overline{)2}(2 \text{ } partial \text{ } quotient$$
$$\underline{2}$$
$$0$$

Each partial quotient is placed under 1.

$$\cfrac{1}{1 + \cfrac{1}{1 + \cfrac{1}{1 + \cfrac{1}{2}}}}$$

Step 3   Place the partial quotients in a row.

1   1   1   2

The first fraction is $\frac{1}{1}$, which is placed under the first partial quotient.

1   1   1   2
$\frac{1}{1}$

The second fraction is $\cfrac{1}{1 + \cfrac{1}{1}}$ or $\cfrac{1}{\cfrac{2}{1}}$ or $\cfrac{1}{2}$.

Place the second fraction under its partial quotient.

$$
\begin{array}{cccc}
1 & 1 & 1 & 2 \\
\dfrac{1}{1} & \dfrac{1}{2} & &
\end{array}
$$

To find the third fraction, multiply the partial quotient, 1, by the *numerator* of the second fraction, 1, which produces 1. Add this to the numerator of the preceding fraction to obtain 2. Go through the same procedure with the *denominator* of the second fraction, 2. This will yield $1 \times 2 + 1$, or 3. Place the third approximate fraction under its partial quotient.

$$
\begin{array}{cccc}
1 & 1 & 1 & 2 \\
\dfrac{1}{1} & \dfrac{1}{2} & \dfrac{2}{3} &
\end{array}
$$

By the same procedure, the fourth quotient will yield a fraction of $\frac{5}{8}$, which is the original fraction. The decimal equivalent of $\frac{1}{2}$ is 0.500. The decimal equivalent of $\frac{2}{3}$ is 0.666.

RULE 15-6 RELATING THE RATIO OF GEARS TO CONTINUED FRACTIONS

To convert the speed ratio to a decimal equivalent, change this into its common fraction and use continued fractions to find other approximately equivalent fractions that may be substituted for the original fraction.

Example 15-7 A set of gears with 50 and 127 teeth, respectively, is unavailable for use. Find other gear tooth combinations that will yield the same speed ratio.

Solution

Step 1 Convert $\frac{50}{127}$ to its decimal equivalent.

$$\frac{50}{127} = 0.3937$$

Step 2 Change 0.3937 into a common fraction.

$$0.3937 = \frac{3937}{10,000}$$

Step 3   Find the continued fraction.

$$3937\overline{)10000(2}$$
$$\underline{7874}$$
$$2126\overline{)3937(1}$$
$$\underline{2126}$$
$$1811\overline{)2126(1}$$
$$\underline{1811}$$
$$315\overline{)1811(5}$$
$$\underline{1575}$$
$$236\overline{)315(1}$$
$$\underline{236}$$
$$79\overline{)236(2}$$
$$\underline{158}$$
$$78\overline{)79(1}$$
$$\underline{78}$$
$$1\overline{)78(78}$$
$$\underline{78}$$
$$0$$

The partial quotients are: 2   1   1   5   1   2   1   78.

The first fraction is $\frac{1}{2}$. The second fraction is

$$\frac{1}{2 + \dfrac{1}{1}}, \quad \text{or} \quad \frac{1}{3}.$$

2	1	1	5	1	2	1	78
$\frac{1}{2}$	$\frac{1}{3}$						

The next fraction is found by multiplying the third partial quotient by the numerator of the preceding fraction and adding the numerator of the preceding fraction. This is $1 \times 1 + 1$, or 2. The same is done with the denominators of the two preceding fractions, yielding $1 \times 3 + 2$, or 5.

The resulting third fraction is $\dfrac{1 \times 1 + 1}{1 \times 3 + 2} = \dfrac{2}{5}$, which is placed under the third partial quotient.

2	1	1	5	1	2	1	78
$\frac{1}{2}$	$\frac{1}{3}$	$\frac{2}{5}$					

The remaining continued fractions are filled in by the same method.

2	1	1	5	1	2	1	78

$$\frac{1}{2} \quad \frac{1}{3} \quad \frac{2}{5} \quad \frac{11}{28} \quad \frac{13}{33} \quad \frac{37}{94} \quad \frac{50}{127}$$

The decimal equivalents of the above fractions are:

0.5 0.3333 0.4000 0.3928 0.3939 0.3936 0.3937

The gear combination of 37 and 94 teeth will give a ratio that is equivalent to 50 : 127 within 0.0001 error. This problem is an example of gear substitution for the standard 50-tooth and 127-tooth "transposing" gears normally required to convert American lathes for cutting *metric* threads.

Example 15-8 A speed ratio calls for a gear train of two gears, one of which has 68 teeth and the other 112 teeth. These gears are unobtainable. Find two other gears that will give an approximate ratio equal to that of the two original gears.

Solution $\dfrac{68}{112} = 0.6071 = \dfrac{6071}{10,000}$

$$
\begin{array}{r}
6071\overline{)10000}(1 \\
\underline{6071} \\
3929\overline{)6071}(1 \\
\underline{3929} \\
2142\overline{)3929}(1 \\
\underline{2124} \\
1787\overline{)2142}(1 \\
\underline{1787} \\
355\overline{)1787}(5 \\
\underline{1775} \\
12\overline{)355}(29 \\
\underline{348} \\
7\overline{)12}(1 \\
\underline{7} \\
5\overline{)7}(1 \\
\underline{5} \\
2\overline{)5}(2 \\
\underline{4} \\
1\overline{)2}(2 \\
\underline{2} \\
0
\end{array}
$$

Partial quotients:

1 1 1 1 5 29 1 1 2 2

Fractions:

$$\frac{1}{1} \quad \frac{1}{2} \quad \frac{2}{3} \quad \frac{3}{5} \quad \frac{17}{28} \quad \frac{496}{817} \quad \frac{513}{845} \quad \frac{1009}{1662} \quad \frac{1552}{2507} \quad \frac{4053}{6676}$$

Decimal equivalents:

$$\frac{1}{1} = 1 \quad \frac{1}{2} = 0.5000 \quad \frac{2}{3} = 0.6667 \quad \frac{3}{5} = 0.6000$$

$$\frac{17}{28} = 0.6071 \quad \frac{496}{817} = 0.6071 \quad \frac{513}{845} = 0.6071$$

$$\frac{1009}{1662} = 0.6071 \quad \frac{1522}{2507} = 0.6071 \quad \frac{4053}{6676} = 0.6071$$

Reviewing the gear combinations above, it becomes obvious that $\frac{17}{28}$ is the last set with a reasonable number of teeth. It may also be apparent that the original number $\frac{68}{112}$ can be factored into $\dfrac{4 \times 17}{4 \times 28}$. However, a table of factors is not always available, nor do all numbers factor into terms practical for gear teeth.

Continued fractions are not merely a mathematical exercise. They have a very real value in solving problems involving gearing, helical milling, and high-number or angular indexing. With a little practice, the process of using continued fractions becomes easy.

Exercises 15-3

By continued fractions, find the gear combinations that best approximate the given speed ratios. Keep the number of teeth in both pinion and gear between 13 and 100 teeth.

	Speed Ratio	Decimal Equivalent	Teeth in Pinion	Teeth in Gear	Actual Ratio
1.	$2\frac{3}{4}:1$	0.3636			
2.	$4.2:1$	0.2380			
3.	$2.1:1$	0.4762			
4.	$1.5:1$	0.6667			
5.	$1.08:1$	0.9259			

15-4 Metric Module System

OBJECTIVES

After completing this section, you should be able to:

1. Find the module of a gear.
2. Find the diametral-pitch equivalent of a metric module.
3. Find the outside diameter of a metric gear.

Most countries that have adopted the metric system utilize the **module** system for gear design. American manufacturers are not likely to change completely from diametral-pitch or circular-pitch systems. This is one of the areas in metrication where, as in the case of the unified thread, a double standard will continue to exist. However, for export and for replacement of imported metric gears, it is necessary to know something of the European module system.

The module system is based on a *direct* measurement. You will recall that diametral pitch for gears is a ratio between the size of the pitch diameter and the number of teeth that can be placed evenly upon it.

RULE 15-7   FINDING THE MODULE OF A GEAR

$$\text{Module of a gear} = \frac{\text{pitch diameter in millimeters}}{\text{number of teeth}}$$

Thus, a gear with a pitch diameter of 60 millimeters and 20 teeth will have a module of 3 millimeters of pitch diameter for each tooth.

Rule 15-8 gives a formula for converting a metric module to diametral pitch.

RULE 15-8   CONVERTING A METRIC MODULE TO DIAMETRAL PITCH

$$\text{Diametral pitch} = \frac{25.4}{\text{module}}$$

Example 15-9   What is the diametral pitch of a 10-module gear?

Solution $\dfrac{25.4}{10} = 2.54$

The nearest equivalent is a diametral pitch of $2\frac{1}{2}$.

In most cases, the number resulting from the formula should be translated into an equivalent circular-pitch gear. Rule 15-9 gives a formula to do this.

RULE 15-9   CONVERTING DIAMETRAL PITCH TO CIRCULAR PITCH

$$P = \frac{\pi}{D.P.}$$

where P is the circular pitch and $D.P.$ is the diametral pitch.

Example 15-10   Find the circular-pitch equivalent of a 2.54 diametral-pitch gear.

Solution $\frac{3.1416}{2.54}$, or a 1.237 circular-pitch gear

The outside diameter of a metric module gear can be calculated by using the formula given in Rule 15-10.

RULE 15-10   FINDING THE OUTSIDE DIAMETER OF A METRIC MODULE GEAR

$O.D._{mm} = (N + 2) \times$ module

Example 15-11   What is the outside diameter of a 4-module gear with 30 teeth?

Solution $O.D._{mm} = (30 + 2)4 = 128$ mm

Exercises 15-4

Complete the columns below, using the formulas given in this section.

	Gear Module	No. of Teeth	P.D. (mm)	Diametral Pitch	Circular Pitch (in.)	O.D. in mm	O.D. in Inches
1.	1	20	20				
2.	3.25	10	32.5				
3.	8	22	176				
4.	1.5	18	24				
5.	3	20	60				
6.	4.6	26	120				

15-5 Indexing

OBJECTIVE

After completing this section, you should be able to:

1. Find the number of turns and spaces for a division on a workpiece.

Indexing is an operation performed with a dividing head for the purpose of making a definite number of cuts on a workpiece. Figure 15-4 shows a dividing head (top) and a setup for milling a spur gear (bottom). The dividing head is mounted on the milling machine bed and rotates about its axis.

Figure 15-4   *(Courtesy of Cincinnati Milacron)*

Plain indexing is the type most often called for. **Angular indexing** extends the range of possible cuts beyond the capabilities of the plates normally used with a dividing head. Differential indexing was formerly used to do the same thing, but is rarely employed now because of the possibilities of human error in calculations.

Most dividing heads have a single-threaded worm meshing with a 40-tooth worm gear, as shown in Figure 15-5. The worm must be cranked 40 complete turns to rotate the worm gear, spindle, and workpiece through 360°. This is a ratio of 40 : 1. Turning the crank 20 times rotates the workpiece through 180°, and so forth. The formula that expresses this relationship is stated in Rule 15-11.

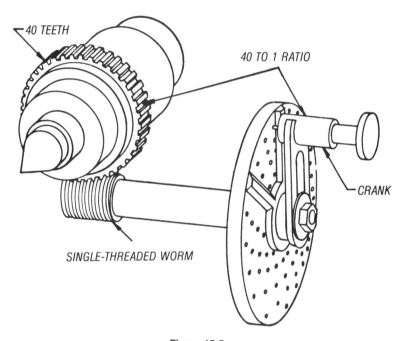

40 TEETH

40 TO 1 RATIO

CRANK

SINGLE-THREADED WORM

Figure 15-5

RULE 15-11 FINDING THE NUMBER OF TURNS OF THE
CRANK FOR INDEXING

$$T = \frac{40}{N}$$

where T is the number of complete and partial crank turns and N is the number of divisions desired.

Example 15-12 How many turns of the crank are needed for work in a dividing head to be indexed 8 equal times?

Solution $T = \dfrac{40}{8} = 5$ complete turns for each division

Where the desired number of divisions divides evenly into 40, there are no *fractional turns*. When a third, a half, or two-thirds of a turn is needed, the two **sector arms** and **index plates** come into use (see Figure 15-6).

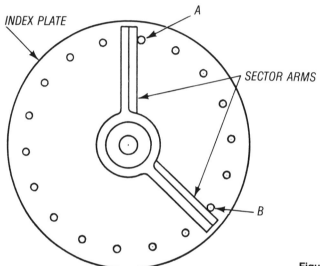

Figure 15-6

Use Figure 15-6 for a typical example of fractional turning. In Figure 15-6, the index plate contains an 18-hole circle. The two sector arms can be adjusted so that they will include any desired angle between them. The length of the crank can also be adjusted to enter the holes on the 18-hole circle.

Suppose 12 teeth are to be cut on a gear blank. The crank must rotate $3\frac{1}{3}$ turns for each tooth. This is determined by the formula as follows:

$$T = \frac{40}{12} = 3\frac{1}{3}$$

Three turns of the crank are easily made, but $\frac{1}{3}$ of a turn requires the use of an index plate and sector arms.

The 18-hole circle shown in Figure 15-6 can be used. If the pin is in hole *A*, it is removed from this hole and the crank is rotated clockwise through *3 turns*. The sector arms remain in the same position and, at the end of the third turn, the pin is at hole *A*. In addition, the crank must be turned through $\frac{1}{3}$ of 18 or 6 spaces. The pin is allowed to fall into hole *B*, since there are six spaces between holes *A* and *B*. From *A* to *B* there are *7 holes but only 6 spaces*. This is an important point to remember. *Spaces determine all indexing divisions; holes are simply devices to separate spaces and to retain the pin.*

It was not necessary to select a circle with 18 holes to index through $3\frac{1}{3}$ turns. Any circle with a number of holes evenly divisible by 3 would serve the purpose. Some indexing attachments have three index plates with circles of holes as follows:

No. 1   15   16   17   18   19   20
No. 2   21   23   27   29   31   33
No. 3   37   39   41   43   47   49

Other indexing attachments can be obtained with one plate drilled on both sides with the following number of holes in each circle:

Front   24   25   28   30   34   37   38   39   41   42   43
Rear   46   47   49   51   53   54   57   58   59   62   66

Example 15-13   Index to cut a gear with 26 teeth.

Solution $T = \dfrac{40}{N} = \dfrac{40}{26} = \dfrac{20}{13} = 1\dfrac{7}{13}$

Select a circle of holes so that 13, the denominator of the fractional part of a turn, will divide into the number of holes in the circle. In this case, select the 39-hole circle. Then rotate the crank *1 turn plus $\frac{7}{13}$ of 39 or 21 spaces on the 39-hole circle for each tooth of the gear.*

Example 15-14   Index for 57 equal divisions.

Solution $T = \dfrac{40}{N} = \dfrac{40}{57}$

Select the 57-hole circle and rotate the crank through *40 spaces* for each division.

Exercises 15-5

List the circle, number of turns, and spaces for the following divisions on a workpiece.

1. 32	**2.** 16	**3.** 60	**4.** 20
5. 40	**6.** 7	**7.** 11	**8.** 27
9. 25	**10.** 114	**11.** 102	**12.** 19
13. 34	**14.** 35	**15.** 75	**16.** 13

15-6 Angular Indexing

OBJECTIVE

After completing this section, you should be able to:

1. Index for degrees, minutes, and seconds.

The dividing head is a versatile machine attachment. One of the unique tasks it is capable of performing is angular indexing. There are many times in a toolroom when accurate spacing of slots or holes in terms of degrees and minutes is called for on blueprints. Examples of such work would be detents for locking devices or cams for interrupted motion.

There are times when a shop does not possess a set of "high-number" indexing plates, which would ordinarily be used for divisions beyond the range of standard plates. Even these plates will not index certain numbers, such as 225, 275, 325, or 375. Often, angular indexing will take the place of these.

Each complete turn of the crank of a dividing head revolves the spindle and workpiece through $\frac{1}{40}$ of 360°, or 9°. Therefore, two holes on the 18-hole circle will equal 1°. Likewise, if the 27-hole circle is used, each three holes equal 1°.

Example 15-15 Use the 18-hole or 27-hole circle for angles having the following fractional parts of a degree: (a) $\frac{1}{2}$, (b) $\frac{1}{3}$, (c) $\frac{2}{3}$.

Solution (a) $\dfrac{5\frac{1}{2}°}{9} = \dfrac{5\frac{1}{2} \times 2}{9 \times 2} = \dfrac{11 \text{ holes}}{18\text{-hole circle}}$

(b) $\dfrac{12\frac{1}{3}°}{9} = \dfrac{12\frac{1}{3} \times 3}{9 \times 3} = \dfrac{37}{27} = 1 \text{ and } \dfrac{10 \text{ holes}}{27\text{-hole circle}}$

(c) $\dfrac{6\frac{2}{3}°}{9} = \dfrac{6\frac{2}{3} \times 3}{9 \times 3} = \dfrac{20 \text{ holes}}{27\text{-hole circle}}$

The 1 in (b) represents one *full* turn of the crank.

Each hole in any circle is equivalent to a certain number of minutes. When two holes are used in an 18-hole circle to arrive at 1°, each hole is worth 30'.

RULE 15-12 FINDING THE VALUE OF ANY SINGLE-HOLE MOVE IN A GIVEN CIRCLE

Change 9° (one complete turn) into minutes and divide by the total number of holes in the circle selected.

Example 15-16 Find the value of a single-hole move in circles with (a) 20, (b) 30, and (c) 43 holes.

Solution $9° = 9 \times 60 = 540'$

(a) One hole in a 20-hole circle $= \dfrac{540}{20} = 27'$.

(b) One hole in a 30-hole circle $= \dfrac{540}{30} = 18'$.

(c) One hole in a 43-hole circle $= \dfrac{540}{43} = 12.56'$ or $12'34''$.

In (c), the decimal part of a minute was multiplied by 60 to convert to seconds of a degree.

Any angle expressed in minutes can be converted to an indexing on a hole circle using continued fractions. The indexing error, in most cases, will be less than plus or minus $\frac{1}{2}°$, which is standard tolerance for angular measurement.

Example 15-17   Convert $6°24'$ to an indexing on a hole circle.

Solution   The first step is to convert degrees and minutes to minutes. The resulting figure is divided by $540'$.

$$6°24' = 6 \times 60 + 24 = 384'; \quad \frac{384}{540} = 0.7111$$

The result is a fractional part of one turn. Convert the decimal 0.7111 to a common fraction and use continued fractions.

```
7111)10000(1
     7111
     2889)7111(2
          5778
          1333)2889(2
               2666
                233)1333(5
                    1115
                     218)223(1
                         218
                           5)218(43
                             215
                               3)5(1
                                 3
                                 2)3(1
                                   2
                                   1)2(2
                                     2
                                     0
```

Partial quotients:

1   2   2   5   1   43   1   1   2

Fractions:

$$\frac{1}{1} \quad \frac{2}{3} \quad \frac{5}{7} \quad \frac{27}{38} \quad \frac{32}{45} \quad \frac{1403}{1973}$$

There is no need to find equivalent fractions past the partial quotient of 43, since the resulting numbers are too high for any hole plate.

Decimal equivalents:

$$\frac{2}{3} = 0.6667 \quad \frac{5}{7} = 0.7143 \quad \frac{27}{38} = 0.7105 \quad \frac{32}{45} = 0.7111$$

The combination of 32 holes in a 45-hole circle gives the desired indexing of 6°24'.

Example 15-18 Find an indexing on a hole circle for 12°18'.

Solution $12°18' = 12 \times 60 + 18 = 738'$; $\dfrac{738}{540} = 1.3666$

One complete turn and a fractional part of a turn are indicated. Do not include the 1 in 1.3666, since this is a whole turn.

```
3666)10000(2
     7332
     2668)3666(1
          2668
          998)2668(2
              1996
              672)998(1
                  672
                  326)672(2
                      652
                      20)326(16
                         320
                         6)20(3
                           18
                           2)6(3
                             6
                             0
```

Partial quotients:

2   1   2   1   2   16   3   3

Fractions:

$$\frac{1}{2} \quad \frac{1}{3} \quad \frac{3}{8} \quad \frac{4}{11} \quad \frac{11}{30} \quad \frac{180}{491} \quad \frac{551}{1503} \quad \frac{1833}{5000}$$

Decimal equivalents:

$$\frac{1}{2} = 0.5 \quad \frac{1}{3} = 0.3333 \quad \frac{3}{8} = 0.375$$

$$\frac{4}{11} = 0.3636 \quad \frac{11}{30} = 0.3667$$

One turn plus 11 holes in a 30-hole circle gives an indexing of 12°18′.

Indexing for seconds of a degree is figured by continued fractions, using the same steps as with conversion of minutes. The only new step is the conversion of degrees and minutes to *seconds*. The number of seconds in one complete turn of the dividing head crank is 9 × 60 × 60 or 32,400 seconds. This figure becomes the denominator of any fractional turn.

Example 15-19   Determine the indexing for 7°15′30″ by continued fractions.

   Solution   7°15′30″ = (7 × 60 × 60) + (15 × 60) + 30
                 = 26,130″

$$\frac{26,130}{32,400} = 0.8065 = \frac{8065}{10,000}$$

```
8065)10000(1
     8065
     1935)8065(4
          7740
          325)1935(5
              1625
              310)325(1
                  310
                  15)310(20
                     300
                     10)15(1
                        10
                        5)10(2
                          10
                          0
```

Partial quotients:

1 4 5 1 20 1 2

Fractions:

$$\frac{1}{1} \quad \frac{4}{5} \quad \frac{21}{26} \quad \frac{25}{31} \quad \frac{521}{646} \quad \frac{546}{677} \quad \frac{1613}{2000}$$

Decimal equivalents:

$$\frac{4}{5} = 0.8000 \quad \frac{21}{26} = 0.8077 \quad \frac{25}{31} = 0.80645$$

$$\frac{521}{646} = 0.8065 \quad \frac{546}{677} = 0.8065 \quad \frac{1613}{2000} = 0.8065$$

The best indexing for 7°15′30″ is 25 holes in the 31-hole circle. The error is $(0.8065 - 0.80645) \times 32,400 = 1.62''$ in an arc of 7°15′30″.

It was said earlier in this chapter that there are certain divisions that are difficult to obtain, even with the high-division plates. One of these divisions mentioned was 275. It is possible, however, to index for very high divisions by means of the angular approach.

Example 15-20 A set of high-division plates is available for indexing but it does not contain the spacing for 275. Arrive at the hole circle that will provide this spacing.

Solution First, 275 must be divided into 360°. This gives an angle of 1°18′32″. If this angle is converted to seconds of a degree, it gives a total of 4712″.

$$\frac{4712}{32,400} = 0.1454, \text{ or } \frac{1454}{10,000}$$

$$
\begin{array}{l}
1454\overline{)10000}(6 \\
\quad \underline{8724} \\
\quad 1276\overline{)1454}(1 \\
\qquad \underline{1276} \\
\qquad 178\overline{)1276}(7 \\
\qquad\quad \underline{1246} \\
\qquad\quad 30\overline{)178}(5 \\
\qquad\qquad \underline{150} \\
\qquad\qquad 28\overline{)30}(1 \\
\qquad\qquad\quad \underline{28} \\
\qquad\qquad\quad 2\overline{)28}(14 \\
\qquad\qquad\qquad \underline{28} \\
\qquad\qquad\qquad \overline{}0
\end{array}
$$

Partial quotients:

6   1   7   5   1   14

Fractions:

$$\frac{1}{6} \quad \frac{1}{7} \quad \frac{8}{55} \quad \frac{41}{282} \quad \frac{49}{337} \quad \frac{727}{5000}$$

Decimal equivalents:

0.1667   0.1429   0.1455   0.1454   0.1454   0.1454

The combination of 8 holes in the 55-hole circle gives a spacing of 275 holes with an error of 3.24″ in an arc of 1°18′32″.

Exercises 15-6

Fill in the column for spacing and circle of holes to be used. The first line is filled in to help you get started.

	Degrees	Spacing			Degrees	Spacing
1.	5	$\frac{10}{18}$ or $\frac{15}{27}$		**13.**	8	
2.	$5\frac{1}{3}$			**14.**	$8\frac{1}{3}$	
3.	$5\frac{1}{2}$			**15.**	$8\frac{1}{2}$	
4.	$5\frac{2}{3}$			**16.**	$8\frac{2}{3}$	
5.	6			**17.**	9	
6.	$6\frac{1}{3}$			**18.**	$9\frac{1}{3}$	
7.	$6\frac{1}{2}$			**19.**	$9\frac{1}{2}$	
8.	$6\frac{2}{3}$			**20.**	$9\frac{2}{3}$	
9.	7			**21.**	10	
10.	$7\frac{1}{3}$			**22.**	$10\frac{1}{3}$	
11.	$7\frac{1}{2}$			**23.**	$10\frac{1}{2}$	
12.	$7\frac{2}{3}$			**24.**	$10\frac{2}{3}$	

25. Find the value of a single-hole move in circles with 33, 47, and 62 holes.

26. Convert 10°15′ to an indexing on a hole circle.

27. Find the indexing for 8°13'25".

28. The high-division plate you are using does not have the spacing for 325. What other hole circle will have this spacing?

Unit 15 Review Exercises

1. Find the outside diameter of a 16-pitch gear with a pitch diameter of $8\frac{1}{4}$ in.

2. An 8-diametral-pitch gear has an outside diameter of 2.875 in. How many teeth will this gear have?

3. Find the center-to-center distance of a pair of 5-pitch gears that have 45 teeth and 62 teeth, respectively.

4. What is the diametral pitch of a 4-module metric gear?

5. Find the outside diameter of a 10-module metric gear with 18 teeth.

6. Find the circular pitch of two gears that are to be operated on shafts $7\frac{5}{16}$ in. apart. The pinion has 42 teeth and the gear has 75 teeth.

7. A set of gears with 40 teeth and 18 teeth, respectively, are unavailable for use. Find other gear tooth combinations that will yield the same speed ratio.

8. Using the formula for indexing, how many turns of the crank are needed to index 10 equal divisions? 8 divisions? 5 divisions?

9. What would be the indexing on a 40 to 1 ratio indexing head for 28 divisions? What hole circle would you choose? How many turns of the crank will be needed? How many spaces?

10. Calculate the indexing for 1°24"15'.

Reference Tables

REFERENCE TABLE I—Diameter of Number, Letter, and Fractional Drills in the Order of Size

Number, Letter, and Fractional Drills	Diameter of Drills in Inches	Number, Letter, and Fractional Drills	Diameter of Drills in Inches	Number, Letter, and Fractional Drills	Diameter of Drills in Inches	Number, Letter, and Fractional Drills	Diameter of Drills in Inches
80	0.01350	43	0.08900	8	0.19900	$\frac{25}{64}$	0.39062
79	0.01450	42	0.09350	7	0.20100	X	0.39700
$\frac{1}{64}$	0.01562	$\frac{3}{32}$	0.09375	$\frac{13}{64}$	0.20312	Y	0.40400
78	0.01600	41	0.09600	6	0.20400	$\frac{13}{32}$	0.40625
77	0.01800	40	0.09800	5	0.20550	Z	0.41300
76	0.02000	39	0.09950	4	0.20900	$\frac{27}{64}$	0.42187
75	0.02100	38	0.10150	3	0.21300	$\frac{7}{16}$	0.43750
74	0.02250	37	0.10400	$\frac{7}{32}$	0.21875	$\frac{29}{64}$	0.45312
73	0.02400	36	0.10650	2	0.22100	$\frac{15}{32}$	0.46875
72	0.02500	$\frac{7}{64}$	0.10937	1	0.22800	$\frac{31}{64}$	0.48437
71	0.02600	35	0.11000	A	0.23400	$\frac{1}{2}$	0.50000
70	0.02800	34	0.11100	$\frac{15}{64}$	0.23437	$\frac{33}{64}$	0.51562
69	0.02925	33	0.11300	B	0.23800	$\frac{17}{32}$	0.53125
68	0.03100	32	0.11600	C	0.24200	$\frac{35}{64}$	0.54687
$\frac{1}{32}$	0.03125	31	0.12000	D	0.24600	$\frac{9}{16}$	0.56250
67	0.03200	$\frac{1}{8}$	0.12500	E	0.25000	$\frac{37}{64}$	0.57812
66	0.03300	30	0.12850	F	0.25700	$\frac{19}{32}$	0.59375
65	0.03500	29	0.13600	G	0.26100	$\frac{39}{64}$	0.60937
64	0.03600	28	0.14050	$\frac{17}{64}$	0.26562	$\frac{5}{8}$	0.62500
63	0.03700	$\frac{9}{64}$	0.14062	H	0.26600	$\frac{41}{64}$	0.64062

REFERENCE TABLE I—continued

Number, Letter, and Fractional Drills	Diameter of Drills in Inches	Number, Letter, and Fractional Drills	Diameter of Drills in Inches	Number, Letter, and Fractional Drills	Diameter of Drills in Inches	Number, Letter, and Fractional Drills	Diameter of Drills in Inches
62	0.03800	27	0.14400	I	0.27200	$\frac{21}{32}$	0.65625
61	0.03900	26	0.14700	J	0.27700	$\frac{43}{64}$	0.67187
60	0.04000	25	0.14950	K	0.28100	$\frac{11}{16}$	0.68750
59	0.04100	24	0.15200	$\frac{9}{32}$	0.28125	$\frac{45}{64}$	0.70312
58	0.04200	23	0.15400	L	0.29000	$\frac{23}{32}$	0.71875
57	0.04300	$\frac{5}{32}$	0.15625	M	0.29500	$\frac{47}{64}$	0.73437
56	0.04650	22	0.15700	$\frac{19}{64}$	0.29687	$\frac{3}{4}$	0.75000
$\frac{3}{64}$	0.04687	21	0.15900	N	0.30200	$\frac{49}{64}$	0.76562
55	0.05200	20	0.16100	$\frac{5}{16}$	0.31250	$\frac{25}{32}$	0.78125
54	0.05500	19	0.16600	O	0.31600	$\frac{51}{64}$	0.79687
53	0.05950	18	0.16950	P	0.32300	$\frac{13}{16}$	0.81250
$\frac{1}{16}$	0.06250	$\frac{11}{64}$	0.17187	$\frac{21}{64}$	0.32812	$\frac{53}{64}$	0.82812
52	0.06350	17	0.17300	Q	0.33200	$\frac{27}{32}$	0.84375
51	0.06700	16	0.17700	R	0.33900	$\frac{55}{64}$	0.85937
50	0.07000	15	0.18000	$\frac{11}{32}$	0.34375	$\frac{7}{8}$	0.87500
49	0.07300	14	0.18200	S	0.34800	$\frac{57}{64}$	0.89062
48	0.07600	13	0.18500	T	0.35800	$\frac{29}{32}$	0.90625
$\frac{5}{64}$	0.07812	$\frac{3}{16}$	0.18750	$\frac{23}{64}$	0.35937	$\frac{59}{64}$	0.92187
47	0.07850	12	0.18900	U	0.36800	$\frac{15}{16}$	0.93750
46	0.08100	11	0.19100	$\frac{3}{8}$	0.37500	$\frac{61}{64}$	0.95312
45	0.08200	10	0.19350	V	0.37700	$\frac{31}{32}$	0.96875
44	0.08600	9	0.19600	W	0.38600	$\frac{63}{64}$	0.98437

REFERENCE TABLE II—Metric ISO* Fine Threads—German DIN**

FINE THREADS COMMONLY USED ON FASTENERS				OTHER FINE THREADS AND SPECIAL THREADS			
Dia. Maximum mm	Pitch mm	Approx. Threads per inch	Recommended Drills mm	Dia. Maximum mm	Pitch mm	Approx. Threads per inch	Recommended Drills mm
M 1	0.20	126	0.80	M 1.5	0.30	85	1.20
M 1.2	0.20	126	1.00	M 1.6	0.35	73	1.45
M 1.4	0.20	126	1.20	M 2.2	0.45	56	1.75
M 1.7	0.20	126	1.50	M 2.5	0.45	56	2.05
M 2	0.25	102	1.75	M 3	0.60	42	2.40
M 2.3	0.25	102	2.05	M 4	0.75	34	3.25
M 2.6	0.35	73	2.25	M 5	0.75	34	4.25
M 3	0.35	73	2.65	M 5	0.90	29	4.10
M 3.5	0.35	73	3.15	M 7	0.50	51	6.50
M 4	0.35	73	3.65	M 9	0.75	34	8.25
M 4	0.50	51	3.50	M 10	0.75	34	9.25
M 4.5	0.50	51	4.00	M 10	1.25	20	8.75
M 5	0.50	51	4.50	M 11	0.75	34	10.25
M 6	0.50	51	5.50	M 11	1.25	20	9.75
M 6	0.75	34	5.25	M 12	1.25	20	10.75
M 7	0.75	34	6.25	M 12	1.00	25	11.00
M 8	0.75	34	7.25	M 14	1.25	20	12.75
M 8	1.00	25	7.00	M 14	1.00	25	13.00
M 9	1.00	25	8.00	M 16	1.25	20	14.75
M 10	1.00	25	9.00	M 16	1.00	25	15.00
M 11	1.00	25	10.00	M 18	1.00	20	17.00
M 12	1.50	17	10.50	M 20	1.00	25	19.00
M 14	1.50	17	12.50	M 20	1.25	20	18.75
M 16	1.50	17	14.50	M 22	1.00	25	21.00
M 18	1.50	17	16.50	M 25	1.00	25	24.00

REFERENCE TABLE II—*continued*

FINE THREADS COMMONLY USED ON FASTENERS				OTHER FINE THREADS AND SPECIAL THREADS			
Dia. Maximum mm	Pitch mm	Approx. Threads per inch	Recom- mended Drills mm	Dia. Maximum mm	Pitch mm	Approx. Threads per inch	Recom- mended Drills mm
M 20	1.50	17	18.50	M 25	1.50	17	23.50
M 20	2.00	$12\frac{1}{2}$	18.00	M 25	2.00	$12\frac{1}{2}$	23.00
M 22	1.50	17	20.50	M 26	1.50	17	24.50
M 22	2.00	$12\frac{1}{2}$	20.00	M 27	1.00	25	26.00
M 24	1.50	17	22.50	M 28	1.00	25	27.00
M 24	2.00	$12\frac{1}{2}$	22.00	M 28	1.50	17	26.50
M 27	1.50	17	25.50	M 28	2.00	$12\frac{1}{2}$	26.00
M 27	2.00	$12\frac{1}{2}$	25.00	M 30	1.00	25	29.00
M 30	1.50	17	28.50				
M 30	2.00	$12\frac{1}{2}$	28.00				

* International Standards Organization
** Deutsche Ingenieur Norm

REFERENCE TABLE III—ISO-DIN Metric Threads

Tapping and Drilling Information		ISO Standard Thread (Germ. Din 13)	
Recommended Drills mm	Approx. Threads per inch	Pitch mm	Dia. Maximum mm
0.75	102	0.25	M 1
0.95	102	0.25	M 1.2
1.10	85	0.30	M 1.4
1.30	73	0.35	M 1.7
1.60	63	0.40	M 2
1.90	63	0.40	M 2.3
2.15	56	0.45	M 2.6
2.50	51	0.50	M 3
2.90	42	0.60	M 3.5
3.30	36	0.70	M 4
4.20	32	0.80	M 5
5.00	25	1.00	M 6
6.00	25	1.00	M 7
6.75	20	1.25	M 8
8.50	17	1.50	M 10
10.25	$14\frac{1}{2}$	1.75	M 12
12.00	$12\frac{1}{2}$	2.00	M 14
14.00	$12\frac{1}{2}$	2.00	M 16
15.50	10	2.50	M 18
17.50	10	2.50	M 20
19.50	10	2.50	M 22
21.00	$8\frac{1}{2}$	3.00	M 24
24.00	$8\frac{1}{2}$	3.00	M 27
26.50	$7\frac{1}{3}$	3.50	M 30

Slight variation of drill sizes depending on type of material.

REFERENCE TABLE IV—Trigonometric Functions

Sine-Cosine

′	0° Sine	0° Cosine	1° Sine	1° Cosine	2° Sine	2° Cosine	3° Sine	3° Cosine	4° Sine	4° Cosine	′
0	.00000	1.	.01745	.99985	.03490	.99939	.05234	.99863	.06976	.99756	60
1	.00029	1.	.01774	.99984	.03519	.99938	.05263	.99861	.07005	.99754	59
2	.00058	1.	.01803	.99984	.03548	.99937	.05292	.99860	.07034	.99752	58
3	.00087	1.	.01832	.99983	.03577	.99936	.05321	.99858	.07063	.99750	57
4	.00116	1.	.01862	.99983	.03606	.99935	.05350	.99857	.07092	.99748	56
5	.00145	1.	.01891	.99982	.03635	.99934	.05379	.99855	.07121	.99746	55
6	.00175	1.	.01920	.99982	.03664	.99933	.05408	.99854	.07150	.99744	54
7	.00204	1.	.01949	.99981	.03693	.99932	.05437	.99852	.07179	.99742	53
8	.00233	1.	.01978	.99980	.03723	.99931	.05466	.99851	.07208	.99740	52
9	.00262	1.	.02007	.99980	.03752	.99930	.05495	.99849	.07237	.99738	51
10	.00291	1.	.02036	.99979	.03781	.99929	.05524	.99847	.07266	.99736	50
11	.00320	.99999	.02065	.99979	.03810	.99927	.05553	.99846	.07295	.99734	49
12	.00349	.99999	.02094	.99978	.03839	.99926	.05582	.99844	.07324	.99731	48
13	.00378	.99999	.02123	.99977	.03868	.99925	.05611	.99842	.07353	.99729	47
14	.00407	.99999	.02152	.99977	.03897	.99924	.05640	.99841	.07382	.99727	46
15	.00436	.99999	.02181	.99976	.03926	.99923	.05669	.99839	.07411	.99725	45
16	.00465	.99999	.02211	.99976	.03955	.99922	.05698	.99838	.07440	.99723	44
17	.00495	.99999	.02240	.99975	.03984	.99921	.05727	.99836	.07469	.99721	43
18	.00524	.99999	.02269	.99974	.04013	.99919	.05756	.99834	.07498	.99719	42
19	.00553	.99998	.02298	.99974	.04042	.99918	.05785	.99833	.07527	.99716	41
20	.00582	.99998	.02327	.99973	.04071	.99917	.05814	.99831	.07556	.99714	40
21	.00611	.99998	.02356	.99972	.04100	.99916	.05844	.99829	.07585	.99712	39
22	.00640	.99998	.02385	.99972	.04129	.99915	.05873	.99827	.07614	.99710	38
23	.00669	.99998	.02414	.99971	.04159	.99913	.05902	.99826	.07643	.99708	37
24	.00698	.99998	.02443	.99970	.04188	.99912	.05931	.99824	.07672	.99705	36
25	.00727	.99997	.02472	.99969	.04217	.99911	.05960	.99822	.07701	.99703	35
26	.00756	.99997	.02501	.99969	.04246	.99910	.05989	.99821	.07730	.99701	34
27	.00785	.99997	.02530	.99968	.04275	.99909	.06018	.99819	.07759	.99699	33
28	.00814	.99997	.02560	.99967	.04304	.99907	.06047	.99817	.07788	.99696	32
29	.00844	.99996	.02589	.99966	.04333	.99906	.06076	.99815	.07817	.99694	31
30	.00873	.99996	.02618	.99966	.04362	.99905	.06105	.99813	.07846	.99692	30
31	.00902	.99996	.02647	.99965	.04391	.99904	.06134	.99812	.07875	.99689	29
32	.00931	.99996	.02676	.99964	.04420	.99902	.06163	.99810	.07904	.99687	28
33	.00960	.99995	.02705	.99963	.04449	.99901	.06192	.99808	.07933	.99685	27
34	.00989	.99995	.02734	.99963	.04478	.99900	.06221	.99806	.07962	.99683	26
35	.01018	.99995	.02763	.99962	.04507	.99898	.06250	.99804	.07991	.99680	25
36	.01047	.99995	.02792	.99961	.04536	.99897	.06279	.99803	.08020	.99678	24
37	.01076	.99994	.02821	.99960	.04565	.99896	.06308	.99801	.08049	.99676	23
38	.01105	.99994	.02850	.99959	.04594	.99894	.06337	.99799	.08078	.99673	22
39	.01134	.99994	.02879	.99959	.04623	.99893	.06366	.99797	.08107	.99671	21
40	.01164	.99993	.02908	.99958	.04653	.99892	.06395	.99795	.08136	.99668	20
41	.01193	.99993	.02938	.99957	.04682	.99890	.06424	.99793	.08165	.99666	19
42	.01222	.99993	.02967	.99956	.04711	.99889	.06453	.99792	.08194	.99664	18
43	.01251	.99992	.02996	.99955	.04740	.99888	.06482	.99790	.08223	.99661	17
44	.01280	.99992	.03025	.99954	.04769	.99886	.06511	.99788	.08252	.99659	16
45	.01309	.99991	.03054	.99953	.04798	.99885	.06540	.99786	.08281	.99657	15
46	.01338	.99991	.03083	.99952	.04827	.99883	.06569	.99784	.08310	.99654	14
47	.01367	.99991	.03112	.99952	.04856	.99882	.06598	.99782	.08339	.99652	13
48	.01396	.99990	.03141	.99951	.04885	.99881	.06627	.99780	.08368	.99649	12
49	.01425	.99990	.03170	.99950	.04914	.99879	.06656	.99778	.08397	.99647	11
50	.01454	.99989	.03199	.99949	.04943	.99878	.06685	.99776	.08426	.99644	10
51	.01483	.99989	.03228	.99948	.04972	.99876	.06714	.99774	.08455	.99642	9
52	.01513	.99989	.03257	.99947	.05001	.99875	.06743	.99772	.08484	.99639	8
53	.01542	.99988	.03286	.99946	.05030	.99873	.06773	.99770	.08513	.99637	7
54	.01571	.99988	.03316	.99945	.05059	.99872	.06802	.99768	.08542	.99635	6
55	.01600	.99987	.03345	.99944	.05088	.99870	.06831	.99766	.08571	.99632	5
56	.01629	.99987	.03374	.99942	.05117	.99869	.06860	.99764	.08600	.99630	4
57	.01658	.99986	.03403	.99942	.05146	.99867	.06889	.99762	.08629	.99627	3
58	.01687	.99986	.03432	.99941	.05175	.99866	.06918	.99760	.08658	.99625	2
59	.01716	.99985	.03461	.99940	.05205	.99864	.06947	.99758	.08687	.99622	1
60	.01745	.99985	.03490	.99939	.05234	.99863	.06976	.99756	.08716	.99619	0
′	Cosine	Sine	Cosine	Sine	Cosine	Sine	Cosine	Sine	Cosine	Sine	′
	89°		88°		87°		86°		85°		

	5°		6°		7°		8°		9°		
	Sine	Cosine	Sine	Cosine	Sine	Cosine	Sine	Cosine	Sine	Cosine	
0	.08716	.99619	.10453	.99452	.12187	.99255	.13917	.99027	.15643	.98769	60
1	.08745	.99617	.10482	.99449	.12216	.99251	.13946	.99023	.15672	.98764	59
2	.08774	.99614	.10511	.99446	.12245	.99248	.13975	.99019	.15701	.98760	58
3	.08803	.99612	.10540	.99443	.12274	.99244	.14004	.99015	.15730	.98755	57
4	.08831	.99609	.10569	.99440	.12302	.99240	.14033	.99011	.15758	.98751	56
5	.08860	.99607	.10597	.99437	.12331	.99237	.14061	.99006	.15787	.98746	55
6	.08889	.99604	.10626	.99434	.12360	.99233	.14090	.99002	.15816	.98741	54
7	.08918	.99602	.10655	.99431	.12389	.99230	.14119	.98998	.15845	.98737	53
8	.08947	.99599	.10684	.99428	.12418	.99226	.14148	.98994	.15873	.98732	52
9	.08976	.99596	.10713	.99424	.12447	.99222	.14177	.98990	.15902	.98728	51
10	.09005	.99594	.10742	.99421	.12476	.99219	.14205	.98986	.15931	.98723	50
11	.09034	.99591	.10771	.99418	.12504	.99215	.14234	.98982	.15959	.98718	49
12	.09063	.99588	.10800	.99415	.12533	.99211	.14263	.98978	.15988	.98714	48
13	.09092	.99586	.10829	.99412	.12562	.99208	.14292	.98973	.16017	.98709	47
14	.09121	.99583	.10858	.99409	.12591	.99204	.14320	.98969	.16046	.98704	46
15	.09150	.99580	.10887	.99406	.12620	.99200	.14349	.98965	.16074	.98700	45
16	.09179	.99578	.10916	.99402	.12649	.99197	.14378	.98961	.16103	.98695	44
17	.09208	.99575	.10945	.99399	.12678	.99193	.14407	.98957	.16132	.98690	43
18	.09237	.99572	.10973	.99396	.12706	.99189	.14436	.98953	.16160	.98686	42
19	.09266	.99570	.11002	.99393	.12735	.99186	.14464	.98948	.16189	.98681	41
20	.09295	.99567	.11031	.99390	.12764	.99182	.14493	.98944	.16218	.98676	40
21	.09324	.99564	.11060	.99386	.12793	.99178	.14522	.98940	.16246	.98671	39
22	.09353	.99562	.11089	.99383	.12822	.99175	.14551	.98936	.16275	.98667	38
23	.09382	.99559	.11118	.99380	.12851	.99171	.14580	.98931	.16304	.98662	37
24	.09411	.99556	.11147	.99377	.12880	.99167	.14608	.98927	.16333	.98657	36
25	.09440	.99553	.11176	.99374	.12908	.99163	.14637	.98923	.16361	.98652	35
26	.09469	.99551	.11205	.99370	.12937	.99160	.14666	.98919	.16390	.98648	34
27	.09498	.99548	.11234	.99367	.12966	.99156	.14695	.98914	.16419	.98643	33
28	.09527	.99545	.11263	.99364	.12995	.99152	.14723	.98910	.16447	.98638	32
29	.09556	.99542	.11291	.99360	.13024	.99148	.14752	.98906	.16476	.98633	31
30	.09585	.99540	.11320	.99357	.13053	.99144	.14781	.98902	.16505	.98629	30
31	.09614	.99537	.11349	.99354	.13081	.99141	.14810	.98897	.16533	.98624	29
32	.09642	.99534	.11378	.99351	.13110	.99137	.14838	.98893	.16562	.98619	28
33	.09671	.99531	.11407	.99347	.13139	.99133	.14867	.98889	.16591	.98614	27
34	.09700	.99528	.11436	.99344	.13168	.99129	.14896	.98884	.16620	.98609	26
35	.09729	.99526	.11465	.99341	.13197	.99125	.14925	.98880	.16648	.98604	25
36	.09758	.99523	.11494	.99337	.13226	.99122	.14954	.98876	.16677	.98600	24
37	.09787	.99520	.11523	.99334	.13254	.99118	.14982	.98871	.16706	.98595	23
38	.09816	.99517	.11552	.99331	.13283	.99114	.15011	.98867	.16734	.98590	22
39	.09845	.99514	.11580	.99327	.13312	.99110	.15040	.98863	.16763	.98585	21
40	.09874	.99511	.11609	.99324	.13341	.99106	.15069	.98858	.16792	.98580	20
41	.09903	.99508	.11638	.99320	.13370	.99102	.15097	.98854	.16820	.98575	19
42	.09932	.99506	.11667	.99317	.13399	.99098	.15126	.98849	.16849	.98570	18
43	.09961	.99503	.11696	.99314	.13427	.99094	.15155	.98845	.16878	.98565	17
44	.09990	.99500	.11725	.99310	.13456	.99091	.15184	.98841	.16906	.98561	16
45	.10019	.99497	.11754	.99307	.13485	.99087	.15212	.98836	.16935	.98556	15
46	.10048	.99494	.11783	.99303	.13514	.99083	.15241	.98832	.16964	.98551	14
47	.10077	.99491	.11812	.99300	.13543	.99079	.15270	.98827	.16992	.98546	13
48	.10106	.99488	.11840	.99297	.13572	.99075	.15299	.98823	.17021	.98541	12
49	.10135	.99485	.11869	.99293	.13600	.99071	.15327	.98818	.17050	.98536	11
50	.10164	.99482	.11898	.99290	.13629	.99067	.15356	.98814	.17078	.98531	10
51	.10192	.99479	.11927	.99286	.13658	.99063	.15385	.98809	.17107	.98526	9
52	.10221	.99476	.11956	.99283	.13687	.99059	.15414	.98805	.17136	.98521	8
53	.10250	.99473	.11985	.99279	.13716	.99055	.15442	.98800	.17164	.98516	7
54	.10279	.99470	.12014	.99276	.13744	.99051	.15471	.98796	.17193	.98511	6
55	.10308	.99467	.12043	.99272	.13773	.99047	.15500	.98791	.17222	.98506	5
56	.10337	.99464	.12071	.99269	.13802	.99043	.15529	.98787	.17250	.98501	4
57	.10366	.99461	.12100	.99265	.13831	.99039	.15557	.98782	.17279	.98496	3
58	.10395	.99458	.12129	.99262	.13860	.99035	.15586	.98778	.17308	.98491	2
59	.10424	.99455	.12158	.99258	.13889	.99031	.15615	.98773	.17336	.98486	1
60	.10453	.99452	.12187	.99255	.13917	.99027	.15643	.98769	.17365	.98481	0
	Cosine	Sine	Cosine	Sine	Cosine	Sine	Cosine	Sine	Cosine	Sine	
	84°		83°		82°		81°		80°		

′	10° Sine	Cosine	11° Sine	Cosine	12° Sine	Cosine	13° Sine	Cosine	14° Sine	Cosine	′
0	.17365	.98481	.19081	.98163	.20791	.97815	.22495	.97437	.24192	.97030	60
1	.17393	.98476	.19109	.98157	.20820	.97809	.22523	.97430	.24220	.97023	59
2	.17422	.98471	.19138	.98152	.20848	.97803	.22552	.97424	.24249	.97015	58
3	.17451	.98466	.19167	.98146	.20877	.97797	.22580	.97417	.24277	.97008	57
4	.17479	.98461	.19195	.98140	.20905	.97791	.22608	.97411	.24305	.97001	56
5	.17508	.98455	.19224	.98135	.20933	.97784	.22637	.97404	.24333	.96994	55
6	.17537	.98450	.19252	.98129	.20962	.97778	.22665	.97398	.24362	.96987	54
7	.17565	.98445	.19281	.98124	.20990	.97772	.22693	.97391	.24390	.96980	53
8	.17594	.98440	.19309	.98118	.21019	.97766	.22722	.97384	.24418	.96973	52
9	.17623	.98435	.19338	.98112	.21047	.97760	.22750	.97378	.24446	.96966	51
10	.17651	.98430	.19366	.98107	.21076	.97754	.22778	.97371	.24474	.96959	50
11	.17680	.98425	.19395	.98101	.21104	.97748	.22807	.97365	.24503	.96952	49
12	.17708	.98420	.19423	.98096	.21132	.97742	.22835	.97358	.24531	.96945	48
13	.17737	.98414	.19452	.98090	.21161	.97735	.22863	.97351	.24559	.96937	47
14	.17766	.98409	.19481	.98084	.21189	.97729	.22892	.97345	.24587	.96930	46
15	.17794	.98404	.19509	.98079	.21218	.97723	.22920	.97338	.24615	.96923	45
16	.17823	.98399	.19538	.98073	.21246	.97717	.22948	.97331	.24644	.96916	44
17	.17852	.98394	.19566	.98067	.21275	.97711	.22977	.97325	.24672	.96909	43
18	.17880	.98389	.19595	.98061	.21303	.97705	.23005	.97318	.24700	.96902	42
19	.17909	.98383	.19623	.98056	.21331	.97698	.23033	.97311	.24728	.96894	41
20	.17937	.98378	.19652	.98050	.21360	.97692	.23062	.97304	.24756	.96887	40
21	.17966	.98373	.19680	.98044	.21388	.97686	.23090	.97298	.24784	.96880	39
22	.17995	.98368	.19709	.98039	.21417	.97680	.23118	.97291	.24813	.96873	38
23	.18023	.98362	.19737	.98033	.21445	.97673	.23146	.97284	.24841	.96866	37
24	.18052	.98357	.19766	.98027	.21474	.97667	.23175	.97278	.24869	.96858	36
25	.18081	.98352	.19794	.98021	.21502	.97661	.23203	.97271	.24897	.96851	35
26	.18109	.98347	.19823	.98016	.21530	.97655	.23231	.97264	.24925	.96844	34
27	.18138	.98341	.19851	.98010	.21559	.97648	.23260	.97257	.24954	.96837	33
28	.18166	.98336	.19880	.98004	.21587	.97642	.23288	.97251	.24982	.96829	32
29	.18195	.98331	.19908	.97998	.21616	.97636	.23316	.97244	.25010	.96822	31
30	.18224	.98325	.19937	.97992	.21644	.97630	.23345	.97237	.25038	.96815	30
31	.18252	.98320	.19965	.97987	.21672	.97623	.23373	.97230	.25066	.96807	29
32	.18281	.98315	.19994	.97981	.21701	.97617	.23401	.97223	.25094	.96800	28
33	.18309	.98310	.20022	.97975	.21729	.97611	.23429	.97217	.25122	.96793	27
34	.18338	.98304	.20051	.97969	.21758	.97604	.23458	.97210	.25151	.96786	26
35	.18367	.98299	.20079	.97963	.21786	.97598	.23486	.97203	.25179	.96778	25
36	.18395	.98294	.20108	.97958	.21814	.97592	.23514	.97196	.25207	.96771	24
37	.18424	.98288	.20136	.97952	.21843	.97585	.23542	.97189	.25235	.96764	23
38	.18452	.98283	.20165	.97946	.21871	.97579	.23571	.97182	.25263	.96756	22
39	.18481	.98277	.20193	.97940	.21899	.97573	.23599	.97176	.25291	.96749	21
40	.18509	.98272	.20222	.97934	.21928	.97566	.23627	.97169	.25320	.96742	20
41	.18538	.98267	.20250	.97928	.21956	.97560	.23656	.97162	.25348	.96734	19
42	.18567	.98261	.20279	.97922	.21985	.97553	.23684	.97155	.25376	.96727	18
43	.18595	.98256	.20307	.97916	.22013	.97547	.23712	.97148	.25404	.96719	17
44	.18624	.98250	.20336	.97910	.22041	.97541	.23740	.97141	.25432	.96712	16
45	.18652	.98245	.20364	.97905	.22070	.97534	.23769	.97134	.25460	.96705	15
46	.18681	.98240	.20393	.97899	.22098	.97528	.23797	.97127	.25488	.96697	14
47	.18710	.98234	.20421	.97893	.22126	.97521	.23825	.97120	.25516	.96690	13
48	.18738	.98229	.20450	.97887	.22155	.97515	.23853	.97113	.25545	.96682	12
49	.18767	.98223	.20478	.97881	.22183	.97508	.23882	.97106	.25573	.96675	11
50	.18795	.98218	.20507	.97875	.22212	.97502	.23910	.97100	.25601	.96667	10
51	.18824	.98212	.20535	.97869	.22240	.97496	.23938	.97093	.25629	.96660	9
52	.18852	.98207	.20563	.97863	.22268	.97489	.23966	.97086	.25657	.96653	8
53	.18881	.98201	.20592	.97857	.22297	.97483	.23995	.97079	.25685	.96645	7
54	.18910	.98196	.20620	.97851	.22325	.97476	.24023	.97072	.25713	.96638	6
55	.18938	.98190	.20649	.97845	.22353	.97470	.24051	.97065	.25741	.96630	5
56	.18967	.98185	.20677	.97839	.22382	.97463	.24079	.97058	.25769	.96623	4
57	.18995	.98179	.20706	.97833	.22410	.97457	.24108	.97051	.25798	.96615	3
58	.19024	.98174	.20734	.97827	.22438	.97450	.24136	.97044	.25826	.96608	2
59	.19052	.98168	.20763	.97821	.22467	.97444	.24164	.97037	.25854	.96600	1
60	.19081	.98163	.20791	.97815	.22495	.97437	.24192	.97030	.25882	.96593	0
′	Cosine	Sine	Cosine	Sine	Cosine	Sine	Cosine	Sine	Cosine	Sine	′
	79°		78°		77°		76°		75°		

′	15° Sine	15° Cosine	16° Sine	16° Cosine	17° Sine	17° Cosine	18° Sine	18° Cosine	19° Sine	19° Cosine	′
0	.25882	.96593	.27564	.96126	.29237	.95630	.30902	.95106	.32557	.94552	60
1	.25910	.96585	.27592	.96118	.29265	.95622	.30929	.95097	.32584	.94542	59
2	.25938	.96578	.27620	.96110	.29293	.95613	.30957	.95088	.32612	.94533	58
3	.25966	.96570	.27648	.96102	.29321	.95605	.30985	.95079	.32639	.94523	57
4	.25994	.96562	.27676	.96094	.29348	.95596	.31012	.95070	.32667	.94514	56
5	.26022	.96555	.27704	.96086	.29376	.95588	.31040	.95061	.32694	.94504	55
6	.26050	.96547	.27731	.96078	.29404	.95579	.31068	.95052	.32722	.94495	54
7	.26079	.96540	.27759	.96070	.29432	.95571	.31095	.95043	.32749	.94485	53
8	.26107	.96532	.27787	.96062	.29460	.95562	.31123	.95033	.32777	.94476	52
9	.26135	.96524	.27815	.96054	.29487	.95554	.31151	.95024	.32804	.94466	51
10	.26163	.96517	.27843	.96046	.29515	.95545	.31178	.95015	.32832	.94457	50
11	.26191	.96509	.27871	.96037	.29543	.95536	.31206	.95006	.32859	.94447	49
12	.26219	.96502	.27899	.96029	.29571	.95528	.31233	.94997	.32887	.94438	48
13	.26247	.96494	.27927	.96021	.29599	.95519	.31261	.94988	.32914	.94428	47
14	.26275	.96486	.27955	.96013	.29626	.95511	.31289	.94979	.32942	.94418	46
15	.26303	.96479	.27983	.96005	.29654	.95502	.31316	.94970	.32969	.94409	45
16	.26331	.96471	.28011	.95997	.29682	.95493	.31344	.94961	.32997	.94399	44
17	.26359	.96463	.28039	.95989	.29710	.95485	.31372	.94952	.33024	.94390	43
18	.26387	.96456	.28067	.95981	.29737	.95476	.31399	.94943	.33051	.94380	42
19	.26415	.96448	.28095	.95972	.29765	.95467	.31427	.94933	.33079	.94370	41
20	.26443	.96440	.28123	.95964	.29793	.95459	.31454	.94924	.33106	.94361	40
21	.26471	.96433	.28150	.95956	.29821	.95450	.31482	.94915	.33134	.94351	39
22	.26500	.96425	.28178	.95948	.29849	.95441	.31510	.94906	.33161	.94342	38
23	.26528	.96417	.28206	.95940	.29876	.95433	.31537	.94897	.33189	.94332	37
24	.26556	.96410	.28234	.95931	.29904	.95424	.31565	.94888	.33216	.94322	36
25	.26584	.96402	.28262	.95923	.29932	.95415	.31593	.94878	.33244	.94313	35
26	.26612	.96394	.28290	.95915	.29960	.95407	.31620	.94869	.33271	.94303	34
27	.26640	.96386	.28318	.95907	.29987	.95398	.31648	.94860	.33298	.94293	33
28	.26668	.96379	.28346	.95898	.30015	.95389	.31675	.94851	.33326	.94284	32
29	.26696	.96371	.28374	.95890	.30043	.95380	.31703	.94842	.33353	.94274	31
30	.26724	.96363	.28402	.95882	.30071	.95372	.31730	.94832	.33381	.94264	30
31	.26752	.96355	.28429	.95874	.30098	.95363	.31758	.94823	.33408	94254	29
32	.26780	.96347	.28457	.95865	.30125	.95354	.31786	.94814	.33436	.94245	28
33	.26808	.96340	.28485	.95857	.30154	.95345	.31813	.94805	.33463	.94235	27
34	.26836	.96332	.28513	.95849	.30182	.95337	.31841	.94795	.33490	.94225	26
35	.26864	.96324	.28541	.95841	.30209	.95328	.31868	.94786	.33518	.94215	25
36	.26892	.96316	.28569	.95832	.30237	.95319	.31896	.94777	.33545	.94206	24
37	.26920	.96308	.28597	.95824	.30265	.95310	.31923	.94768	.33573	.94196	23
38	.26948	.96301	.28625	.95816	.30292	.95301	.31951	.94758	.33600	.94186	22
39	.26976	.96293	.28652	.95807	.30320	.95293	.31979	.94749	.33627	.94176	21
40	.27004	.96285	.28680	.95799	.30348	.95284	.32006	.94740	.33655	.94167	20
41	.27032	.96277	.28708	.95791	.30376	.95275	.32034	.94730	.33682	.94157	19
42	.27060	.96269	.28736	.95782	.30403	.95266	.32061	.94721	.33710	.94147	18
43	.27088	.96261	.28764	.95774	.30431	.95257	.32089	.94712	.33737	.94137	17
44	.27116	.96253	.28792	.95766	.30459	.95248	.32116	.94702	.33764	.94127	16
45	.27144	.96246	.28820	.95757	.30486	.95240	.32144	.94693	.33792	.94118	15
46	.27172	.96238	.28847	.95749	.30514	.95231	.32171	.94684	.33819	.94108	14
47	.27200	.96230	.28875	.95740	.30542	.95222	.32199	.94674	.33846	.94098	13
48	.27228	.96222	.28903	.95732	.30570	.95213	.32227	.94665	.33874	.94088	12
49	.27256	.96214	.28931	.95724	.30597	.95204	.32254	.94656	.33901	.94078	11
50	.27284	.96206	.28959	.95715	.30625	.95195	.32282	.94646	.33929	.94068	10
51	.27312	.96198	.28987	.95707	.30653	.95186	.32309	.94637	.33956	.94058	9
52	.27340	.96190	.29015	.95698	.30680	.95177	.32337	.94627	.33983	.94049	8
53	.27368	.96182	.29042	.95690	.30708	.95168	.32364	.94618	.34011	.94039	7
54	.27396	.96174	.29070	.95681	.30736	.95159	.32392	.94609	.34038	.94029	6
55	.27424	.96166	.29098	.95673	.30763	.95150	.32419	.94599	.34065	.94019	5
56	.27452	.96158	.29126	.95664	.30791	.95142	.32447	.94590	.34093	.94009	4
57	.27480	.96150	.29154	.95656	.30819	.95133	.32474	.94580	.34120	.93999	3
58	.27508	.96142	.29182	.95647	.30846	.95124	.32502	.94571	.34147	.93989	2
59	.27536	.96134	.29209	.95639	.30874	.95115	.32529	.94561	.34175	.93979	1
60	.27564	.96126	.29237	.95630	.30902	.95106	.32557	.94552	.34202	.93969	0
′	74° Cosine	74° Sine	73° Cosine	73° Sine	72° Cosine	72° Sine	71° Cosine	71° Sine	70° Cosine	70° Sine	′

382 SHOP MATHEMATICS

′	20° Sine	20° Cosine	21° Sine	21° Cosine	22° Sine	22° Cosine	23° Sine	23° Cosine	24° Sine	24° Cosine	′
0	.34202	.93969	.35837	.93358	.37461	.92718	.39073	.92050	.40674	.91355	60
1	.34229	.93959	.35864	.93348	.37488	.92707	.39100	.92039	.40700	.91343	59
2	.34257	.93949	.35891	.93337	.37515	.92697	.39127	.92028	.40727	.91331	58
3	.34284	.93939	.35918	.93327	.37542	.92686	.39153	.92016	.40753	.91319	57
4	.34311	.93929	.35945	.93316	.37569	.92675	.39180	.92005	.40780	.91307	56
5	.34339	.93919	.35973	.93306	.37595	.92664	.39207	.91994	.40806	.91295	55
6	.34366	.93909	.36000	.93295	.37622	.92653	.39234	.91982	.40833	.91283	54
7	.34393	.93899	.36027	.93285	.37649	.92642	.39260	.91971	.40860	.91272	53
8	.34421	.93889	.36054	.93274	.37676	.92631	.39287	.91959	.40886	.91260	52
9	.34448	.93879	.36081	.93264	.37703	.92620	.39314	.91948	.40913	.91248	51
10	.34475	.93869	.36108	.93253	.37730	.92609	.39341	.91936	.40939	.91236	50
11	.34503	.93859	.36135	.93243	.37757	.92598	.39367	.91925	.40966	.91224	49
12	.34530	.93849	.36162	.93232	.37784	.92587	.39394	.91914	.40992	.91212	48
13	.34557	.93839	.36190	.93222	.37811	.92576	.39421	.91902	.41019	.91200	47
14	.34584	.93829	.36217	.93211	.37838	.92565	.39448	.91891	.41045	.91188	46
15	.34612	.93819	.36244	.93201	.37865	.92554	.39474	.91879	.41072	.91176	45
16	.34639	.93809	.36271	.93190	.37892	.92543	.39501	.91868	.41098	.91164	44
17	.34666	.93799	.36298	.93180	.37919	.92532	.39528	.91856	.41125	.91152	43
18	.34694	.93789	.36325	.93169	.37946	.92521	.39555	.91845	.41151	.91140	42
19	.34721	.93779	.36352	.93159	.37973	.92510	.39581	.91833	.41178	.91128	41
20	.34748	.93769	.36379	.93148	.37999	.92499	.39608	.91822	.41204	.91116	40
21	.34775	.93759	.36406	.93137	.38026	.92488	.39635	.91810	.41231	.91104	39
22	.34803	.93748	.36434	.93127	.38053	.92477	.39661	.91799	.41257	.91092	38
23	.34830	.93738	.36461	.93116	.38080	.92466	.39688	.91787	.41284	.91080	37
24	.34857	.93728	.36488	.93106	.38107	.92455	.39715	.91775	.41310	.91068	36
25	.34884	.93718	.36515	.93095	.38134	.92444	.39741	.91764	.41337	.91056	35
26	.34912	.93708	.36542	.93084	.38161	.92432	.39768	.91752	.41363	.91044	34
27	.34939	.93698	.36569	.93074	.38188	.92421	.39795	.91741	.41390	.91032	33
28	.34966	.93688	.36596	.93063	.38215	.92410	.39822	.91729	.41416	.91020	32
29	.34993	.93677	.36623	.93052	.38241	.92399	.39848	.91718	.41443	.91008	31
30	.35021	.93667	.36650	.93042	.38268	.92388	.39875	.91706	.41469	.90996	30
31	.35048	.93657	.36677	.93031	.38295	.92377	.39902	.91694	.41496	.90984	29
32	.35075	.93647	.36704	.93020	.38322	.92366	.39928	.91683	.41522	.90972	28
33	.35102	.93637	.36731	.93010	.38349	.92355	.39955	.91671	.41549	.90960	27
34	.35130	.93626	.36758	.92999	.38376	.92343	.39982	.91660	.41575	.90948	26
35	.35157	.93616	.36785	.92988	.38403	.92332	.40008	.91648	.41602	.90936	25
36	.35184	.93606	.36812	.92978	.38430	.92321	.40035	.91636	.41628	.90924	24
37	.35211	.93596	.36839	.92967	.38456	.92310	.40062	.91625	.41655	.90911	23
38	.35239	.93585	.36867	.92956	.38483	.92299	.40088	.91613	.41681	.90899	22
39	.35266	.93575	.36894	.92945	.38510	.92287	.40115	.91601	.41707	.90887	21
40	.35293	.93565	.36921	.92935	.38537	.92276	.40141	.91590	.41734	.90875	20
41	.35320	.93555	.36948	.92924	.38564	.92265	.40168	.91578	.41760	.90863	19
42	.35347	.93544	.36975	.92913	.38591	.92254	.40195	.91566	.41787	.90851	18
43	.35375	.93534	.37002	.92902	.38617	.92243	.40221	.91555	.41813	.90839	17
44	.35402	.93524	.37029	.92892	.38644	.92231	.40248	.91543	.41840	.90826	16
45	.35429	.93514	.37056	.92881	.38671	.92220	.40275	.91531	.41866	.90814	15
46	.35456	.93503	.37083	.92870	.38698	.92209	.40301	.91519	.41892	.90802	14
47	.35484	.93493	.37110	.92859	.38725	.92198	.40328	.91508	.41919	.90790	13
48	.35511	.93483	.37137	.92849	.38752	.92186	.40355	.91496	.41945	.90778	12
49	.35538	.93472	.37164	.92838	.38778	.92175	.40381	.91484	.41972	.90766	11
50	.35565	.93462	.37191	.92827	.38805	.92164	.40408	.91472	.41998	.90753	10
51	.35592	.93452	.37218	.92816	.38832	.92152	.40434	.91461	.42024	.90741	9
52	.35619	.93441	.37245	.92805	.38859	.92141	.40461	.91449	.42051	.90729	8
53	.35647	.93431	.37272	.92794	.38886	.92130	.40488	.91437	.42077	.90717	7
54	.35674	.93420	.37299	.92784	.38912	.92119	.40514	.91425	.42104	.90704	6
55	.35701	.93410	.37326	.92773	.38939	.92107	.40541	.91414	.42130	.90692	5
56	.35728	.93400	.37353	.92762	.38966	.92096	.40567	.91402	.42156	.90680	4
57	.35755	.93389	.37380	.92751	.38993	.92085	.40594	.91390	.42183	.90668	3
58	.35782	.93379	.37407	.92740	.39020	.92073	.40621	.91378	.42209	.90655	2
59	.35810	.93368	.37434	.92729	.39046	.92062	.40647	.91366	.42235	.90643	1
60	.35837	.93358	.37461	.92718	.39073	.92050	.40674	.91355	.42262	.90631	0
′	Cosine	Sine	Cosine	Sine	Cosine	Sine	Cosine	Sine	Cosine	Sine	′
	69°		68°		67°		66°		65°		

′	25° Sine	Cosine	26° Sine	Cosine	27° Sine	Cosine	28° Sine	Cosine	29° Sine	Cosine	′
0	.42262	.90631	.43837	.89879	.45399	.89101	.46947	.88295	.48481	.87462	60
1	.42288	.90618	.43863	.89867	.45425	.89087	.46973	.88281	.48506	.87448	59
2	.42315	.90606	.43889	.89854	.45451	.89074	.46999	.88267	.48532	.87434	58
3	.42341	.90594	.43916	.89841	.45477	.89061	.47024	.88254	.48557	.87420	57
4	.42367	.90582	.43942	.89828	.45503	.89048	.47050	.88240	.48583	.87406	56
5	.42394	.90569	.43968	.89816	.45529	.89035	.47076	.88226	.48608	.87391	55
6	.42420	.90557	.43994	.89803	.45554	.89021	.47101	.88213	.48634	.87377	54
7	.42446	.90545	.44020	.89790	.45580	.89008	.47127	.88199	.48659	.87363	53
8	.42473	.90532	.44046	.89777	.45606	.88995	.47153	.88185	.48684	.87349	52
9	.42499	.90520	.44072	.89764	.45632	.88981	.47178	.88172	.48710	.87335	51
10	.42525	.90507	.44098	.89752	.45658	.88968	.47204	.88158	.48735	.87321	50
11	.42552	.90495	.44124	.89739	.45684	.88955	.47229	.88144	.48761	.87306	49
12	.42578	.90483	.44151	.89726	.45710	.88942	.47255	.88130	.48786	.87292	48
13	.42604	.90470	.44177	.89713	.45736	.88928	.47281	.88117	.48811	.87278	47
14	.42631	.90458	.44203	.89700	.45762	.88915	.47306	.88103	.48837	.87264	46
15	.42657	.90446	.44229	.89687	.45787	.88902	.47332	.88089	.48862	.87250	45
16	.42683	.90433	.44255	.89674	.45813	.88888	.47358	.88075	.48888	.87235	44
17	.42709	.90421	.44281	.89662	.45839	.88875	.47383	.88062	.48913	.87221	43
18	.42736	.90408	.44307	.89649	.45865	.88862	.47409	.88048	.48938	.87207	42
19	.42762	.90396	.44333	.89636	.45891	.88848	.47434	.88034	.48964	.87193	41
20	.42788	.90383	.44359	.89623	.45917	.88835	.47460	.88020	.48989	.87178	40
21	.42815	.90371	.44385	.89610	.45942	.88822	.47486	.88006	.49014	.87164	39
22	.42841	.90358	.44411	.89597	.45968	.88808	.47511	.87993	.49040	.87150	38
23	.42867	.90346	.44437	.89584	.45994	.88795	.47537	.87979	.49065	.87136	37
24	.42894	.90334	.44464	.89571	.46020	.88782	.47562	.87965	.49090	.87121	36
25	.42920	.90321	.44490	.89558	.46046	.88768	.47588	.87951	.49116	.87107	35
26	.42946	.90309	.44516	.89545	.46072	.88755	.47614	.87937	.49141	.87093	34
27	.42972	.90296	.44542	.89532	.46097	.88741	.47639	.87923	.49166	.87079	33
28	.42999	.90284	.44568	.89519	.46123	.88728	.47665	.87909	.49192	.87064	32
29	.43025	.90271	.44594	.89506	.46149	.88715	.47690	.87896	.49217	.87050	31
30	.43051	.90259	.44620	.89493	.46175	.88701	.47716	.87882	.49242	.87036	30
31	.43077	.90246	.44646	89480	.46201	.88688	.47741	.87868	.49268	.87021	29
32	.43104	.90233	.44672	.89467	.46226	.88674	.47767	.87854	.49293	.87007	28
33	.43130	.90221	.44698	.89454	.46252	.88661	.47793	.87840	.49318	.86993	27
34	.43156	.90208	.44724	.89441	.46278	.88647	.47818	.87826	.49344	.86978	26
35	.43182	.90196	.44750	.89428	.46304	.88634	.47844	.87812	.49369	.86964	25
36	.43209	.90183	.44776	.89415	.46330	.88620	.47869	.87798	.49394	.86949	24
37	.43235	.90171	.44802	.89402	.46355	.88607	.47895	.87784	.49419	.86935	23
38	.43261	.90158	.44828	.89389	.46381	.88593	.47920	.87770	.49445	.86921	22
39	.43287	.90146	.44854	.89376	.46407	.88580	.47946	.87756	.49470	.86906	21
40	.43313	.90133	.44880	.89363	.46433	.88566	.47971	.87743	.49495	.86892	20
41	.43340	.90120	.44906	.89350	.46458	.88553	.47997	.87729	.49521	.86878	19
42	.43366	.90108	.44932	.89337	.46484	.88539	.48022	.87715	.49546	.86863	18
43	.43392	.90095	.44958	.89324	.46510	.88526	.48048	.87701	.49571	.86849	17
44	.43418	.90082	.44984	.89311	.46536	.88512	.48073	.87687	.49596	.86834	16
45	.43445	.90070	.45010	.89298	.46561	.88499	.48099	.87673	.49622	.86820	15
46	.43471	.90057	.45036	.89285	.46587	.88485	.48124	.87659	.49647	.86805	14
47	.43497	.90045	.45062	.89272	.46613	.88472	.48150	.87645	.49672	.86791	13
48	.43523	.90032	.45088	.89259	.46639	.88458	.48175	.87631	.49697	.86777	12
49	.43549	.90019	.45114	.89245	.46664	.88445	.48201	.87617	.49723	.86762	11
50	.43575	.90007	.45140	.89232	.46690	.88431	.48226	.87603	.49748	.86748	10
51	.43602	.89994	.45166	.89219	.46716	.88417	.48252	.87589	.49773	.86733	9
52	.43628	.89981	.45192	.89206	.46742	.88404	.48277	.87575	.49798	.86719	8
53	.43654	.89968	.45218	.89193	.46767	.88390	.48303	.87561	.49824	.86704	7
54	.43680	.89956	.45243	.89180	.46793	.88377	.48328	.87546	.49849	.86690	6
55	.43706	.89943	.45269	.89167	.46819	.88363	.48354	.87532	.49874	.86675	5
56	.43733	.89930	.45295	.89153	.46844	.88349	.48379	.87518	.49899	.86661	4
57	.43759	.89918	.45321	.89140	.46870	.88336	.48405	.87504	.49924	.86646	3
58	.43785	.89905	.45347	.89127	.46896	.88322	.48430	.87490	.49950	.86632	2
59	.43811	.89892	.45373	.89114	.46921	.88308	.48456	.87476	.49975	.86617	1
60	.43837	.89879	.45399	.89101	.46947	.88295	.48481	.87462	.50000	.86603	0
′	Cosine	Sine	Cosine	Sine	Cosine	Sine	Cosine	Sine	Cosine	Sine	′
	64°		63°		62°		61°		60°		

′	30° Sine	30° Cosine	31° Sine	31° Cosine	32° Sine	32° Cosine	33° Sine	33° Cosine	34° Sine	34° Cosine	′
0	.50000	.86603	.51504	.85717	.52992	.84805	.54464	.83867	.55919	.82904	60
1	.50025	.86588	.51529	.85702	.53017	.84789	.54488	.83851	.55943	.82887	59
2	.50050	.86573	.51554	.85687	.53041	.84774	.54513	.83835	.55968	.82871	58
3	.50076	.86559	.51579	.85672	.53066	.84759	.54537	.83819	.55992	.82855	57
4	.50101	.86544	.51604	.85657	.53091	.84743	.54561	.83804	.56016	.82839	56
5	.50126	.86530	.51628	.85642	.53115	.84728	.54586	.83788	.56040	.82822	55
6	.50151	.86515	.51653	.85627	.53140	.84712	.54610	.83772	.56064	.82806	54
7	.50176	.86501	.51678	.85612	.53164	.84697	.54635	.83756	.56088	.82790	53
8	.50201	.86486	.51703	.85597	.53189	.84681	.54659	.83740	.56112	.32773	52
9	.50227	.86471	.51728	.85582	.53214	.84666	.54683	.83724	.56136	.82757	51
10	.50252	.86457	.51753	.85567	.53238	.84650	.54708	.83708	.56160	.82741	50
11	.50277	.86442	.51778	.85551	.53263	.84635	.54732	.83692	.56184	.82724	49
12	.50302	.86427	.51803	.85536	.53288	.84619	.54756	.83676	.56208	.82708	48
13	.50327	.86413	.51828	.85521	.53312	.84604	.54781	.83660	.56232	.82692	47
14	.50352	.86398	.51852	.85506	.53337	.84588	.54805	.83645	.56256	.82675	46
15	.50377	.86384	.51877	.85491	.53361	.84573	.54829	.83629	.56280	.82659	45
16	.50403	.86369	.51902	.85476	.53386	.84557	.54854	.83613	.56305	.82643	44
17	.50428	.86354	.51927	.85461	.53411	.84542	.54878	.83597	.56329	.82626	43
18	.50453	.86340	.51952	.85446	.53435	.84526	.54902	.83581	.56353	.82610	42
19	.50478	.86325	.51977	.85431	.53460	.84511	.54927	.83565	.56377	.82593	41
20	.50503	.86310	.52002	.85416	.53484	.84495	.54951	.83549	.56401	.82577	40
21	.50528	.86295	.52026	.85401	.53509	.84480	.54975	.83533	.56425	.82561	39
22	.50553	.86281	.52051	.85385	.53534	.84464	.54999	.83517	.56449	.82544	38
23	.50578	.86266	.52076	.85370	.53558	.84448	.55024	.83501	.56473	.82528	37
24	.50603	.86251	.52101	.85355	.53583	.84433	.55048	.83485	.56497	.82511	36
25	.50628	.86237	.52126	.85340	.53607	.84417	.55072	.83469	.56521	.82495	35
26	.50654	.86222	.52151	.85325	.53632	.84402	.55097	.83453	.56545	.82478	34
27	.50679	.86207	.52175	.85310	.53656	.84386	.55121	.83437	.56569	.82462	33
28	.50704	.86192	.52200	.85294	.53681	.84370	.55145	.83421	.56593	.82446	32
29	.50729	.86178	.52225	.85279	.53705	.84355	.55169	.83405	.56617	.82429	31
30	.50754	.86163	.52250	.85264	.53730	.84339	.55194	.83389	.56641	.82413	30
31	.50779	.86148	.52275	.85249	.53754	.84324	.55218	.83373	.56665	.82396	29
32	.50804	.86133	.52299	.85234	.53779	.84308	.55242	.83356	.56689	.82380	28
33	.50829	.86119	.52324	.85218	.53804	.84292	.55266	.83340	.56713	.82363	27
34	.50854	.86104	.52349	.85203	.53828	.84277	.55291	.83324	.56736	.82347	26
35	.50879	.86089	.52374	.85188	.53853	.84261	.55315	.83308	.56760	.82330	25
36	.50904	.86074	.52399	.85173	.53877	.84245	.55339	.83292	.56784	.82314	24
37	.50929	.86059	.52423	.85157	.53902	.84230	.55363	.83276	.56808	.82297	23
38	.50954	.86045	.52448	.85142	.53926	.84214	.55388	.83260	.56832	.82281	22
39	.50979	.86030	.52473	.85127	.53951	.84198	.55412	.83244	.56856	.82264	21
40	.51004	.86015	.52498	.85112	.53975	.84182	.55436	.83228	.56880	.82248	20
41	.51029	.86000	.52522	.85096	.54000	.84167	.55460	.83212	.56904	.82231	19
42	.51054	.85985	.52547	.85081	.54024	.84151	.55484	.83195	.56928	.82214	18
43	.51079	.85970	.52572	.85066	.54049	.84135	.55509	.83179	.56952	.82198	.17
44	.51104	.85956	.52597	.85051	.54073	.84120	.55533	.83163	.56976	.82181	16
45	.51129	.85941	.52621	.85035	.54097	.84104	.55557	.83147	.57000	.82165	15
46	.51154	.85926	.52646	.85020	.54122	.84088	.55581	.83131	.57024	.82148	14
47	.51179	.85911	.52671	.85005	.54146	.84072	.55605	.83115	.57047	.82132	13
48	.51204	.85896	.52696	.84989	.54171	.84057	.55630	.83098	.57071	.82115	12
49	.51229	.85881	.52720	.84974	.54195	.84041	.55654	.83082	.57095	.82098	11
50	.51254	.85866	.52745	.84959	.54220	.84025	.55678	.83066	.57119	.82082	10
51	.51279	.85851	.52770	.84943	.54244	.84009	.55702	.83050	.57143	.82065	9
52	.51304	.85836	.52794	.84928	.54269	.83994	.55726	.83034	.57167	.82048	8
53	.51329	.85821	.52819	.84913	.54293	.83978	.55750	.83017	.57191	.82032	7
54	.51354	.85806	.52844	.84897	.54317	.83962	.55775	.83001	.57215	.82015	6
55	.51379	.85792	.52869	.84882	.54342	.83946	.55799	.82985	.57238	.81999	5
56	.51404	.85777	.52893	.84866	.54366	.83930	.55823	.82969	.57262	.81982	4
57	.51429	.85762	.52918	.84851	.54391	.83915	.55847	.82953	.57286	.81965	3
58	.51454	.85747	.52943	.84836	.54415	.83899	.55871	.82936	.57310	.81949	2
59	.51479	.85732	.52967	.84820	.54440	.83883	.55895	.82920	.57334	.81932	1
60	.51504	.85717	.52992	.84805	.54464	.83867	.55919	.82904	.57358	.81915	0
′	Cosine	Sine	Cosine	Sine	Cosine	Sine	Cosine	Sine	Cosine	Sine	′
	59°		58°		57°		56°		55°		

′	35° Sine	Cosine	36° Sine	Cosine	37° Sine	Cosine	38° Sine	Cosine	39° Sine	Cosine	′
0	.57358	.81915	.58779	.80902	.60182	.79864	.61566	.78801	.62932	.77715	60
1	.57381	.81899	.58802	.80885	.60205	.79846	.61589	.78783	.62955	.77696	59
2	.57405	.81882	.58826	.80867	.60228	.79829	.61612	.78765	.62977	.77678	58
3	.57429	.81865	.58849	.80850	.60251	.79811	.61635	.78747	.63000	.77660	57
4	.57453	.81848	.58873	.80833	.60274	.79793	.61658	.78729	.63022	.77641	56
5	.57477	.81832	.58896	.80816	.60298	.79776	.61681	.78711	.63045	.77623	55
6	.57501	.81815	.58920	.80799	.60321	.79758	.61704	.78694	.63068	.77605	54
7	.57524	.81798	.58943	.80782	.60344	.79741	.61726	.78676	.63090	.77586	53
8	.57548	.81782	.58967	.80765	.60367	.79723	.61749	.78658	.63113	.77568	52
9	.57572	.81765	.58990	.80748	.60390	.79706	.61772	.78640	.63135	.77550	51
10	.57596	.81748	.59014	.80730	.60414	.79688	.61795	.78622	.63158	.77531	50
11	.57619	.81731	.59037	.80713	.60437	.79671	.61818	.78604	.63180	.77513	49
12	.57643	.81714	.59061	.80696	.60460	.79653	.61841	.78586	.63203	.77494	48
13	.57667	.81698	.59034	.80679	.60483	.79635	.61864	.78568	.63225	.77476	47
14	.57691	.81681	.59108	.80662	.60506	.79618	.61887	.78550	.63248	.77458	46
15	.57715	.81664	.59131	.80644	.60529	.79600	.61909	.78532	.63271	.77439	45
16	.57738	.81647	.59154	.80627	.60553	.79583	.61932	.78514	.63293	.77421	44
17	.57762	.81631	.59178	.80610	.60576	.79565	.61955	.78496	.63316	.77402	43
18	.57786	.81614	.59201	.80593	.60599	.79547	.61978	.78478	.63338	.77384	42
19	.57810	.81597	.59225	.80576	.60622	.79530	.62001	.78460	.63361	.77366	41
20	.57833	.81580	.59248	.80558	.60645	.79512	.62024	.78442	.63383	.77347	40
21	.57857	.81563	.59272	.80541	.60668	.79494	.62046	.78424	.63406	.77329	39
22	.57881	.81546	.59295	.80524	.60691	.79477	.62069	.78405	.63428	.77310	38
23	.57904	.81530	.59318	.80507	.60714	.79459	.62092	.78387	.63451	.77292	37
24	.57928	.81513	.59342	.80489	.60738	.79441	.62115	.78369	.63473	.77273	36
25	.57952	.81496	.59365	.80472	.60761	.79424	.62138	.78351	.63496	.77255	35
26	.57976	.81479	.59389	.80455	.60784	.79406	.62160	.78333	.63518	.77236	34
27	.57999	.81462	.59412	.80438	.60807	.79388	.62183	.78315	.63540	.77218	33
28	.58023	.81445	.59436	.80420	.60830	.79371	.62206	.78297	.63563	.77199	32
29	.58047	.81428	.59459	.80403	.60853	.79353	.62229	.78279	.63585	.77181	31
30	.58070	.81412	.59482	.80386	.60876	.79335	.62251	.78261	.63608	.77162	30
31	.58094	.81395	.59506	.80368	.60899	.79318	.62274	.78243	.63630	.77144	29
32	.58118	.81378	.59529	.80351	.60922	.79300	.62297	.78225	.63653	.77125	28
33	.58141	.81361	.59552	.80334	.60945	.79282	.62320	.78206	.63675	.77107	27
34	.58165	.81344	.59576	.80316	.60968	.79264	.62342	.78188	.63698	.77088	26
35	.58189	.81327	.59599	.80299	.60991	.79247	.62365	.78170	.63720	.77070	25
36	.58212	.81310	.59622	.80282	.61015	.79229	.62388	.78152	.63742	.77051	24
37	.58236	.81293	.59646	.80264	.61038	.79211	.62411	.78134	.63765	.77033	23
38	.58260	.81276	.59669	.80247	.61061	.79193	.62433	.78116	.63787	.77014	22
39	.58283	.81259	.59693	.80230	.61084	.79176	.62456	.78098	.63810	.76996	21
40	.58307	.81242	.59716	.80212	.61107	.79158	.62479	.78079	.63832	.76977	20
41	.58330	.81225	.59739	.80195	.61130	.79140	.62502	.78061	.63854	.76959	19
42	.58354	.81208	.59763	.80178	.61153	.79122	.62524	.78043	.63877	.76940	18
43	.58378	.81191	.59786	.80160	.61176	.79105	.62547	.78025	.63899	.76921	17
44	.38401	.81174	.59809	.80143	.61199	.79087	.62570	.78007	.63922	.76903	16
45	.58425	.81157	.59832	.80125	.61222	.79069	.62592	.77988	.63944	.76884	15
46	.58449	.81140	.59856	.80108	.61245	.79051	.62615	.77970	.63966	.76866	14
47	.58472	.81123	.59879	.80091	.61268	.79033	.62638	.77952	.63989	.76847	13
48	.58496	.81106	.59902	.80073	.61291	.79016	.62660	.77934	.64011	.76828	12
49	.58519	.81089	.59926	.80056	.61314	.78998	.62683	.77916	.64033	.76810	11
50	.58543	.81072	.59949	.80038	.61337	.78980	.62706	.77897	.64056	.76791	10
51	.58567	.81055	.59972	.80021	.61360	.78962	.62728	.77879	.64078	.76772	9
52	.58590	.81038	.59995	.80003	.61383	.78944	.62751	.77861	.64100	.76754	8
53	.58614	.81021	.60019	.79986	.61406	.78926	.62774	.77843	.64123	.76735	7
54	.58637	.81004	.60042	.79968	.61429	.78908	.62796	.77824	.64145	.76717	6
55	.58661	.80987	.60065	.79951	.61451	.78891	.62819	.77806	.64167	.76698	5
56	.58684	.80970	.60089	.79934	.61474	.78873	.62842	.77788	.64190	.76679	4
57	.58708	.80953	.60112	.79916	.61497	.78855	.62864	.77769	.64212	.76661	3
58	.58731	.80936	.60135	.79899	.61520	.78837	.62887	.77751	.64234	.76642	2
59	.58755	.80919	.60158	.79881	.61543	.78819	.62909	.77733	.64256	.76623	1
60	.58779	.80902	.60182	.79864	.61566	.78801	.62932	.77715	.64279	.76604	0
′	Cosine	Sine	Cosine	Sine	Cosine	Sine	Cosine	Sine	Cosine	Sine	′

| | 54° | | 53° | | 52° | | 51° | | 50° | | |

′	40° Sine	40° Cosine	41° Sine	41° Cosine	42° Sine	42° Cosine	43° Sine	43° Cosine	44° Sine	44° Cosine	′
0	.64279	.76604	.65606	.75471	.66913	.74314	.68200	.73135	.69466	.71934	60
1	.64301	.76586	.65628	.75452	.66935	.74295	.68221	.73116	.69487	.71914	59
2	.64323	.76567	.65650	.75433	.66956	.74276	.68242	.73096	.69508	.71894	58
3	.64346	.76548	.65672	.75414	.66978	.74256	.68264	.73076	.69529	.71873	57
4	.64368	.76530	.65694	.75395	.66999	.74237	.68285	.73056	.69549	.71853	56
5	.64390	.76511	.65716	.75375	.67021	.74217	.68306	.73036	.69570	.71833	55
6	.64412	.76492	.65738	.75356	.67043	.74198	.68327	.73016	.69591	.71813	54
7	.64435	.76473	.65759	.75337	.67064	.74178	.68349	.72996	.69612	.71792	53
8	.64457	.76455	.65781	.75318	.67086	.74159	.68370	.72976	.69633	.71772	52
9	.64479	.76436	.65803	.75299	.67107	.74139	.68391	.72957	.69654	.71752	51
10	.64501	.76417	.65825	.75280	.67129	.74120	.68412	.72937	.69675	.71732	50
11	.64524	.76398	.65847	.75261	.67151	.74100	.68434	.72917	.69696	.71711	49
12	.64546	.76380	.65869	.75241	.67172	.74080	.68455	.72897	.69717	.71691	48
13	.64568	.76361	.65891	.75222	.67194	.74061	.68476	.72877	.69737	.71671	47
14	.64590	.76342	.65913	.75203	.67215	.74041	.68497	.72857	.69758	.71650	46
15	.64612	.76323	.65935	.75184	.67237	.74022	.68518	.72837	.69779	.71630	45
16	.64635	.76304	.65956	.75165	.67258	.74002	.68539	.72817	.69800	.71610	44
17	.64657	.76286	.65978	.75146	.67280	.73983	.68561	.72797	.69821	.71590	43
18	.64679	.76267	.66000	.75126	.67301	.73963	.68582	.72777	.69842	.71569	42
19	.64701	.76248	.66022	.75107	.67323	.73944	.68603	.72757	.69862	.71549	41
20	.64723	.76229	.66044	.75088	.67344	.73924	.68624	.72737	.69883	.71529	40
21	.64746	.76210	.66066	.75069	.67366	.73904	.68645	.72717	.69904	.71508	39
22	.64768	.76192	.66088	.75050	.67387	.73885	.68666	.72697	.69925	.71488	38
23	.64790	.76173	.66109	.75030	.67409	.73865	.68688	.72677	.69946	.71468	37
24	.64812	.76154	.66131	.75011	.67430	.73846	.68709	.72657	.69966	.71447	36
25	.64834	.76135	.66153	.74992	.67452	.73826	.68730	.72637	.69987	.71427	35
26	.64856	.76116	.66175	.74973	.67473	.73806	.68751	.72617	.70008	.71407	34
27	.64878	.76097	.66197	.74953	.67495	.73787	.68772	.72597	.70029	.71386	33
28	.64901	.76078	.66218	.74934	.67516	.73767	.68793	.72577	.70049	.71366	32
29	.64923	.76059	.66240	.74915	.67538	.73747	.68814	.72557	.70070	.71345	31
30	.64945	.76041	.66262	.74896	.67559	.73728	.68835	.72537	.70091	.71325	30
31	.64967	.76022	.66284	.74876	.67580	.73708	.68857	.72517	.70112	.71305	29
32	.64989	.76003	.66306	.74857	.67602	.73688	.68878	.72497	.70132	.71284	28
33	.65011	.75984	.66327	.74838	.67623	.73669	.68899	.72477	.70153	.71264	27
34	.65033	.75965	.66349	.74818	.67645	.73649	.68920	.72457	.70174	.71243	26
35	.65055	.75946	.66371	.74799	.67666	.73629	.68941	.72437	.70195	.71223	25
36	.65077	.75927	.66393	.74780	.67688	.73610	.68962	.72417	.70215	.71203	24
37	.65100	.75908	.66414	.74760	.67709	.73590	.68983	.72397	.70236	.71182	23
38	.65122	.75889	.66436	.74741	.67730	.73570	.69004	.72377	.70257	.71162	22
39	.65144	.75870	.66458	.74722	.67752	.73551	.69025	.72357	.70277	.71141	21
40	.65166	.75851	.66480	.74703	.67773	.73531	.69046	.72337	.70298	.71121	20
41	.65188	.75832	.66501	.74683	.67795	.73511	.69067	.72317	.70319	.71100	19
42	.65210	.75813	.66523	.74664	.67816	.73491	.69088	.72297	.70339	.71080	18
43	.65232	.75794	.66545	.74644	.67837	.73472	.69109	.72277	.70360	.71059	17
44	.65254	.75775	.66566	.74625	.67859	.73452	.69130	.72257	.70381	.71039	16
45	.65276	.75756	.66588	.74606	.67880	.73432	.69151	.72236	.70401	.71019	15
46	.65298	.75738	.66610	.74586	.67901	.73413	.69172	.72216	.70422	.70998	14
47	.65320	.75719	.66632	.74567	.67923	.73393	.69193	.72196	.70443	.70978	13
48	.65342	.75700	.66653	.74548	.67944	.73373	.69214	.72176	.70463	.70957	12
49	.65364	.75680	.66675	.74528	.67965	.73353	.69235	.72156	.70484	.70937	11
50	.65386	.75661	.66697	.74509	.67987	.73333	.69256	.72136	.70505	.70916	10
51	.65408	.75642	.66718	.74489	.68008	.73314	.69277	.72116	.70525	.70896	9
52	.65430	.75623	.66740	.74470	.68029	.73294	.69298	.72095	.70546	.70875	8
53	.65452	.75604	.66762	.74451	.68051	.73274	.69319	.72075	.70567	.70855	7
54	.65474	.75585	.66783	.74431	.68072	.73254	.69340	.72055	.70587	.70834	6
55	.65496	.75566	.66805	.74412	.68093	.73234	.69361	.72035	.70608	.70813	5
56	.65518	.75547	.66827	.74392	.68115	.73215	.69382	.72015	.70628	.70793	4
57	.65540	.75528	.66848	.74373	.68136	.73195	.69403	.71995	.70649	.70772	3
58	.65562	.75509	.66870	.74353	.68157	.73175	.69424	.71974	.70670	.70752	2
59	.65584	.75490	.66891	.74334	.68179	.73155	.69445	.71954	.70690	.70731	1
60	.65606	.75471	.66913	.74314	.68200	.73135	.69466	.71934	.70711	.70711	0
′	Cosine	Sine	Cosine	Sine	Cosine	Sine	Cosine	Sine	Cosine	Sine	′
	49°		48°		47°		46°		45°		

Tangent-Cotangent

,	0° Tang	0° Cotang	1° Tang	1° Cotang	2° Tang	2° Cotang	3° Tang	3° Cotang	4° Tang	4° Cotang	,
0	.00000	Infin.	.01746	57.2900	.03492	28.6363	.05241	19.0811	.06993	14.3007	60
1	.00029	3437.75	.01775	56.3506	.03521	28.3994	.05270	18.9755	.07022	14.2411	59
2	.00058	1718.87	.01804	55.4415	.03550	28.1664	.05299	18.8711	.07051	14.1821	58
3	.00087	1145.92	.01833	54.5613	.03579	27.9372	.05328	18.7678	.07080	14.1235	57
4	.00116	859.436	.01862	53.7086	.03609	27.7117	.05357	18.6656	.07110	14.0655	56
5	.00145	687.549	.01891	52.8821	.03638	27.4899	.05387	18.5645	.07139	14.0079	55
6	.00175	572.957	.01920	52.0807	.03667	27.2715	.05416	18.4645	.07168	13.9507	54
7	.00204	491.106	.01949	51.3032	.03696	27.0566	.05445	18.3655	.07197	13.8940	53
8	.00233	429.718	.01978	50.5485	.03725	26.8450	.05474	18.2677	.07227	13.8378	52
9	.00262	381.971	.02007	49.8157	.03754	26.6367	.05503	18.1708	.07256	13.7821	51
10	.00291	343.774	.02036	49.1039	.03783	26.4316	.05533	18.0750	.07285	13.7267	50
11	.00320	312.521	.02066	48.4121	.03812	26.2296	.05562	17.9802	.07314	13.6719	49
12	.00349	286.478	.02095	47.7395	.03842	26.0307	.05591	17.8863	.07344	13.6174	48
13	.00378	264.441	.02124	47.0853	.03871	25.8348	.05620	17.7934	.07373	13.5634	47
14	.00407	245.552	.02153	46.4489	.03900	25.6418	.05649	17.7015	.07402	13.5098	46
15	.00436	229.182	.02182	45.8294	.03929	25.4517	.05678	17.6106	.07431	13.4566	45
16	.00465	214.858	.02211	45.2261	.03958	25.2644	.05708	17.5205	.07461	13.4039	44
17	.00495	202.219	.02240	44.6386	.03987	25.0798	.05737	17.4314	.07490	13.3515	43
18	.00524	190.984	.02269	44.0661	.04016	24.8978	.05766	17.3432	.07519	13.2996	42
19	.00553	180.932	.02298	43.5081	.04046	24.7185	.05795	17.2558	.07548	13.2480	41
20	.00582	171.885	.02328	42.9641	.04075	24.5418	.05824	17.1693	.07578	13.1969	40
21	.00611	163.700	.02357	42.4335	.04104	24.3675	.05854	17.0837	.07607	13.1461	39
22	.00640	156.259	.02386	41.9158	.04133	24.1957	.05883	16.9990	.07636	13.0958	38
23	.00669	149.465	.02415	41.4106	.04162	24.0263	.05912	16.9150	.07665	13.0458	37
24	.00698	143.237	.02444	40.9174	.04191	23.8593	.05941	16.8319	.07695	12.9962	36
25	.00727	137.507	.02473	40.4358	.04220	23.6945	.05970	16.7496	.07724	12.9469	35
26	.00756	132.219	.02502	39.9655	.04250	23.5321	.05999	16.6681	.07753	12.8981	34
27	.00785	127.321	.02531	39.5059	.04279	23.3718	.06029	16.5874	.07782	12.8496	33
28	.00815	122.774	.02560	39.0568	.04308	23.2137	.06058	16.5075	.07812	12.8014	32
29	.00844	118.540	.02589	38.6177	.04337	23.0577	.06087	16.4283	.07841	12.7536	31
30	.00873	114.589	.02619	38.1885	.04366	22.9038	.06116	16.3499	.07870	12.7062	30
31	.00902	110.892	.02648	37.7686	.04395	22.7519	.06145	16.2722	.07899	12.6591	29
32	.00931	107.426	.02677	37.3579	.04424	22.6020	.06175	16.1952	.07929	12.6124	28
33	.00960	104.171	.02706	36.9560	.04454	22.4541	.06204	16.1190	.07958	12.5660	27
34	.00989	101.107	.02735	36.5627	.04483	22.3081	.06233	16.0435	.07987	12.5199	26
35	.01018	98.2179	.02764	36.1776	.04512	22.1640	.06262	15.9687	.08017	12.4742	25
36	.01047	95.4895	.02793	35.8006	.04541	22.0217	.06291	15.8945	.08046	12.4288	24
37	.01076	92.9085	.02822	35.4313	.04570	21.8813	.06321	15.8211	.08075	12.3838	23
38	.01105	90.4633	.02851	35.0695	.04599	21.7426	.06350	15.7483	.08104	12.3390	22
39	.01135	88.1436	.02881	34.7151	.04628	21.6056	.06379	15.6762	.08134	12.2946	21
40	.01164	85.9398	.02910	34.3678	.04658	21.4704	.06408	15.6048	.08163	12.2505	20
41	.01193	83.8435	.02939	34.0273	.04687	21.3369	.06437	15.5340	.08192	12.2067	19
42	.01222	81.8470	.02968	33.6935	.04716	21.2049	.06467	15.4638	.08221	12.1632	18
43	.01251	79.9434	.02997	33.3662	.04745	21.0747	.06496	15.3943	.08251	12.1201	17
44	.01280	78.1263	.03026	33.0452	.04774	20.9460	.06525	15.3254	.08280	12.0772	16
45	.01309	76.3900	.03055	32.7303	.04803	20.8188	.06554	15.2571	.08309	12.0346	15
46	.01338	74.7292	.03084	32.4213	.04833	20.6932	.06584	15.1893	.08339	11.9923	14
47	.01367	73.1390	.03114	32.1181	.04862	20.5691	.06613	15.1222	.08368	11.9504	13
48	.01396	71.6151	.03143	31.8205	.04891	20.4465	.06642	15.0557	.08397	11.9087	12
49	.01425	70.1533	.03172	31.5284	.04920	20.3253	.06671	14.9898	.08427	11.8673	11
50	.01455	68.7501	.03201	31.2416	.04949	20.2056	.06700	14.9244	.08456	11.8262	10
51	.01484	67.4019	.03230	30.9599	.04978	20.0872	.06730	14.8596	.08485	11.7853	9
52	.01513	66.1055	.03259	30.6833	.05007	19.9702	.06759	14.7954	.08514	11.7448	8
53	.01542	64.8580	.03288	30.4116	.05037	19.8546	.06788	14.7317	.08544	11.7045	7
54	.01571	63.6567	.03317	30.1446	.05066	19.7403	.06817	14.6685	.08573	11.6645	6
55	.01600	62.4992	.03346	29.8823	.05095	19.6273	.06847	14.6059	.08602	11.6248	5
56	.01629	61.3829	.03376	29.6245	.05124	19.5156	.06876	14.5438	.08632	11.5853	4
57	.01658	60.3058	.03405	29.3711	.05153	19.4051	.06905	14.4823	.08661	11.5461	3
58	.01687	59.2659	.03434	29.1220	.05182	19.2959	.06934	14.4212	.08690	11.5072	2
59	.01716	58.2612	.03463	28.8771	.05212	19.1879	.06963	14.3607	.08720	11.4685	1
60	.01746	57.2900	.03492	28.6363	.05241	19.0811	.06993	14.3007	.08749	11.4301	0
,	Cotang	Tang	Cotang	Tang	Cotang	Tang	Cotang	Tang	Cotang	Tang	
	89°		88°		87°		86°		85°		

′	5°		6°		7°		8°		9°		′
	Tang	Cotang	Tang	Cotang	Tang	Cotang	Tang	Cotang	Tang	Cotang	
0	.08749	11.4301	.10510	9.51436	.12278	8.14435	.14054	7.11537	.15838	6.31375	60
1	.08778	11.3919	.10540	9.48781	.12308	8.12481	.14084	7.10038	.15868	6.30189	59
2	.08807	11.3540	.10569	9.46141	.12338	8.10536	.14113	7.08546	.15898	6.29007	58
3	.08837	11.3163	.10599	9.43515	.12367	8.08600	.14143	7.07059	.15928	6.27829	57
4	.08866	11.2789	.10628	9.40904	.12397	8.06674	14173	7.05579	.15958	6.26655	56
5	.08895	11.2417	.10657	9.38307	.12426	8.04756	.14202	7.04105	.15988	6.25486	55
6	.08925	11.2048	.10687	9.35724	.12456	8.02848	.14232	7.02637	.16017	6.24321	54
7	.08954	11.1681	.10716	9.33155	.12485	8.00948	.14262	7.01174	.16047	6.23160	53
8	.08983	11.1316	.10746	9.30599	.12515	7.99058	.14291	6.99718	.16077	6.22003	52
9	.09013	11.0954	.10775	9.28058	.12544	7.97176	.14321	6.98268	.16107	6.20851	51
10	.09042	11.0594	.10805	9.25530	.12574	7.95302	.14351	6.96823	.16137	6.19703	50
11	.09071	11.0237	.10834	9.23016	.12603	7.93438	.14381	6.95385	.16167	6.18559	49
12	.09101	10.9882	.10863	9.20516	.12633	7.91582	.14410	6.93952	.16196	6.17419	48
13	.09130	10.9529	.10893	9.18028	.12662	7.89734	.14440	6.92525	.16226	6.16283	47
14	.09159	10.9178	.10922	9.15554	.12692	7.87895	.14470	6.91104	.16256	6.15151	46
15	.09189	10.8829	.10952	9.13093	.12722	7.86064	.14499	6.89688	.16286	6.14023	45
16	.09218	10.8483	.10981	9.10646	.12751	7.84242	.14529	6.88278	.16316	6.12899	44
17	.09247	10.8139	.11011	9.08211	.12781	7.82428	.14559	6.86874	.16346	6.11779	43
18	.09277	10.7797	.11040	9.05789	.12810	7.80622	.14588	6.85475	.16376	6.10664	42
19	.09306	10.7457	.11070	9.03379	.12840	7.78825	.14618	6.84082	.16405	6.09552	41
20	.09335	10.7119	.11099	9.00983	.12869	7.77035	.14648	6.82694	.16435	6.08444	40
21	.09365	10.6783	.11128	8.98598	.12899	7.75254	.14678	6.81812	.16465	6.07340	39
22	.09394	10.6450	.11158	8.96227	.12929	7.73480	.14707	6.79936	.16495	6.06240	38
23	.09423	10.6118	.11187	8.93867	.12958	7.71715	.14737	6.78564	.16525	6.05143	37
24	.09453	10.5789	.11217	8.91520	.12988	7.69957	.14767	6.77199	.16555	6.04051	36
25	.09482	10.5462	.11246	8.89185	.13017	7.68208	.14796	6.75838	.16585	6.02962	35
26	.09511	10.5136	.11276	8.86862	.13047	7.66466	.14826	6.74483	.16615	6.01878	34
27	.09541	10.4813	.11305	8.84551	.13076	7.64732	.14856	6.73133	.16645	6.00797	33
28	.09570	10.4491	.11335	8.82252	.13106	7.63005	.14886	6.71789	.16674	5.99720	32
29	.09600	10.4172	.11364	8.79964	.13136	7.61287	.14915	6.70450	.16704	5.98646	31
30	.09629	10.3854	.11394	8.77689	.13165	7.59575	.14945	6.69116	.16734	5.97576	30
31	.09658	10.3538	.11423	8.75425	.13195	7.57872	.14975	6.67787	.16764	5.96510	29
32	.09688	10.3224	.11452	8.73172	.13224	7.56176	.15005	6.66463	.16794	5.95448	28
33	.09717	10.2913	.11482	8.70931	.13254	7.54487	.15034	6.65144	.16824	5.94390	27
34	.09746	10.2602	.11511	8.68701	.13284	7.52806	.15064	6.63831	.16854	5.93335	26
35	.09776	10.2294	.11541	8.66482	.13313	7.51132	.15094	6.62523	.16884	5.92283	25
36	.09805	10.1988	.11570	8.64275	.13343	7.49465	.15124	6.61219	.16914	5.91236	24
37	.09834	10.1683	.11600	8.62078	.13372	7.47806	.15153	6.59921	.16944	5.90191	23
38	.09864	10.1381	.11629	8.59893	.13402	7.46154	.15183	6.58627	.16974	5.89151	22
39	.09893	10.1080	.11659	8.57718	.13432	7.44509	.15213	6.57339	.17004	5.88114	21
40	.09923	10.0780	.11688	8.55555	.13461	7.42871	.15243	6.56055	.17033	5.87080	20
41	.09952	10.0483	.11718	8.53402	.13491	7.41240	.15272	6.54777	.17063	5.86051	19
42	.09981	10.0187	.11747	8.51259	.13521	7.39616	.15302	6.53503	.17093	5.85024	18
43	.10011	9.98931	.11777	8.49128	.13550	7.37999	.15332	6.52234	.17123	5.84001	17
44	.10040	9.96007	.11806	8.47007	.13580	7.36389	.15362	6.50970	.17153	5.82982	16
45	.10069	9.93101	.11836	8.44896	.13609	7.34786	.15391	6.49710	.17183	5.81966	15
46	.10099	9.90211	.11865	8.42795	.13639	7.33190	.15421	6.48456	.17213	5.80953	14
47	.10128	9.87338	.11895	8.40705	.13669	7.31600	.15451	6.47206	.17243	5.79944	13
48	.10158	9.84482	.11924	8.38625	.13698	7.30018	.15481	6.45961	.17273	5.78938	12
49	.10187	9.81641	.11954	8.36555	.13728	7.28442	.15511	6.44720	.17303	5.77936	11
50	.10216	9.78817	.11983	8.34496	.13758	7.26873	.15540	6.43484	.17333	5.76937	10
51	.10246	9.76009	.12013	8.32446	.13787	7.25310	.15570	6.42253	.17363	5.75941	9
52	.10275	9.73217	.12042	8.30406	.13817	7.23754	.15600	6.41026	.17393	5.74949	8
53	.10305	9.70441	.12072	8.28376	.13846	7.22204	.15630	6.39804	.17423	5.73960	7
54	.10334	9.67680	.12101	8.26355	.13876	7.20661	.15660	6.38587	.17453	5.72974	6
55	.10363	9.64935	.12131	8.24345	.13906	7.19125	.15689	6.37374	.17483	5.71992	5
56	.10393	9.62205	.12160	8.22344	.13935	7.17594	.15719	6.36165	.17513	5.71013	4
57	.10422	9.59490	.12190	8.20352	.13965	7.16071	.15749	6.34961	.17543	5.70037	3
58	.10452	9.56791	.12219	8.18370	.13995	7.14553	.15779	6.33761	.17573	5.69064	2
59	.10481	9.54106	.12249	8.16398	.14024	7.13042	.15809	6.32566	.17603	5.68094	1
60	.10510	9.51436	.12278	8.14435	.14054	7.11537	.15838	6.31375	.17633	5.67128	0
′	Cotang	Tang	Cotang	Tang	Cotang	Tang	Cotang	Tang	Cotang	Tang	′
	84°		83°		82°		81°		80°		

′	10° Tang	10° Cotang	11° Tang	11° Cotang	12° Tang	12° Cotang	13° Tang	13° Cotang	14° Tang	14° Cotang	′
0	.17633	5.67128	.19438	5.14455	.21256	4.70463	.23087	4.33148	.24933	4.01078	60
1	.17663	5.66165	.19468	5.13658	.21286	4.69791	.23117	4.32573	.24964	4.00582	59
2	.17693	5.65205	.19498	5.12862	.21316	4.69121	.23148	4.32001	.24995	4.00086	58
3	.17723	5.64248	.19529	5.12069	.21347	4.68452	.23179	4.31430	.25026	3.99592	57
4	.17753	5.63295	.19559	5.11279	.21377	4.67786	.23209	4.30860	.25056	3.99099	56
5	.17783	5.62344	.19589	5.10490	.21408	4.67121	.23240	4.30291	.25087	3.98607	55
6	.17813	5.61397	.19619	5.09704	.21438	4.66458	.23271	4.29724	.25118	3.98117	54
7	.17843	5.60452	.19649	5.08921	.21469	4.65797	.23301	4.29159	.25149	3.97627	53
8	.17873	5.59511	.19680	5.08139	.21499	4.65138	.23332	4.28595	.25180	3.97139	52
9	.17903	5.58573	.19710	5.07360	.21529	4.64480	.23363	4.28032	.25211	3.96651	51
10	.17933	5.57638	.19740	5.06584	.21560	4.63825	.23393	4.27471	.25242	3.96165	50
11	.17963	5.56706	.19770	5.05809	.21590	4.63171	.23424	4.26911	.25273	3.95680	49
12	.17993	5.55777	.19801	5.05037	.21621	4.62518	.23455	4.26352	.25304	3.95196	48
13	.18023	5.54851	.19831	5.04267	.21651	4.61868	.23485	4.25795	.25335	3.94713	47
14	.18053	5.53927	.19861	5.03499	.21682	4.61219	.23516	4.25239	.25366	3.94232	46
15	.18083	5.53007	.19891	5.02734	.21712	4.60572	.23547	4.24685	.25397	3.93751	45
16	.18113	5.52090	.19921	5.01971	.21743	4.59927	.23578	4.24132	.25428	3.93271	44
17	.18143	5.51176	.19952	5.01210	.21773	4.59283	.23608	4.23580	.25459	3.92793	43
18	.18173	5.50264	.19982	5.00451	.21804	4.58641	.23639	4.23030	.25490	3.92316	42
19	.18203	5.49356	.20012	4.99695	.21834	4.58001	.23670	4.22481	.25521	3.91839	41
20	.18233	5.48451	.20042	4.98940	.21864	4.57363	.23700	4.21933	.25552	3.91364	40
21	.18263	5.47548	.20073	4.98188	.21895	4.56726	.23731	4.21387	.25583	3.90890	39
22	.18293	5.46648	.20103	4.97438	.21925	4.56091	.23762	4.20842	.25614	3.90417	38
23	.18323	5.45751	.20133	4.96690	.21956	4.55458	.23793	4.20298	.25645	3.89945	37
24	.18353	5.44857	.20164	4.95945	.21986	4.54826	.23823	4.19756	.25676	3.89474	36
25	.18384	5.43966	.20194	4.95201	.22017	4.54196	.23854	4.19215	.25707	3.89004	35
26	.18414	5.43077	.20224	4.94460	.22047	4.53568	.23885	4.18675	.25738	3.88536	34
27	.18444	5.42192	.20254	4.93721	.22078	4.52941	.23916	4.18137	.25769	3.88068	33
28	.18474	5.41309	.20285	4.92984	.22108	4.52316	.23946	4.17600	.25800	3.87601	32
29	.18504	5.40429	.20315	4.92249	.22139	4.51693	.23977	4.17064	.25831	3.87136	31
30	.18534	5.39552	.20345	4.91516	.22169	4.51071	.24008	4.16530	.25862	3.86671	30
31	.18564	5.38677	.20376	4.90785	.22200	4.50451	.24039	4.15997	.25893	3.86208	29
32	.18594	5.37805	.20406	4.90056	.22231	4.49832	.24069	4.15465	.25924	3.85745	28
33	.18624	5.36936	.20436	4.89330	.22261	4.49215	.24100	4.14934	.25955	3.85284	27
34	.18654	5.36070	.20466	4.88605	.22292	4.48600	.24131	4.14405	.25986	3.84824	26
35	.18684	5.35206	.20497	4.87882	.22322	4.47986	.24162	4.13877	.26017	3.84364	25
36	.18714	5.34345	.20527	4.87162	.22353	4.47374	.24193	4.13350	.26048	3.83906	24
37	.18745	5.33487	.20557	4.86444	.22383	4.46764	.24223	4.12825	.26079	3.83449	23
38	.18775	5.32631	.20588	4.85727	.22414	4.46155	.24254	4.12301	.26110	3.82992	22
39	.18805	5.31778	.20618	4.85013	.22444	4.45548	.24285	4.11778	.26141	3.82537	21
40	.18835	5.30928	.20648	4.84300	.22475	4.44942	.24316	4.11256	.26172	3.82083	20
41	.18865	5.30080	.20679	4.83590	.22505	4.44338	.24347	4.10736	.26203	3.81630	19
42	.18895	5.29235	.20709	4.82882	.22536	4.43735	.24377	4.10216	.26235	3.81177	18
43	.18925	5.28393	.20739	4.82175	.22567	4.43134	.24408	4.09699	.26266	3.80726	17
44	.18955	5.27553	.20770	4.81471	.22597	4.42534	.24439	4.09182	.26297	3.80276	16
45	.18986	5.26715	.20800	4.80769	.22628	4.41936	.24470	4.08666	.26328	3.79827	15
46	.19016	5.25880	.20830	4.80068	.22658	4.41340	.24501	4.08152	.26359	3.79378	14
47	.19046	5.25048	.20861	4.79370	.22689	4.40745	.24532	4.07639	.26390	3.78931	13
48	.19076	5.24218	.20891	4.78673	.22719	4.40152	.24562	4.07127	.26421	3.78485	12
49	.19106	5.23391	.20921	4.77978	.22750	4.39560	.24593	4.06616	.26452	3.78040	11
50	.19136	5.22566	.20952	4.77286	.22781	4.38969	.24624	4.06107	.26483	3.77595	10
51	.19166	5.21744	.20982	4.76595	.22811	4.38381	.24655	4.05599	.26515	3.77152	9
52	.19197	5.20925	.21013	4.75906	.22842	4.37793	.24686	4.05092	.26546	3.76709	8
53	.19227	5.20107	.21043	4.75219	.22872	4.37207	.24717	4.04586	.26577	3.76268	7
54	.19257	5.19293	.21073	4.74534	.22903	4.36623	.24747	4.04081	.26608	3.75828	6
55	.19287	5.18480	.21104	4.73851	.22934	4.36040	.24778	4.03578	.26639	3.75388	5
56	.19317	5.17671	.21134	4.73170	.22964	4.35459	.24809	4.03076	.26670	3.74950	4
57	.19347	5.16863	.21164	4.72490	.22995	4.34879	.24840	4.02574	.26701	3.74512	3
58	.19378	5.16058	.21195	4.71813	.23026	4.34300	.24871	4.02074	.26733	3.74075	2
59	.19408	5.15256	.21225	4.71137	.23056	4.33723	.24902	4.01576	.26764	3.73640	1
60	.19438	5.14455	.21256	4.70463	.23087	4.33148	.24933	4.01078	.26795	3.73205	0
′	Cotang	Tang	Cotang	Tang	Cotang	Tang	Cotang	Tang	Cotang	Tang	′
	79°		78°		77°		76°		75°		

′	15° Tang	Cotang	16° Tang	Cotang	17° Tang	Cotang	18° Tang	Cotang	19° Tang	Cotang	′
0	.26795	3.73205	.28675	3.48741	.30573	3.27085	.32492	3.07768	.34433	2.90421	60
1	.26826	3.72771	.28706	3.48359	.30605	3.26745	.32524	3.07464	.34465	2.90147	59
2	.26857	3.72338	.28738	3.47977	.30637	3.26406	32556	3.07160	.34498	2.89873	58
3	.26888	3.71907	.28769	3.47596	.30669	3.26067	.32588	3.06857	.34530	2.89600	57
4	.26920	3.71476	.28800	3.47216	.30700	3.25729	.32621	3.06554	.34563	2.89327	56
5	.26951	3.71046	.28832	3.46837	.30732	3.25392	.32653	3.06252	.34596	2.89055	55
6	.26982	3.70616	.28864	3.46458	.30764	3.25055	.32685	3.05950	.34628	2.88783	54
7	.27013	3.70188	.28895	3.46080	.30796	3.24719	.32717	3.05649	.34661	2.88511	53
8	.27044	3.69761	.28927	3.45703	.30828	3.24383	.32749	3.05349	.34693	2.88240	52
9	.27076	3.69335	.28958	3.45327	.30860	3.24049	.32782	3.05049	.34726	2.87970	51
10	.27107	3.68909	.28990	3.44951	.30891	3.23714	.32814	3.04749	.34758	2.87700	50
11	.27138	3.68485	.29021	3.44576	.30923	3.23381	.32846	3.04450	.34791	2.87430	49
12	.27169	3.68061	.29053	3.44202	.30955	3.23048	.32878	3.04152	.34824	2.87161	48
13	.27201	3.67638	.29084	3.43829	.30987	3.22715	.32911	3.03854	.34856	2.86892	47
14	.27232	3.67217	.29116	3.43456	.31019	3.22384	.32943	3.03556	.34889	2.86624	46
15	.27263	3.66796	.29147	3.43084	.31051	3.22053	.32975	3.03260	.34922	2.86356	45
16	.27294	3.66376	.29179	3.42713	.31083	3.21722	.33007	3.02963	.34954	2.86089	44
17	.27326	3.65957	.29210	3.42343	.31115	3.21392	.33040	3.02667	.34987	2.85822	43
18	.27357	3.65538	.29242	3.41973	.31147	3.21063	.33072	3.02372	.35020	2.85555	42
19	.27388	3.65121	.29274	3.41604	.31178	3.20734	.33104	3.02077	.35052	2.85289	41
20	.27419	3.64705	.29305	3.41236	.31210	3.20406	.33136	3.01783	.35085	2.85023	40
21	.27451	3.64289	.29337	3.40869	.31242	3.20079	.33169	3.01489	.35118	2.84758	39
22	.27482	3.63874	.29368	3.40502	.31274	3.19752	.33201	3.01196	.35150	2.84494	38
23	.27513	3.63461	.29400	3.40136	.31306	3.19426	.33233	3.00903	.35183	2.84229	37
24	.27545	3.63048	.29432	3.39771	.31338	3.19100	.33266	3.00611	.35216	2.83965	36
25	.27576	3.62636	.29463	3.39406	.31370	3.18775	.33298	3.00319	.35248	2.83702	35
26	.27607	3.62224	.29495	3.39042	.31402	3.18451	.33330	3.00028	.35281	2.83439	34
27	.27638	3.61814	.29526	3.38679	.31434	3.18127	.33363	2.99738	.35314	2.83176	33
28	.27670	3.61405	.29558	3.38317	.31466	3.17804	.33395	2.99447	.35346	2.82914	32
29	.27701	3.60996	.29590	3.37955	.31498	3.17481	.33427	2.99158	.35379	2.82653	31
30	.27732	3.60588	.29621	3.37594	.31530	3.17159	.33460	2.98868	.35412	2.82391	30
31	.27764	3.60181	.29653	3.37234	.31562	3.16838	.33492	2.98580	.35445	2.82130	29
32	.27795	3.59775	.29685	3.36875	.31594	3.16517	.33524	2.98292	.35477	2.81870	28
33	.27826	3.59370	.29716	3.36516	.31626	3.16197	.33557	2.98004	.35510	2.81610	27
34	.27858	3.58966	.29748	3.36158	.31658	3.15877	33589	2.97717	.35543	2.81350	26
35	.27889	3.58562	.29780	3.35800	.31690	3.15558	.33621	2.97430	.35576	2.81091	25
36	.27921	3.58160	.29811	3.35443	.31722	3.15240	.33654	2.97144	.35608	2.80833	24
37	.27952	3.57758	.29843	3.35087	.31754	3.14922	.33686	2.96858	.35641	2.80574	23
38	.27983	3.57357	.29875	3.34732	.31786	3.14605	.33718	2.96573	.35674	2.80316	22
39	.28015	3.56957	.29906	3.34377	.31818	3.14288	.33751	2.96288	.35707	2.80059	21
40	.28046	3.56557	.29938	3.34023	.31850	3.13972	.33783	2.96004	.35740	2.79802	20
41	.28077	3.56159	.29970	3.33670	.31882	3.13656	.33816	2.95721	.35772	2.79545	19
42	.28109	3.55761	.30001	3.33317	.31914	3.13341	.33848	2.95437	.35805	2.79289	18
43	.28140	3.55364	.30033	3.32965	.31946	3.13027	.33881	2.95155	.35838	2.79033	17
44	.28172	3.54968	.30065	3.32614	.31978	3.12713	.33913	2.94872	.35871	2.78778	16
45	.28203	3.54573	.30097	3.32264	.32010	3.12400	.33945	2.94591	.35904	2.78523	15
46	.28234	3.54179	.30128	3.31914	.32042	3.12087	.33978	2.94309	.35937	2.78269	14
47	.28266	3.53785	.30160	3.31565	.32074	3.11775	.34010	2.94028	.35969	2.78014	13
48	.28297	3.53393	.30192	3.31216	.32106	3.11464	.34043	2.93748	.36002	2.77761	12
49	.28329	3.53001	.30224	3.30868	.32139	3.11153	.34075	2.93468	.36035	2.77507	11
50	.28360	3.52609	.30255	3.30521	.32171	3.10842	.34108	2.93189	.36068	2.77254	10
51	.28391	3.52219	.30287	3.30174	.32203	3.10532	.34140	2.92910	.36101	2.77002	9
52	.28423	3.51829	.30319	3.29829	.32235	3.10223	.34173	2.92632	.36134	2.76750	8
53	.28454	3.51441	.30351	3.29483	.32267	3.09914	.34205	2.92354	.36167	2.76498	7
54	.28486	3.51053	.30382	3.29139	.32299	3.09606	.34238	2.92076	.36199	2.76247	6
55	.28517	3.50666	.30414	3.28795	.32331	3.09298	.34270	2.91799	.36232	2.75996	5
56	.28549	3.50279	.30446	3.28452	.32363	3.08991	.34303	2.91523	.36265	2.75746	4
57	.28580	3.49894	.30478	3.28109	.32396	3.08685	.34335	2.91246	.36298	2.75496	3
58	.28612	3.49509	.30509	3.27767	.32428	3.08379	.34368	2.90971	.36331	2.75246	2
59	.28643	3.49125	.30541	3.27426	.32460	3.08073	.34400	2.90696	.36364	2.74997	1
60	.28675	3.48741	.30573	3.27085	.32492	3.07768	.34433	2.90421	.36397	2.74748	0
′	Cotang	Tang	Cotang	Tang	Cotang	Tang	Cotang	Tang	Cotang	Tang	′
	74°		73°		72°		71°		70°		

′	20° Tang	Cotang	21° Tang	Cotang	22° Tang	Cotang	23° Tang	Cotang	24° Tang	Cotang	′
0	.36397	2.74748	.38386	2.60509	.40403	2.47509	.42447	2.35585	.44523	2.24604	60
1	.36430	2.74499	.38420	2.60283	.40436	2.47302	.42482	2.35395	.44558	2.24428	59
2	.36463	2.74251	.38453	2.60057	.40470	2.47095	.42516	2.35205	.44593	2.24252	58
3	.36496	2.74004	.38487	2.59831	.40504	2.46888	.42551	2.35015	.44627	2.24077	57
4	.36529	2.73756	.38520	2.59606	.40538	2.46682	.42585	2.34825	.44662	2.23902	56
5	.36562	2.73509	.38553	2.59381	.40572	2.46476	.42619	2.34636	.44697	2.23727	55
6	.36595	2.73263	.38587	2.59156	.40606	2.46270	.42654	2.34447	.44732	2.23553	54
7	.36628	2.73017	.38620	2.58932	.40640	2.46065	.42688	2.34258	.44767	2.23378	53
8	.36661	2.72771	.38654	2.58708	.40674	2.45860	.42722	2.34069	.44802	2.23204	52
9	.36694	2.72526	.38687	2.58484	.40707	2.45655	.42757	2.33881	.44837	2.23030	51
10	.36727	2.72281	.38721	2.58261	.40741	2.45451	.42791	2.33693	.44872	2.22857	50
11	.36760	2.72036	.38754	2.58038	.40775	2.45246	.42826	2.33505	.44907	2.22683	49
12	.36793	2.71792	.38787	2.57815	.40809	2.45043	.42860	2.33317	.44942	2.22510	48
13	.36826	2.71548	.38821	2.57593	.40843	2.44839	.42894	2.33130	.44977	2.22337	47
14	.36859	2.71305	.38854	2.57371	.40877	2.44636	.42929	2.32943	.45012	2.22164	46
15	.36892	2.71062	.38888	2.57150	.40911	2.44433	.42963	2.32756	.45047	2.21992	45
16	.36925	2.70819	.38921	2.56928	.40945	2.44230	.42998	2.32570	.45082	2.21819	44
17	.36958	2.70577	.38955	2.56707	.40979	2.44027	.43032	2.32383	.45117	2.21647	43
18	.36991	2.70335	.38988	2.56487	.41013	2.43825	.43067	2.32197	.45152	2.21475	42
19	.37024	2.70094	.39022	2.56266	.41047	2.43623	.43101	2.32012	.45187	2.21304	41
20	.37057	2.69853	.39055	2.56046	.41081	2.43422	.43136	2.31826	.45222	2.21132	40
21	.37090	2.69612	.39089	2.55827	.41115	2.43220	.43170	2.31641	.45257	2.20961	39
22	.37123	2.69371	.39122	2.55608	.41149	2.43019	.43205	2.31456	.45292	2.20790	38
23	.37157	2.69131	.39156	2.55389	.41183	2.42819	.43239	2.31271	.45327	2.20619	37
24	.37190	2.68892	.39190	2.55170	.41217	2.42618	.43274	2.31086	.45362	2.20449	36
25	.37223	2.68653	.39223	2.54952	.41251	2.42418	.43308	2.30902	.45397	2.20278	35
26	.37256	2.68414	.39257	2.54734	.41285	2.42218	.43343	2.30718	.45432	2.20108	34
27	.37289	2.68175	.39290	2.54516	.41319	2.42019	.43378	2.30534	.45467	2.19938	33
28	.37322	2.67937	.39324	2.54299	.41353	2.41819	.43412	2.30351	.45502	2.19769	32
29	.37355	2.67700	.39357	2.54082	.41387	2.41620	.43447	2.30167	.45538	2.19599	31
30	.37388	2.67462	.39391	2.53865	.41421	2.41421	.43481	2.29984	.45573	2.19430	30
31	.37422	2.67225	.39425	2.53648	.41455	2.41223	.43516	2.29801	.45608	2.19261	29
32	.37455	2.66989	.39458	2.53432	.41490	2.41025	.43550	2.29619	.45643	2.19092	28
33	.37488	2.66752	.39492	2.53217	.41524	2.40827	.43585	2.29437	.45678	2.18923	27
34	.37521	2.66516	.39526	2.53001	.41558	2.40629	.43620	2.29254	.45713	2.18755	26
35	.37554	2.66281	.39559	2.52786	.41592	2.40432	.43654	2.29073	.45748	2.18587	25
36	.37588	2.66046	.39593	2.52571	.41626	2.40235	.43689	2.28891	.45784	2.18419	24
37	.37621	2.65811	.39626	2.52357	.41660	2.40038	.43724	2.28710	.45819	2.18251	23
38	.37654	2.65576	.39660	2.52142	.41694	2.39841	.43758	2.28528	.45854	2.18084	22
39	.37687	2.65342	.39694	2.51929	.41728	2.39645	.43793	2.28348	.45889	2.17916	21
40	.37720	2.65109	.39727	2.51715	.41763	2.39449	.43828	2.28167	.45924	2.17749	20
41	.37754	2.64875	.39761	2.51502	.41797	2.39253	.43862	2.27987	.45960	2.17582	19
42	.37787	2.64642	.39795	2.51289	.41831	2.39058	.43897	2.27806	.45995	2.17416	18
43	.37820	2.64410	.39829	2.51076	.41865	2.38863	.43932	2.27626	.46030	2.17249	17
44	.37853	2.64177	.39862	2.50864	.41899	2.38668	.43966	2.27447	.46065	2.17083	16
45	.37887	2.63945	.39896	2.50652	.41933	2.38473	.44001	2.27267	.46101	2.16917	15
46	.37920	2.63714	.39930	2.50440	.41968	2.38279	.44036	2.27088	.46136	2.16751	14
47	.37953	2.63483	.39963	2.50229	.42002	2.38084	.44071	2.26909	.46171	2.16585	13
48	.37986	2.63252	.39997	2.50018	.42036	2.37891	.44105	2.26730	.46206	2.16420	12
49	.38020	2.63021	.40031	2.49807	.42070	2.37697	.44140	2.26552	.46242	2.16255	11
50	.38053	2.62791	.40065	2.49597	.42105	2.37504	.44175	2.26374	.46277	2.16090	10
51	.38086	2.62561	.40098	2.49386	.42139	2.37311	.44210	2.26196	.46312	2.15925	9
52	.38120	2.62332	.40132	2.49177	.42173	2.37118	.44244	2.26018	.46348	2.15760	8
53	.38153	2.62103	.40166	2.48967	.42207	2.36925	.44279	2.25840	.46383	2.15596	7
54	.38186	2.61874	.40200	2.48758	.42242	2.36733	.44314	2.25663	.46418	2.15432	6
55	.38220	2.61646	.40234	2.48549	.42276	2.36541	.44349	2.25486	.46454	2.15268	5
56	.38253	2.61418	.40267	2.48340	.42310	2.36349	.44384	2.25309	.46489	2.15104	4
57	.38286	2.61190	.40301	2.48132	.42345	2.36158	.44418	2.25132	.46525	2.14940	3
58	.38320	2.60963	.40335	2.47924	.42379	2.35967	.44453	2.24956	.46560	2.14777	2
59	.38353	2.60736	.40369	2.47716	.42413	2.35776	.44488	2.24780	.46595	2.14614	1
60	.38386	2.60509	.40403	2.47509	.42447	2.35585	.44523	2.24604	.46631	2.14451	0
′	Cotang	Tang	Cotang	Tang	Cotang	Tang	Cotang	Tang	Cotang	Tang	′
	69°		68°		67°		66°		65°		

′	25° Tang	25° Cotang	26° Tang	26° Cotang	27° Tang	27° Cotang	28° Tang	28° Cotang	29° Tang	29° Cotang	′
0	.46631	2.14451	.48773	2.05030	.50953	1.96261	.53171	1.88073	.55431	1.80405	60
1	.46666	2.14288	.48809	2.04879	.50989	1.96120	.53208	1.87941	.55469	1.80281	59
2	.46702	2.14125	.48845	2.04728	.51026	1.95979	.53246	1.87809	.55507	1.80158	58
3	.46737	2.13963	.48881	2.04577	.51063	1.95838	.53283	1.87677	.55545	1.80034	57
4	.46772	2.13801	.48917	2.04426	.51099	1.95698	.53320	1.87546	.55583	1.79911	56
5	.46808	2.13639	.48953	2.04276	.51136	1.95557	.53358	1.87415	.55621	1.79788	55
6	.46843	2.13477	.48989	2.04125	.51173	1.95417	.53395	1.87283	.55659	1.79665	54
7	.46879	2.13316	.49026	2.03975	.51209	1.95277	.53432	1.87152	.55697	1.79542	53
8	.46914	2.13154	.49062	2.03825	.51246	1.95137	.53470	1.87021	.55736	1.79419	52
9	.46950	2.12993	.49098	2.03675	.51283	1.94997	.53507	1.86891	.55774	1.79296	51
10	.46985	2.12832	.49134	2.03526	.51319	1.94858	.53545	1.86760	.55812	1.79174	50
11	.47021	2.12671	.49170	2.03376	.51356	1.94718	.53582	1.86630	.55850	1.79051	49
12	.47056	2.12511	.49206	2.03227	.51393	1.94579	.53620	1.86499	.55888	1.78929	48
13	.47092	2.12350	.49242	2.03078	.51430	1.94440	.53657	1.86369	.55926	1.78807	47
14	.47128	2.12190	.49278	2.02929	.51467	1.94301	.53694	1.86239	.55964	1.78685	46
15	.47163	2.12030	.49315	2.02780	.51503	1.94162	.53732	1.86109	.56003	1.78563	45
16	.47199	2.11871	.49351	2.02631	.51540	1.94023	.53769	1.85979	.56041	1.78441	44
17	.47234	2.11711	.49387	2.02483	.51577	1.93885	.53807	1.85850	.56079	1.78319	43
18	.47270	2.11552	.49423	2.02335	.51614	1.93746	.53844	1.85720	.56117	1.78198	42
19	.47305	2.11392	.49459	2.02187	.51651	1.93608	.53882	1.85591	.56156	1.78077	41
20	.47341	2.11233	.49495	2.02039	.51688	1.93470	.53920	1.85462	.56194	1.77955	40
21	.47377	2.11075	.49532	2.01891	.51724	1.93332	.53957	1.85333	.56232	1.77834	39
22	.47412	2.10916	.49568	2.01743	.51761	1.93195	.53995	1.85204	.56270	1.77713	38
23	.47448	2.10758	.49604	2.01596	.51798	1.93057	.54032	1.85075	.56309	1.77592	37
24	.47483	2.10600	.49640	2.01449	.51835	1.92920	.54070	1.84946	.56347	1.77471	36
25	.47519	2.10442	.49677	2.01302	.51872	1.92782	.54107	1.84818	.56385	1.77351	35
26	.47555	2.10284	.49713	2.01155	.51909	1.92645	.54145	1.84689	.56424	1.77230	34
27	.47590	2.10126	.49749	2.01008	.51946	1.92508	.54183	1.84561	.56462	1.77110	33
28	.47626	2.09969	.49786	2.00862	.51983	1.92371	.54220	1.84433	.56501	1.76990	32
29	.47662	2.09811	.49822	2.00715	.52020	1.92235	.54258	1.84305	.56539	1.76869	31
30	.47698	2.09654	.49858	2.00569	.52057	1.92098	.54296	1.84177	.56577	1.76749	30
31	.47733	2.09498	.49894	2.00423	.52094	1.91962	.54333	1.84049	.56616	1.76629	29
32	.47769	2.09341	.49931	2.00277	.52131	1.91826	.54371	1.83922	.56654	1.76510	28
33	.47805	2.09184	.49967	2.00131	.52168	1.91690	.54409	1.83794	.56693	1.76390	27
34	.47840	2.09028	.50004	1.99986	.52205	1.91554	.54446	1.83667	.56731	1.76271	26
35	.47876	2.08872	.50040	1.99841	.52242	1.91418	.54484	1.83540	.56769	1.76151	25
36	.47912	2.08716	.50076	1.99695	.52279	1.91282	.54522	1.83413	.56808	1.76032	24
37	.47948	2.08560	.50113	1.99550	.52316	1.91147	.54560	1.83286	.56846	1.75913	23
38	.47984	2.08405	.50149	1.99406	.52353	1.91012	.54597	1.83159	.56885	1.75794	22
39	.48019	2.08250	.50185	1.99261	.52390	1.90876	.54635	1.83033	.56923	1.75675	21
40	.48055	2.08094	.50222	1.99116	.52427	1.90741	.54673	1.82906	.56962	1.75556	20
41	.48091	2.07939	.50258	1.98972	.52464	1.90607	.54711	1.82780	.57000	1.75437	19
42	.48127	2.07785	.50295	1.98828	.52501	1.90472	.54748	1.82654	.57039	1.75319	18
43	.48163	2.07630	.50331	1.98684	.52538	1.90337	.54786	1.82528	.57078	1.75200	17
44	.48198	2.07476	.50368	1.98540	.52575	1.90203	.54824	1.82402	.57116	1.75082	16
45	.48234	2.07321	.50404	1.98396	.52613	1.90069	.54862	1.82276	.57155	1.74964	15
46	.48270	2.07167	.50441	1.98253	.52650	1.89935	.54900	1.82150	.57193	1.74846	14
47	.48306	2.07014	.50477	1.98110	.52687	1.89801	.54938	1.82025	.57232	1.74728	13
48	.48342	2.06860	.50514	1.97966	.52724	1.89667	.54975	1.81899	.57271	1.74610	12
49	.48378	2.06706	.50550	1.97823	.52761	1.89533	.55013	1.81774	.57309	1.74492	11
50	.48414	2.06553	.50587	1.97681	.52798	1.89400	.55051	1.81649	.57348	1.74375	10
51	.48450	2.06400	.50623	1.97538	.52836	1.89266	.55089	1.81524	.57386	1.74257	9
52	.48486	2.06247	.50660	1.97395	.52873	1.89133	.55127	1.81399	.57425	1.74140	8
53	.48521	2.06094	.50696	1.97253	.52910	1.89000	.55165	1.81274	.57464	1.74022	7
54	.48557	2.05942	.50733	1.97111	.52947	1.88867	.55203	1.81150	.57503	1.73905	6
55	.48593	2.05790	.50769	1.96969	.52985	1.88734	.55241	1.81025	.57541	1.73788	5
56	.48629	2.05637	.50806	1.96827	.53022	1.88602	.55279	1.80901	.57580	1.73671	4
57	.48665	2.05485	.50843	1.96685	.53059	1.88469	.55317	1.80777	.57619	1.73555	3
58	.48701	2.05333	.50879	1.96544	.53096	1.88337	.55355	1.80653	.57657	1.73438	2
59	.48737	2.05182	.50916	1.96402	.53134	1.88205	.55393	1.80529	.57696	1.73321	1
60	.48773	2.05030	.50953	1.96261	.53171	1.88073	.55431	1.80405	.57735	1.73205	0
′	Cotang	Tang	Cotang	Tang	Cotang	Tang	Cotang	Tang	Cotang	Tang	′
	64°		63°		62°		61°		60°		

′	30° Tang	30° Cotang	31° Tang	31° Cotang	32° Tang	32° Cotang	33° Tang	33° Cotang	34° Tang	34° Cotang	′
0	.57735	1.73205	.60086	1.66428	.62487	1.60033	.64941	1.53986	.67451	1.48256	60
1	.57774	1.73089	.60126	1.66318	.62527	1.59930	.64982	1.53888	.67493	1.48163	59
2	.57813	1.72973	.60165	1.66209	.62568	1.59826	.65024	1.53791	.67536	1.48070	58
3	.57851	1.72857	.60205	1.66099	.62608	1.59723	.65065	1.53693	.67578	1.47977	57
4	.57890	1.72741	.60245	1.65990	.62649	1.59620	.65106	1.53595	.67620	1.47885	56
5	.57929	1.72625	.60284	1.65881	.62689	1.59517	.65148	1.53497	.67663	1.47792	55
6	.57968	1.72509	.60324	1.65772	.62730	1.59414	.65189	1.53400	.67705	1.47699	54
7	.58007	1.72393	.60364	1.65663	.62770	1.59311	.65231	1.53302	.67748	1.47607	53
8	.58046	1.72278	.60403	1.65554	.62811	1.59208	.65272	1.53205	.67790	1.47514	52
9	.58085	1.72163	.60443	1.65445	.62852	1.59105	.65314	1.53107	.67832	1.47422	51
10	.58124	1.72047	.60483	1 65337	.62892	1.59002	.65355	1.53010	.67875	1.47330	50
11	.58162	1.71932	.60522	1.65228	.62933	1.58900	.65397	1.52913	.67917	1.47238	49
12	.58201	1.71817	.60562	1.65120	.62973	1.58797	.65438	1.52816	.67960	1.47146	48
13	.58240	1.71702	.60602	1.65011	.63014	1.58695	.65480	1.52719	.68002	1.47053	47
14	.58279	1.71588	.60642	1.64903	.63055	1.58593	.65521	1.52622	.68045	1.46962	46
15	.58318	1.71473	.60681	1.64795	.63095	1.58490	.65563	1.52525	.68088	1.46870	45
16	.58357	1.71358	.60721	1.64687	.63136	1.58388	.65604	1.52429	.68130	1.46778	44
17	.58396	1.71244	.60761	1.64579	.63177	1.58286	.65646	1.52332	.68173	1.46686	43
18	.58435	1.71129	.60801	1.64471	.63217	1.58184	.65688	1.52235	.68215	1.46595	42
19	.58474	1.71015	.60841	1.64363	.63258	1.58083	.65729	1.52139	.68258	1.46503	41
20	.58513	1.70901	.60881	1.64256	.63299	1.57981	.65771	1.52043	.68301	1.46411	40
21	.58552	1.70787	.60921	1.64148	.63340	1.57879	.65813	1.51946	.68343	1.46320	39
22	.58591	1.70673	.60960	1.64041	.63380	1.57778	.65854	1.51850	.68386	1.46229	38
23	.58631	1.70560	.61000	1.63934	.63421	1.57676	.65896	1.51754	.68429	1.46137	37
24	.58670	1.70446	.61040	1.63826	.63462	1.57575	.65938	1.51658	.68471	1.46046	36
25	.58709	1.70332	.61080	1.63719	.63503	1.57474	.65980	1.51562	.68514	1.45955	35
26	.58748	1.70219	.61120	1.63612	.63544	1.57372	.66021	1.51466	.68557	1.45864	34
27	.58787	1.70106	.61160	1.63505	.63584	1.57271	.66063	1.51370	.68600	1.45773	33
28	.58826	1.69992	.61200	1.63398	.63625	1.57170	.66105	1.51275	.68642	1.45682	32
29	.58865	1.69879	.61240	1.63292	.63665	1.57069	.66147	1.51179	.68685	1.45592	31
30	.58905	1.69766	.61280	1.63185	.63707	1.56969	.66189	1.51084	.68728	1.45501	30
31	.58944	1.69653	.61320	1.63079	.63748	1.56868	.66230	1.50988	.68771	1.45410	29
32	.58983	1.69541	.61360	1.62972	.63789	1.56767	.66272	1.50893	.68814	1.45320	28
33	.59022	1.69428	.61400	1.62866	.63830	1.56667	.66314	1.50797	.68857	1.45229	27
34	.59061	1.69316	.61440	1.62760	.63871	1.56566	.66356	1.50702	.68900	1.45139	26
35	.59101	1.69203	.61480	1.62654	.63912	1.56466	.66398	1.50607	.68942	1.45049	25
36	.59149	1.69091	.61520	1.62548	.63953	1.56366	.66440	1.50512	.68985	1.44958	24
37	.59179	1.68979	.61561	1.62442	.63994	1.56265	.66482	1.50417	.69028	1.44868	23
38	.59218	1.68866	.61601	1.62336	.64035	1.56165	.66524	1.50322	.69071	1.44778	22
39	.59258	1.68754	.61641	1.62230	.64076	1.56065	.66566	1.50228	.69114	1.44688	21
40	.59297	1.68643	.61681	1.62125	.64117	1.55966	.66608	1.50133	.69157	1.44598	20
41	.59336	1.68531	.61721	1.62019	.64158	1.55866	.66650	1.50038	.69200	1.44508	19
42	.59376	1.68419	.61761	1.61914	.64199	1.55766	.66692	1.49944	.69243	1.44418	18
43	.59415	1.68308	.61801	1.61808	.64240	1.55666	.66734	1.49849	.69286	1.44329	17
44	.59454	1.68196	.61842	1.61703	.64281	1.55567	.66776	1.49755	.69329	1.44239	16
45	.59494	1.68085	.61882	1.61598	.64322	1.55467	.66818	1.49661	.69372	1.44149	15
46	.59533	1.67974	.61922	1.61493	.64363	1.55368	.66860	1.49566	.69416	1.44060	14
47	.59573	1.67863	.61962	1.61388	.64404	1.55269	.66902	1.49472	.69459	1.43970	13
48	.59612	1.67752	.62003	1.61283	.64446	1.55170	.66944	1.49378	.69502	1.43881	12
49	.59651	1.67641	.62043	1.61179	.64487	1.55071	.66986	1.49284	.69545	1.43792	11
50	.59691	1.67530	.62083	1.61074	.64528	1.54972	.67028	1.49190	.69588	1.43703	10
51	.59730	1.67419	.62124	1.60970	.64569	1.54873	.67071	1.49097	.69631	1.43614	9
52	.59770	1.67309	.62164	1.60865	.64610	1.54774	.67113	1.49003	.69675	1.43525	8
53	.59809	1.67198	.62204	1.60761	.64652	1.54675	.67155	1 48909	.69718	1.43436	7
54	.59849	1.67088	.62245	1.60657	.64693	1.54576	.67197	1.48816	.69761	1.43347	6
55	.59888	1.66978	.62285	1.60553	.64734	1.54478	.67239	1.48722	.69804	1.43258	5
56	.59928	1.66867	.62325	1.60449	.64775	1.54379	.67282	1.48629	.69847	1.43169	4
57	.59967	1.66757	.62366	1.60345	.64817	1.54281	.67324	1.48536	.69891	1.43080	3
58	.60007	1.66647	.62406	1.60241	.64858	1.54183	.67366	1.48442	.69934	1.42992	2
59	.60046	1.66538	.62446	1.60137	.64899	1.54085	.67409	1.48349	.69977	1.42903	1
60	.60086	1.66428	.62487	1.60033	.64941	1.53986	.67451	1.48256	.70021	1.42815	0
′	Cotang	Tang	Cotang	Tang	Cotang	Tang	Cotang	Tang	Cotang	Tang	′
	59°		58°		57°		56°		55°		

′	35° Tang	35° Cotang	36° Tang	36° Cotang	37° Tang	37° Cotang	38° Tang	38° Cotang	39° Tang	39° Cotang	′
0	.70021	1.42815	.72654	1.37638	.75355	1.32704	.78129	1.27994	.80978	1.23490	60
1	.70064	1.42726	.72699	1.37554	.75401	1.32624	.78175	1.27917	.81027	1.23416	59
2	.70107	1.42638	.72743	1.37470	.75447	1.32544	.78222	1.27841	.81075	1.23343	58
3	.70151	1.42550	.72788	1.37386	.75492	1.32464	.78269	1.27764	.81123	1.23270	57
4	.70194	1.42462	.72832	1.37302	.75538	1.32384	.78316	1.27688	.81171	1.23196	56
5	.70238	1.42374	.72877	1.37218	.75584	1.32304	.78363	1.27611	.81220	1.23123	55
6	.70281	1.42286	.72921	1.37134	.75629	1.32224	.78410	1.27535	.81268	1.23050	54
7	.70325	1.42198	.72966	1.37050	.75675	1.32144	.78457	1.27458	.81316	1.22977	53
8	.70368	1.42110	.73010	1.36967	.75721	1.32064	.78504	1.27382	.81364	1.22904	52
9	.70412	1.42022	.73055	1.36883	.75767	1.31984	.78551	1.27306	.81413	1.22831	51
10	.70455	1.41934	.73100	1.36800	.75812	1.31904	.78598	1.27230	.81461	1.22758	50
11	.70499	1.41847	.73144	1.36716	.75858	1.31825	.78645	1.27153	.81510	1.22685	49
12	.70542	1.41759	.73189	1.36633	.75904	1.31745	.78692	1.27077	.81558	1.22612	48
13	.70586	1.41672	.73234	1.36549	.75950	1.31666	.78739	1.27001	.81606	1.22539	47
14	.70629	1.41584	.73278	1.36466	.75996	1.31586	.78786	1.26925	.81655	1.22467	46
15	.70673	1.41497	.73323	1.36383	.76042	1.31507	.78834	1.26849	.81703	1.22394	45
16	.70717	1.41409	.73368	1.36300	.76088	1.31427	.78881	1.26774	.81752	1.22321	44
17	.70760	1.41322	.73413	1.36217	.76134	1.31348	.78928	1.26698	.81800	1.22249	43
18	.70804	1.41235	.73457	1.36134	.76180	1.31269	.78975	1.26622	.81849	1.22176	42
19	.70848	1.41148	.73502	1.36051	.76226	1.31190	.79022	1.26546	.81898	1.22104	41
20	.70891	1.41061	.73547	1.35968	.76272	1.31110	.79070	1.26471	.81946	1.22031	40
21	.70935	1.40974	.73592	1.35885	.76318	1.31031	.79117	1.26395	.81995	1.21959	39
22	.70979	1.40887	.73637	1.35802	.76364	1.30952	.79164	1.26319	.82044	1.21886	38
23	.71023	1.40800	.73681	1.35719	.76410	1.30873	.79212	1.26244	.82092	1.21814	37
24	.71066	1.40714	.73726	1.35637	.76456	1.30795	.79259	1.26169	.82141	1.21742	36
25	.71110	1.40627	.73771	1.35554	.76502	1.30716	.79306	1.26093	.82190	1.21670	35
26	.71154	1.40540	.73816	1.35472	.76548	1.30637	.79354	1.26018	.82238	1.21598	34
27	.71198	1.40454	.73861	1.35389	.76594	1.30558	.79401	1.25943	.82287	1.21526	33
28	.71242	1.40367	.73906	1.35307	.76640	1.30480	.79449	1.25867	.82336	1.21454	32
29	.71285	1.40281	.73951	1.35224	.76686	1.30401	.79496	1.25792	.82385	1.21382	31
30	.71329	1.40195	.73996	1.35142	.76733	1.30323	.79544	1.25717	.82434	1.21310	30
31	.71373	1.40109	.74041	1.35060	.76779	1.30244	.79591	1.25642	.82483	1.21238	29
32	.71417	1.40022	.74086	1.34978	.76825	1.30166	.79639	1.25567	.82531	1.21166	28
33	.71461	1.39936	.74131	1.34896	.76871	1.30087	.79686	1.25492	.82580	1.21094	27
34	.71505	1.39850	.74176	1.34814	.76918	1.30009	.79734	1.25417	.82629	1.21023	26
35	.71549	1.39764	.74221	1.34732	.76964	1.29931	.79781	1.25343	.82678	1.20951	25
36	.71593	1.39679	.74267	1.34650	.77010	1.29853	.79829	1.25268	.82727	1.20879	24
37	.71637	1.39593	.74312	1.34568	.77057	1.29775	.79877	1.25193	.82776	1.20808	23
38	.71681	1.39507	.74357	1.34487	.77103	1.29696	.79924	1.25118	.82825	1.20736	22
39	.71725	1.39421	.74402	1.34405	.77149	1.29618	.79972	1.25044	.82874	1.20665	21
40	.71769	1.39336	.74447	1.34323	.77196	1.29541	.80020	1.24969	.82923	1.20593	20
41	.71813	1.39250	.74492	1.34242	.77242	1.29463	.80067	1.24895	.82972	1.20522	19
42	.71857	1.39165	.74538	1.34160	.77289	1.29385	.80115	1.24820	.83022	1.20451	18
43	.71901	1.39079	.74583	1.34079	.77335	1.29307	.80163	1.24746	.83071	1.20379	17
44	.71946	1.38994	.74628	1.33998	.77382	1.29229	.80211	1.24672	.83120	1.20308	16
45	.71990	1.38909	.74674	1.33916	.77428	1.29152	.80258	1.24597	.83169	1.20237	15
46	.72034	1.38824	.74719	1.33835	.77475	1.29074	.80306	1.24523	.83218	1.20166	14
47	.72078	1.38738	.74764	1.33754	.77521	1.28997	.80354	1.24449	.83268	1.20095	13
48	.72122	1.38653	.74809	1.33673	.77568	1.28919	.80402	1.24375	.83317	1.20024	12
49	.72167	1.38568	.74855	1.33592	.77615	1.28842	.80450	1.24301	.83366	1.19953	11
50	.72211	1.38484	.74900	1.33511	.77661	1.28764	.80498	1.24227	.83415	1.19882	10
51	.72255	1.38399	.74946	1.33430	.77708	1.28687	.80546	1.24153	.83465	1.19811	9
52	.72299	1.38314	.74991	1.33349	.77754	1.28610	.80594	1.24079	.83514	1.19740	8
53	.72344	1.38229	.75037	1.33268	.77801	1.28533	.80642	1.24005	.83564	1.19669	7
54	.72388	1.38145	.75082	1.33187	.77848	1.28456	.80690	1.23931	.83613	1.19599	6
55	.72432	1.38060	.75128	1.33107	.77895	1.28379	.80738	1.23858	.83662	1.19528	5
56	.72477	1.37976	.75173	1.33026	.77941	1.28302	.80786	1.23784	.83712	1.19457	4
57	.72521	1.37891	.75219	1.32946	.77988	1.28225	.80834	1.23710	.83761	1.19387	3
58	.72565	1.37807	.75264	1.32865	.78035	1.28148	.80882	1.23637	.83811	1.19316	2
59	.72610	1.37722	.75310	1.32785	.78082	1.28071	.80930	1.23563	.83860	1.19246	1
60	.72654	1.37638	.75355	1.32704	.78129	1.27994	.80978	1.23490	.83910	1.19175	0

′	Cotang	Tang	Cotang	Tang	Cotang	Tang	Cotang	Tang	Cotang	Tang	′
	54°		53°		52°		51°		50°		

′	40°		41°		42°		43°		44°		′
	Tang	Cotang	Tang	Cotang	Tang	Cotang	Tang	Cotang	Tang	Cotang	
0	.83910	1.19175	.86929	1.15037	.90040	1.11061	.93252	1.07237	.96569	1.03553	60
1	.83960	1.19105	.86980	1.14969	.90093	1.10996	.93306	1.07174	.96625	1.03493	59
2	.84009	1.19035	.87031	1.14902	.90146	1.10931	.93360	1.07112	.96681	1.03433	58
3	.84059	1.18964	.87082	1.14834	.90199	1.10867	.93415	1.07049	.96738	1.03372	57
4	.84108	1.18894	.87133	1.14767	.90251	1.10802	.93469	1.06987	.96794	1.03312	56
5	.84158	1.18824	.87184	1.14699	.90304	1.10737	.93524	1.06925	.96850	1.03252	55
6	.84208	1.18754	.87236	1.14632	.90357	1.10672	.93578	1.06862	.96907	1.03192	54
7	.84258	1.18684	.87287	1.14565	.90410	1.10607	.93633	1.06800	.96963	1.03132	53
8	.84307	1.18614	.87338	1.14498	.90463	1.10543	.93688	1.06738	.97020	1.03072	52
9	.84357	1.18544	.87389	1.14430	.90516	1.10478	.93742	1.06676	.97076	1.03012	51
10	.84407	1.18474	.87441	1.14363	.90569	1.10414	.93797	1.06613	.97133	1.02952	50
11	.84457	1.18404	.87492	1.14296	.90621	1.10349	.93852	1.06551	.97189	1.02892	49
12	.84507	1.18334	.87543	1.14229	.90674	1.10285	.93906	1.06489	.97246	1.02832	48
13	.84556	1.18264	.87595	1.14162	.90727	1.10220	.93961	1.06427	.97302	1.02772	47
14	.84606	1.18194	.87646	1.14095	.90781	1.10156	.94016	1.06365	.97359	1.02713	46
15	.84656	1.18125	.87698	1.14028	.90834	1.10091	.94071	1.06303	.97416	1.02653	45
16	.84706	1.18055	.87749	1.13961	.90887	1.10027	.94125	1.06241	.97472	1.02593	44
17	.84756	1.17986	.87801	1.13894	.90940	1.09963	.94180	1.06179	.97529	1.02533	43
18	.84806	1.17916	.87852	1.13828	.90993	1.09899	.94235	1.06117	.97586	1.02474	42
19	.84856	1.17846	.87904	1.13761	.91046	1.09834	.94290	1.06056	.97643	1.02414	41
20	.84906	1.17777	.87955	1.13694	.91099	1.09770	.94345	1.05994	.97700	1.02355	40
21	.84956	1.17708	.88007	1.13627	.91153	1.09706	.94400	1.05932	.97756	1.02295	39
22	.85006	1.17638	.88059	1.13561	.91206	1.09642	.94455	1.05870	.97813	1.02236	38
23	.85057	1.17569	.88110	1.13494	.91259	1.09578	.94510	1.05809	.97870	1.02176	37
24	.85107	1.17500	.88162	1.13428	.91313	1.09514	.94565	1.05747	.97927	1.02117	36
25	.85157	1.17430	.88214	1.13361	.91366	1.09450	.94620	1.05685	.97984	1.02057	35
26	.85207	1.17361	.88265	1.13295	.91419	1.09386	.94676	1.05624	.98041	1.01998	34
27	.85257	1.17292	.88317	1.13228	.91473	1.09322	.94731	1.05562	.98098	1.01939	33
28	.85308	1.17223	.88369	1.13162	.91526	1.09258	.94786	1.05501	.98155	1.01879	32
29	.85358	1.17154	.88421	1.13096	.91580	1.09195	.94841	1.05439	.98213	1.01820	31
30	.85408	1.17085	.88473	1.13029	.91633	1.09131	.94896	1.05378	.98270	1.01761	30
31	.85458	1.17016	.88524	1.12963	.91687	1.09067	.94952	1.05317	.98327	1.01702	29
32	.85509	1.16947	.88576	1.12897	.91740	1.09003	.95007	1.05255	.98384	1.01642	28
33	.85559	1.16878	.88628	1.12831	.91794	1.08940	.95062	1.05194	.98441	1.01583	27
34	.85609	1.16809	.88680	1.12765	.91847	1.08876	.95118	1.05133	.98499	1.01524	26
35	.85660	1.16741	.88732	1.12699	.91901	1.08813	.95173	1.05072	.98556	1.01465	25
36	.85710	1.16672	.88784	1.12633	.91955	1.08749	.95229	1.05010	.98613	1.01406	24
37	.85761	1.16603	.88836	1.12567	.92008	1.08686	.95284	1.04949	.98671	1.01347	23
38	.85811	1.16535	.88888	1.12501	.92062	1.08622	.95340	1.04888	.98728	1.01288	22
39	.85862	1.16466	.88940	1.12435	.92116	1.08559	.95395	1.04827	.98786	1.01229	21
40	.85912	1.16398	.88992	1.12369	.92170	1.08496	.95451	1.04766	.98843	1.01170	20
41	.85963	1.16329	.89045	1.12303	.92224	1.08432	.95506	1.04705	.98901	1.01112	19
42	.86014	1.16261	.89097	1.12238	.92277	1.08369	.95562	1.04644	.98958	1.01053	18
43	.86064	1.16192	.89149	1.12172	.92331	1.08306	.95618	1.04583	.99016	1.00994	17
44	.86115	1.16124	.89201	1.12106	.92385	1.08243	.95673	1.04522	.99073	1.00935	16
45	.86166	1.16056	.89253	1.12041	.92439	1.08179	.95729	1.04461	.99131	1.00876	15
46	.86216	1.15987	.89306	1.11975	.92493	1.08116	.95785	1.04401	.99189	1.00818	14
47	.86267	1.15919	.89358	1.11909	.92547	1.08053	.95841	1.04340	.99247	1.00759	13
48	.86318	1.15851	.89410	1.11844	.92601	1.07990	.95897	1.04279	.99304	1.00701	12
49	.86368	1.15783	.89463	1.11778	.92655	1.07927	.95952	1.04218	.99362	1.00642	11
50	.86419	1.15715	.89515	1.11713	.92709	1.07864	.96008	1.04158	.99420	1.00583	10
51	.86470	1.15647	.89567	1.11648	.92763	1.07801	.96064	1.04097	.99478	1.00525	9
52	.86521	1.15579	.89620	1.11582	.92817	1.07738	.96120	1.04036	.99536	1.00467	8
53	.86572	1.15511	.89672	1.11517	.92872	1.07676	.96176	1.03976	.99594	1.00408	7
54	.86623	1.15443	.89725	1.11452	.92926	1.07613	.96232	1.03915	.99652	1.00350	6
55	.86674	1.15375	.89777	1.11387	.92980	1.07550	.96288	1.03855	.99710	1.00291	5
56	.86725	1.15308	.89830	1.11321	.93034	1.07487	.96344	1.03794	.99768	1.00233	4
57	.86776	1.15240	.89883	1.11256	.93088	1.07425	.96400	1.03734	.99826	1.00175	3
58	.86827	1.15172	.89935	1.11191	.93143	1.07362	.96457	1.03674	.99884	1.00116	2
59	.86878	1.15104	.89988	1.11126	.93197	1.07299	.96513	1.03613	.99942	1.00058	1
60	.86929	1.15037	.90040	1.11061	.93252	1.07237	.96569	1.03553	1.00000	1.00000	0
′	Cotang	Tang	Cotang	Tang	Cotang	Tang	Cotang	Tang	Cotang	Tang	′
	49°		48°		47°		46°		45°		

REFERENCE TABLE V—Squares, Square Roots, Diameters, and Circumferences

NUMBER N	SQUARE OF NUMBER N^2	SQUARE ROOT OF NUMBER $\sqrt{N}$	AREA OF CIRCLE IF N = DIA.*	CIRCUMFERENCE OF CIRCLE IF N = DIA.*
1	1	1.00000	0.7854	3.1416
2	4	1.4142	3.1416	6.283
3	9	1.7321	7.0686	9.425
4	16	2.0000	12.5664	12.566
5	25	2.2361	19.6350	15.708
6	36	2.4495	28.2743	18.850
7	49	2.6458	38.4845	21.991
8	64	2.8284	50.2655	25.133
9	81	3.0000	63.6173	28.274
10	100	3.1623	78.5398	31.416
11	121	3.3166	95.033	34.558
12	144	3.4641	113.097	37.699
13	169	3.6056	132.732	40.841
14	196	3.7417	153.938	43.982
15	225	3.8730	176.715	47.124
16	256	4.0000	201.062	50.265
17	289	4.1231	226.980	53.407
18	324	4.2426	254.469	56.549
19	361	4.3589	283.529	59.690
20	400	4.4721	314.159	62.832
21	441	4.5826	346.361	65.973
22	484	4.6904	380.133	69.115
23	529	4.7958	415.476	72.257
24	576	4.8990	452.389	75.398
25	625	5.0000	490.874	78.540
26	676	5.0990	530.929	81.681
27	729	5.1962	572.555	84.823
28	784	5.2915	615.752	87.965
29	841	5.3852	660.520	91.106
30	900	5.4772	706.858	94.248
31	961	5.5678	754.768	97.389
32	1024	5.6569	804.248	100.531
33	1089	5.7446	855.299	103.673
34	1156	5.8310	907.920	106.814
35	1225	5.9161	962.113	109.956

*$\dfrac{\pi}{4} = 0.785398$ $\pi = 3.141593$

REFERENCE TABLE V—*continued*

NUMBER N	SQUARE OF NUMBER N^2	SQUARE ROOT OF NUMBER $\sqrt{N}$	AREA OF CIRCLE IF $N = $ DIA.*	CIRCUM- FERENCE OF CIRCLE IF $N = $ DIA.*
36	1296	6.0000	1017.88	113.097
37	1369	6.0828	1075.21	116.239
38	1444	6.1644	1134.11	119.381
39	1521	6.2450	1194.59	122.522
40	1600	6.3246	1256.64	125.664
41	1681	6.4031	1320.25	128.81
42	1764	6.4807	1385.44	131.95
43	1849	6.5574	1452.20	135.09
44	1936	6.6332	1520.53	138.23
45	2025	6.7082	1590.43	141.37
46	2116	6.7823	1661.90	144.51
47	2209	6.8557	1734.94	147.65
48	2304	6.9282	1809.56	150.80
49	2401	7.0000	1885.74	153.94
50	2500	7.0711	1963.50	157.08
51	2601	7.1414	2042.82	160.22
52	2704	7.2111	2123.72	163.36
53	2809	7.2801	2206.18	166.50
54	2916	7.3485	2290.22	169.65
55	3025	7.4162	2375.83	172.79
56	3136	7.4833	2463.01	175.93
57	3249	7.5498	2551.76	179.07
58	3364	7.6158	2642.08	182.21
59	3481	7.6811	2733.97	185.35
60	3600	7.7460	2827.43	188.50
61	3721	7.8102	2922.47	191.64
62	3844	7.8740	3019.07	194.78
63	3969	7.9373	3117.25	197.92
64	4096	8.0000	3216.99	201.06
65	4225	8.0623	3318.31	204.20
66	4356	8.1240	3421.19	207.35
67	4489	8.1854	3525.65	210.49
68	4624	8.2462	3631.68	213.63
69	4761	8.3066	3739.28	216.77
70	4900	8.3666	3848.45	219.91

REFERENCE TABLE V—continued

Number N	Square of Number N²	Square Root of Number √N	Area of Circle if N = Dia.*	Circum- ference of Circle if N = Dia.*
71	5041	8.4261	3959.19	223.05
72	5184	8.4853	4071.50	226.19
73	5329	8.5440	4185.39	229.34
74	5476	8.6023	4300.84	232.48
75	5625	8.6603	4417.86	235.62
76	5776	8.7178	4536.46	238.76
77	5929	8.7750	4656.63	241.90
78	6084	8.8318	4778.36	245.04
79	6241	8.8882	4901.67	248.19
80	6400	8.9443	5026.55	251.33
81	6561	9.0000	5153.00	254.47
82	6724	9.0554	5281.02	257.61
83	6889	9.1104	5410.61	260.75
84	7056	9.1652	5541.77	263.89
85	7225	9.2195	5674.50	267.04
86	7396	9.2736	5808.80	270.18
87	7569	9.3274	5944.68	273.32
88	7744	9.3808	6082.12	276.46
89	7921	9.4340	6221.14	279.60
90	8100	9.4868	6361.73	282.74
91	8281	9.5394	6503.88	285.88
92	8464	9.5917	6647.61	289.03
93	8649	9.6437	6792.91	292.17
94	8836	9.6954	6939.78	295.31
95	9025	9.7468	7088.22	298.45
96	9216	9.7980	7238.23	301.59
97	9409	9.8489	7389.81	304.73
98	9604	9.8995	7542.96	307.88
99	9801	9.9499	7697.69	311.02
100	10000	10.0000	7853.98	314.16

Answers to Odd-Numbered Exercises

Exercises 1-1 (pages 4–5)
1. 74   **3.** 3065   **5.** 73,857   **7.** 870,401
9. 20 + 6   **11.** 100 + 20 + 5   **13.** 1000 + 800 + 20
15. 900,000 + 80,000 + 400 + 60 + 2
17. 4   **19.** 20   **21.** 40   **23.** 800,000
25. 50   **27.** 3700   **29.** 800   **31.** 1800
33. 8000   **35.** 26,000   **37.** 27,700   **39.** 300
41. 19,000   **43.** 32,600

Exercises 1-2 (pages 9–11)
1. 79   **3.** 9979   **5.** 83   **7.** 6027
9. 1343   **11.** 38,243   **13.** 2425   **15.** 1,542,594
17. 4428   **19.** 531   **21.** 37   **23.** 2749
25. 86,359   **27.** 107,514   **29.** 209,456   **31.** 97
33. 58,689   **35.** 8,914,537   **37.** 84 mm   **39.** 224 in.
41. 171 min   **43.** 3388 km   **45.** 22 in.   **47.** 480 h
49. 10,367 h   **51.** 823

Exercises 1-3 (page 13)
1. 774   **3.** 5355   **5.** 18,846   **7.** 220,742
9. 56,742   **11.** 32,907,645   **13.** 1000   **15.** 13,719,000
17. 90,270   **19.** 147,958,475   **21.** 154,420,000   **23.** 1,960,000,000
25. 80 sq m   **27.** 1071 cu yd   **29.** 415 km   **31.** 600 cm

Exercises 1-4 (pages 17–18)
1. 84   **3.** 599 R3   **5.** 85   **7.** 15
9. 81 R11   **11.** 401 R17   **13.** 18   **15.** 96 R426
17. 107 km   **19.** 162 ft   **21.** 17 ft   **23.** 794 cm
25. 21 R1712   **27.** 85   **29.** 6 m   **31.** 103 cm
33. 7 $^m/_{min}$ **35.** 38   **37.** 22; 8 ft

Exercises 1-5 (pages 19–20)

1. 8^2	**3.** 51^3	**5.** $3^1 \times 4^3$	**7.** 200^2
9. $2^2 \times 3^3$	**11.** $14^3 \times 3^3$	**13.** 15^{13}	**15.** 5^2
17. 64	**19.** 27	**21.** 256	**23.** 13
25. 64	**27.** 1600	**29.** 25,934,336	**31.** 1

Exercises 1-6 (pages 21–22)

1. $2^3 \times 3$	**3.** $2^3 \times 3^2$	**5.** 3×19	**7.** $2 \times 3 \times 17$
9. 2×7^2	**11.** 3×5^2	**13.** 3×41	**15.** $3 \times 5 \times 11$
17. 11×13	**19.** $2 \times 3 \times 43$	**21.** 7×43	**23.** 3×59
25. 7×47	**27.** $2^4 \times 37$	**29.** 19×37	**31.** $2 \times 7 \times 47$

Unit 1 Review Exercises (page 22)

1. 4200 h	**3.** 3 min	**5.** 6954 wheels	**7.** 45
9. $3^3 \times 2^4$			

Exercises 2-1 (pages 26–27)

1. $1\frac{1}{2}$	**3.** $19\frac{1}{4}$	**5.** $88\frac{3}{5}$	**7.** $55\frac{5}{12}$
9. $9\frac{3}{10}$	**11.** $1015\frac{5}{8}$	**13.** $7\frac{1}{2}$	**15.** $2\frac{1}{5}$
17. $4\frac{3}{4}$	**19.** $15\frac{7}{8}$	**21.** $2\frac{9}{16}$	**23.** $4\frac{1}{6}$
25. $3\frac{3}{4}$ ft	**27.** $1\frac{1}{8}$ in.		

Exercises 2–2 (page 31)

1. yes	**3.** no	**5.** yes	**7.** yes
9. $\frac{25}{30}$	**11.** $\frac{2}{12}$	**13.** $\frac{42}{45}$	**15.** $\frac{108}{99}$
17. $1\frac{2}{3}$	**19.** $\frac{112}{630}$	**21.** $\frac{1}{3}$	**23.** $\frac{2}{7}$
25. $\frac{5}{3}$ or $1\frac{2}{3}$	**27.** $\frac{2}{1}$ or 2	**29.** $\frac{7}{8}$	**31.** $\frac{4}{9}$
33. $\frac{2}{3}$	**35.** $\frac{17}{27}$	**37.** yes	**39.** yes

Exercises 2-3 (page 33)

1. correct	**3.** not correct	**5.** not correct	**7.** $\frac{3}{4}$ in.
9. $2\frac{5}{32}$ in.	**11.** 1 in.	**13.** $\frac{1}{4}$ in.	**15.** $1\frac{5}{32}$ in.
17. $\frac{1}{2}$ in.	**19.** $1\frac{1}{4}$ in.	**21.** $\frac{7}{16}$ in.	**23.** $\frac{13}{16}$ in.
25. $1\frac{1}{4}$ in.	**27.** $1\frac{9}{16}$ in.	**29.** $\frac{3}{8}$ in.	**31.** $\frac{7}{8}$ in.
33. $1\frac{9}{16}$ in.	**35.** $2\frac{1}{16}$ in.		

Exercises 2-4 (pages 36–37)

1. 4	**3.** 66	**5.** 12	**7.** 45
9. 64	**11.** 16	**13.** $\frac{10}{14}, \frac{9}{14}$	**15.** $\frac{16}{60}, \frac{35}{60}$
17. $\frac{42}{90}, \frac{55}{90}$	**19.** $\frac{45}{50}, \frac{8}{50}$	**21.** $\frac{18}{45}, \frac{5}{45}, \frac{21}{45}$	
23. $\frac{285}{1080}, \frac{350}{1080}, \frac{684}{1080}$		**25.** 32	**27.** $\frac{47}{51}, \frac{45}{51}$

Exercises 2-5 (pages 39–40)

1. true	**3.** false	**5.** true	**7.** false
9. true	**11.** false	**13.** true	**15.** true

17. $\frac{5}{6} > \frac{5}{9}$	**19.** $\frac{5}{7} > \frac{3}{7}$	**21.** $\frac{21}{64} > \frac{10}{32}$	**23.** $8\frac{5}{7} < 8\frac{7}{9}$
25. $\frac{2}{4}, \frac{3}{4}, \frac{7}{8}$	**27.** $\frac{2}{3}, \frac{3}{4}, \frac{5}{6}$	**29.** $6\frac{7}{15}, 6\frac{3}{5}, 6\frac{3}{4}$	**31.** Jan, Mar, Feb
33. $\frac{1}{4}$ in. $< \frac{19}{64}$ in.			

Exercises 2-6 (pages 43–44)

1. $\frac{2}{3}$	**3.** $\frac{7}{10}$	**5.** $1\frac{5}{12}$	**7.** $\frac{23}{30}$
9. $11\frac{1}{2}$	**11.** $18\frac{2}{3}$	**13.** $8\frac{5}{6}$	**15.** $17\frac{7}{24}$
17. $70\frac{25}{99}$	**19.** $983\frac{37}{45}$	**21.** 1	**23.** $\frac{17}{64}$
25. $1\frac{43}{60}$	**27.** $1\frac{43}{96}$	**29.** $1\frac{7}{64}$	**31.** $\frac{11}{64}$ in.
33. $53\frac{974}{1000}$ mm	**35.** $13\frac{47}{64}$ in.	**37.** $4\frac{17}{32}$ in.	

Exercises 2-7 (pages 47–48)

1. $\frac{5}{8}$	**3.** $\frac{1}{2}$	**5.** $\frac{25}{36}$	**7.** $\frac{23}{42}$
9. $4\frac{1}{4}$	**11.** $4\frac{3}{8}$	**13.** $3\frac{1}{7}$	**15.** $2\frac{5}{24}$
17. $6\frac{1}{6}$	**19.** $9\frac{7}{9}$	**21.** $28\frac{9}{20}$	**23.** $168\frac{108}{247}$
25. $\frac{1}{2}$	**27.** $\frac{2}{7}$	**29.** $2\frac{4}{35}$	**31.** $14\frac{17}{32}$ in.
33. $\frac{15}{32}$ in.			

Exercises 2-8 (pages 51–52)

1. $\frac{3}{10}$	**3.** $\frac{5}{12}$	**5.** $\frac{4}{7}$	**7.** $\frac{7}{18}$
9. $\frac{3}{14}$	**11.** $\frac{1}{5}$	**13.** 9	**15.** 3
17. 75	**19.** 174	**21.** $6\frac{1}{3}$	**23.** $41\frac{2}{3}$
25. $2\frac{19}{32}$ in.	**27.** $10\frac{1}{8}$ in.	**29.** 15 gal	**31.** $184\frac{27}{32}$ in.

Exercises 2-9 (pages 55–56)

1. $\frac{7}{8}$	**3.** $\frac{9}{11}$	**5.** $\frac{3}{5}$	**7.** 10
9. $27\frac{1}{2}$	**11.** $\frac{7}{16}$	**13.** $\frac{1}{12}$	**15.** 4
17. $\frac{5}{33}$	**19.** $\frac{15}{26}$	**21.** $\frac{15}{16}$	**23.** $\frac{27}{52}$
25. 46	**27.** $\frac{19}{64}$ in.	**29.** 15	**31.** $2\frac{7}{16}$ in.

Unit 2 Review Exercises (page 56)

1. $\frac{338}{9}$	**3.** $4\frac{1}{52}$	**5.** $1\frac{1}{2}$	**7.** $\frac{33}{64}$ in.
9. $\frac{7}{64}$ oz			

Exercises 3-1 (pages 62–63)

1. twenty-one hundredths
3. one-hundredth
5. four hundred thirty-five ten-thousandths
7. nine and ten thousand thirty-two hundred-thousandths

9. 0.6	**11.** 0.117	**13.** 2.7	**15.** 100.065
17. $\frac{3}{10}$	**19.** $\frac{17}{20}$	**21.** $\frac{3}{8}$	**23.** $\frac{5}{80}$
25. 0.9	**27.** 0.09	**29.** 0.0009	**31.** 0.009
33. 0.0625	**35.** 0.703125	**37.** 0.125	**39.** 0.65625
41. 0.421875	**43.** $\frac{11}{32}$ in.	**45.** 0.875 in.	

Exercises 3-2 (pages 66–67)

1. 0.2	**3.** 0.8	**5.** 16.0	**7.** 99.8
9. 14.79	**11.** 9.99	**13.** 0.17	**15.** 67.97
17. 0.844	**19.** 0.101	**21.** 10.900	**23.** 0.001
25. 3	**27.** 4	**29.** 6	**31.** 7
33. 5	**35.** 2	**37.** 1	**39.** 5
41. 38	**43.** 6183.01	**45.** 8.13	**47.** 1,200
49. 2.750			

Exercises 3-3 (pages 69–72)

1. 4.2	**3.** 17.246	**5.** 8.74	**7.** 12.275
9. 254.981	**11.** 11.3055	**13.** 11.892	**15.** 133.2301
17. 0.08	**19.** 0.09	**21.** 0.4	**23.** 2.9452
25. 2.8839	**27.** 1.2431	**29.** 4.523	**31.** 137.0349
33. 3.1875 in.	**35.** 4.575 in.	**37.** 0.375 in.	**39.** 1.9845 in.
41. 79.375 mm	**43.** 1.750 in.		

Exercises 3-4 (page 75)

1. 0.0502; 0.0590; 0.0800; 0.1000; 1.0000
3. 0.0507; 0.0520; 0.1400; 0.4000; 0.5000; 1.0000
5. 0.0506; 0.0550; 0.1100; 1.0000
7. 0.0503; 0.0560; 0.1300; 0.4000; 1.0000; 2.0000
9. 0.0502; 0.0570; 0.1300; 0.5000
11. 0.0502; 0.0700; 0.5000; 2.0000

Exercises 3–5 (page 77)

1. 9.80224	**3.** 0.05922	**5.** 2.8	**7.** 24.3
9. 0.208	**11.** 0.0126	**13.** 3.944	**15.** 11.637
17. 364.98	**19.** 1.0208	**21.** 383.75766	**23.** 120.38898
25. 1.704	**27.** 76.563	**29.** 1014.288	**31.** 6.582 in.
33. 0.36 in.	**35.** 25.1 in.	**37.** 5.105 in.	**39.** 408.958 mm

Exercises 3-6 (pages 80–81)

1. 1.96	**3.** 0.0006	**5.** 0.08	**7.** 2.35
9. 75	**11.** 0.02	**13.** 0.75	**15.** 0.002
17. 198	**19.** 10.2	**21.** 0.04	**23.** 0.45
25. 1.3	**27.** 0.4	**29.** 0.5	**31.** 0.7854
33. 0.219 in.			

Unit 3 Review Exercises (pages 81–82)

1. 0.640625
3. 39.6
5. 1.688 in.
7. 0.0502; 0.054; 0.100; 0.110; 2.000
9. 19.6 ft

Exercises 4-1 (pages 89–90)

1. 144 in.   **3.** 21.6 in.   **5.** 40 ft   **7.** 13,200 ft
9. 5 yd   **11.** 5984 yd   **13.** 8 pt   **15.** 4.5 pt
17. 46 qt   **19.** 5 qt   **21.** 4 pk   **23.** 3.5 pk
25. 10,000 lb **27.** 2837 lb   **29.** 1542 s   **31.** 6780 s
33. 12 min   **35.** 1440 min   **37.** ⅓ day   **39.** 28 days
41. (a) 8800 yd
       (b) 8.8 mi
43. (a) 15⅝ gal
       (b) 4.5 gal
45. (a) 234 weeks (b) 54 months
47. ⅕
49. 45,000 lb

Exercises 4-2 (pages 94–96)

1. 10,500 m   **3.** 1 m   **5.** 4.8 cm
7. 807,000 cm   **9.** 6.905 km   **11.** 5.0 km
13. 50,200 L   **15.** 4.85 L   **17.** 2300 mL
19. 5,800,000 mL **21.** 0.7 g   **23.** 8920 g
25. 10.58 kg   **27.** 9.1 kg   **29.** 4.38 m
31. 8.52 L   **33.** 4.87 g
35. 9.7 cm; 97 mm; 9 cm 7 mm   **37.** 7.5 cm; 75 mm; 7 cm 5 mm
39. 10.5 cm; 105 mm; 10 cm 5 mm   **41.** 0.832 kg
43. × 10; ÷ 100; ÷ 100,000   **45.** 1200 L

Exercises 4-3 (pages 100–101)

1. 0.7366 mm   **3.** 10.32 mm   **5.** 15.88 mm   **7.** 3.49 cm
9. 15.2 cm   **11.** 68.6 cm   **13.** 45.7 cm   **15.** 114 cm
17. 259 cm   **19.** 0.762 m   **21.** 1.83 m   **23.** 4.42 m
25. 1.21 km   **27.** 515 km   **29.** 226.8 g   **31.** 85.05 g
33. 141.8 g   **35.** 3629 g   **37.** 9072 g   **39.** 3175 g
41. 90.72 kg   **43.** 1134 kg   **45.** 0.9072 kg   **47.** 7.57 L
49. 37.9 L   **51.** 14.2 L   **53.** 3.31 L   **55.** 19.0 ft
57. 325 in.   **59.** 350 yd   **61.** 5 oz   **63.** 6400 lb
65. 8.19 pt   **67.** 10.6 qt.   **69.** 0.697 in.

Exercises 4-4 (pages 104–105)

1. 15 ft 8 in.   **3.** 125 km 151 m   **5.** 9 qt 1 pt
7. 16 da 4 h 33 min   **9.** 34 km 5 hm 5 dam   **11.** 6 yd 3 in.
13. 4 mi 1132 yd   **15.** 1 lb 6 oz   **17.** 5 gal 2 qt
19. 7 cm 4 mm   **21.** 22 yd   **23.** 197 mi 935 yd
25. 80 sq m   **27.** 1071 cu yd   **29.** 1 h 13 min
31. 3 yd 2 ft   **33.** 3 L 4 dL   **35.** 8 lb 3 oz
37. 170 cm   **39.** 162 g 9 dg 6 cg   **41.** 30 cm 2.25 mm,
                                                         or 302.25 mm

Unit 4 Review Exercises (page 106)
1. 144 in.; 56 in.; 21.6 in.
3. 338 parts   **5.** 18   **7.** 4.07 gal   **9.** 14 lb 4⅝ oz

Exercises 5-1 (page 110)
1. 11   **3.** 36.86   **5.** 13⁵⁄₂₄   **7.** 812
9. 69.342   **11.** 5¼   **13.** 4   **15.** 2.8
17. 28.408

Exercises 5-2 (pages 112–113)
1. 5   **3.** 7   **5.** 12   **7.** 85
9. 8.4   **11.** 10   **13.** 10.4   **15.** 3.72
17. 1   **19.** 49.22

Exercises 5-3 (pages 115–116)
1. 20   **3.** 10.25   **5.** 0.350   **7.** 0.5625
9. 100   **11.** 0.390625   **13.** 12¹⁷⁷⁄₂₅₆   **15.** 5.641
17. 3.1416   **19.** 1.767   **21.** 13.0   **23.** 40
25. 25.75   **27.** 0.042   **29.** 0.056   **31.** 200
33. 200

Exercises 5-4 (pages 118–119)
1. $A = 180° - (B+C); B = 180° - (A+C)$
3. $D = M + 1.5155p - 3G$
5. $D = d + \frac{1}{N}$ **7.** $c = W - 2a$
9. $b = c \cos A$ **11.** $S = \frac{ts}{T}$
13. $a = \frac{2A}{b}$ **15.** $P = \frac{2D}{1.299}$
17. $a = \frac{2A}{b + b'}$

Exercises 5-5 (page 124–125)
1. 11   **3.** 9   **5.** 22   **7.** 19
9. 7.211   **11.** 5.292   **13.** 9.899   **15.** 7.937
17. 9.220   **19.** 4.359   **21.** 2.828   **23.** 5.196
25. 8.485   **27.** 7.348   **29.** 14.142   **31.** 22.045
33. 17.889   **35.** 19.053   **37.** 15.81 ft   **39.** 18 in.
41. 3.5 ft

Exercises 5-6 (page 129)
1. 5   **3.** 36   **5.** 52   **7.** 77
9. 1508   **11.** 1.50   **13.** 8.40   **15.** 1.70
17. 88.50   **19.** 5.20   **21.** 11.18   **23.** 0.35
25. 0.18

Unit 5 Review Exercises (page 130)
1. 144.02   **3.** 673.08   **5.** 3.1416   **7.** 15.2 ft
9. 3.728 in.

Exercises 6-1 (page 134)

1. 7/9   3. 4/11   5. 1/3   7. 2/3
9. 4/3   11. 2/7   13. 9/5   15. 2/3
17. 2/3   19. 1/4   21. 5/3   23. 2/7
25. 3/8   27. 1/2   29. 840 rpm   31. 437.5 mi/h
33. (a) 4/3 (b) 3/4 (c) 3/7   35. 5/16   37. 1/72 in.

Exercises 6-2 (pages 136–137)

1. false   3. true   5. false   7. true
9. 15   11. 5   13. 81   15. 80
17. 2 1/7   19. 45.9   21. 36   23. 26
25. 64

Exercises 6-3 (pages 140–141)

1. 3 1/3 ft   3. 8 1/3 ft   5. 1 in.
7. 1/12 in.   9. 1 in. = 8 ft   11. 5 in.
13. 4 in.   15. 1 in. = 1.33 in.

Exercises 6-4 (pages 145–146)

1. 7 1/2   3. 0.825   5. 6.48   7. 24
9. 1   11. 96.75   13. 3 min   15. 75 lb/sq in.
17. directly   19. inversely

Exercises 6-5 (pages 148–150)

1. 0.21   3. 2.58   5. 0.065   7. 0.0062
9. 0.053   11. 3/10   13. 3/20   15. 7/20
17. 1 7/8   19. 4 4/5   21. 67%   23. 23.3%
25. 9.2%   27. 122%   29. 800%   31. 61%
33. 725%   35. 60%   37. 11 1/9%   39. 0.5%
41. (b) 0.25; 25%   (c) 0.375; 37.5%   (d) 0.5; 50%   (e) 0.625; 62.5%
   (f) 0.75; 75%   (g) 0.875; 87.5%
43. (a) 0.01; 1%   (b) 0.02; 2%   (c) 0.04; 4%   (d) 0.05; 5%   (e) 0.0625;
   6.25%   (f) 0.08 1/3; 8 1/3%

Exercises 6-6 (pages 151–152)

1. 51.3   3. $5760   5. 7.56   7. 98.6
9. 175.5   11. 10,845   13. 0.12   15. 0.4
17. 50%   19. 87.5%   21. 6%   23. 20%
25. 200%   27. 100%   29. 1.75%   31. 77.8%
33. 20   35. 500   37. 1871   39. 70
41. 3200   43. 73,500   45. 68.5   47. $140

Exercises 6-7 (page 154)

1. 24.3   3. 84   5. 80
7. 48.4   9. 5.4   11. 49.68

13. 6.75   **15.** $62.37   **17.** 5/12
19. 161.7 lb   **21.** 347 board ft   **23.** 134 lb copper; 66 lb
                                                    zinc

Exercises 6-8 (pages 155–157)
1. 44%   **3.** 25%   **5.** 0.5%
7. 4.4%   **9.** 180%   **11.** 5%
13. 450%   **15.** 5%   **17.** 28.6%
19. 90%   **21.** 0.093%   **23.** 1.04%
25. Vance; Maria
27. Piece 1 is medium-carbon steel; Piece 2 is high-carbon steel; Piece 3 is
wrought iron; Piece 4 is cast iron; Piece 5 is low-carbon steel

Unit 6 Review Exercises (page 158)
1. ½   **3.** 30.155   **5.** 75.36   **7.** 60%
9. 3600 rpm

Exercises 7-1 (pages 162–163)
1. 0.106   **3.** 0.218   **5.** 0.008   **7.** 0.247
9. 0.075   **11.** 0.125 in.   **13.** 0.115 in   **15.** 0.143
17. 0.099   **19.** 0.008

Exercises 7-2 (pages 165–166)
1. 0.1624   **3.** 0.0551   **5.** 0.1002   **7.** 0.2440
9. 0.0248   **11.** 0.1413

Exercises 7-3 (page 170)
1. 1.083   **3.** 3.001   **5.** 1.997   **7.** 0.141
9. 0.006   **11.** 0.364   **13.** 0.297   **15.** 4.016

Exercises 7-4 (pages 172–173)
1. 15° 20′   **3.** 23° 5′   **5.** 89° 50′
7. 5° 25′   **9.** 36° 35′   **11.** 27° 0′
13. (a) 45° 10′   (b) 37° 0°   (c) 55° 5′

Exercises 7-5 (page 176)
1. 1.150; 1.154; 1.150; 0.004   **3.** 0.875; 0.876; 0.873; 0.003
5. 0.500; 0.5015; 0.500; 0.0015   **7.** 5/8; 41/64; 39/64; 1/32
9. 0.8125; 0.8130; 0.8120; 0.001

Unit 7 Review Exercises (page 177)
1. 40   **3.** 0.131 in.   **5.** 0.1624   **7.** 5.266
9. to the right of 10°; sixth

Exercises 8-1 (pages 183–184)
1. 8°19′10″   **3.** 31°24′34″   **5.** 41°39′   **7.** 57°41′24″

9. 48°29′19″   **11.** 79°5′51″   **13.** 142°32′   **15.** 82°12′

17. 10°22′36″   **19.** 25.75°   **21.** 16.17°   **23.** 36.2°

25. 27.6°   **27.** 16°20′13″   **29.** 31°45′18″   **31.** 54°32′42″

33. 19°37′30″   **35.** 67°48′43″   **37.** 4°58′30″

39. Wrong, should be 2°50′; 5⅔° = 5°40′, 5°40′ = 4°100′, 4°100′ ÷ 2 = 2°50′.

Exercises 8-2 (pages 186–188)

1. Group 1 *GOB, RPS, POX, EPO, STY, CTF, OYD, TYH*

3. Group 3 *JZY, DZX, SRA, PRI*

5. Group 5 *TSX, YXZ, OXS, PSR*

7. 131°40′   **9.** 40°   **11.** 121°   **13.** 156°

15. 21°20′   **17.** 146°41′

Exercises 8-3 (pages 193–196)

1. 105°   **3.** 92°40′   **5.** 40°   **7.** 90°, 71°18′

9. 13.454   **11.** 8.485   **13.** 24.432   **15.** 5.998

17. 12.775   **19.** 4 in.   **21.** 14.151 in.   **23.** 9.22 ft

25. 12.021 in.   **27.** (a) 11.314 in.   (b) 45°, 45°, 90°   **29.** 51.96 ft

31. 4⅞ in.   **33.** 15 in.   **35.** 45°   **37.** 0.144 in.

Exercises 8-4 (pages 200–201)

1. 135°, 135°; 45°   **3.** 8.5 in.   **5.** 1³¹⁄₃₂ in.

7. 2.598 in.   **9.** 23.1 in.   **11.** 3.46 in.

13. 1.191 in.   **15.** 2.53 in.

Exercises 8-5 (pages 205–207)

1. 1.625 in.   **3.** I.D. is 1⅝ in.; O.D. is 3¹⁄₂₀ in.

5. 15 ft   **7.** ¹⁷⁄₆₄ in.   **9.** 0.333 in.   **11.** 0.2813 in.

13. 5.510 mm   **15.** 6.970 mm

Unit 8 Review Exercises (pages 207–208)

1. 37.5°   **3.** 120°45′37″   **5.** 15.811   **7.** 8.662 in.

9. 0.0414 in.

Exercises 9-1 (pages 212–214)

1. 32 in.   **3.** 19 ft   **5.** 18 in.   **7.** 45.2 m

9. 28.1   **11.** 36.4 m   **13.** 66 ft   **15.** 2.1 yd

17. 31½ in.   **19.** 131.9 mm   **21.** 113.1 in.   **23.** 8 ft 6¼ in.

25. 3.1416 ft

27. 17.14 in., or about 17⁵⁄₃₂ in.

29. 6.534 in., or about 6¹⁷⁄₃₂ in.

Exercises 9-2 (pages 219–221)

1. 29.1 sq in.   **3.** 32.1 sq cm   **5.** 33.83 sq in.   **7.** 8.06 sq ft

9. 42.79 sq in.   **11.** 46.87 sq ft   **13.** 14.67 sq ft   **15.** 40 sq in.

17. 19.5 sq cm   **19.** 10.63 sq m   **21.** 15.87 in.; 71.43 sq in.

23. s = 23.34 in.

25. (a) 100 ft   (b) 7500 sq ft
27. 15 sq ft   **29.** 4.16 sq yd   **31.** 420 sq in.   **33.** 92 sq in.
35. 40.88 sq ft   **37.** 36.6 sq ft   **39.** 11.616 sq in.

Exercises 9-3 (pages 224–226)
1. 28.274 sq in.   **3.** 30.680 sq ft   **5.** 30.943 ft   **7.** 26°51′27″
9. 3.927 in.   **11.** 19.67 mm   **13.** 5.236 in.   **15.** 2.405 sq in.
17. 0.000325 in.

Exercises 9-4 (pages 232–235)
1. 3.25 cu yd   **3.** 68.6 cu ft   **5.** 498.96 cu in
7. 424 cu in.   **9.** 11.0 cu ft   **11.** 21.56 cu yd
13. 360.50 cu in.   **15.** 1.224 in.   **17.** 4.539 in.
19. 78.07 lb   **21.** 128.65 lb   **23.** 2.53 cu ft
25. 42.67 cu in.   **27.** 654.50 cu in.   **29.** 1813.75 cu in.
31. 7.3 cu yd   **33.** 1017.88 sq in.
35. sphere: 523.6 cu in.; cube: 1000 cu in.
37. 3619.1 cu ft   **39.** 114.14 cu in.

Exercises 9-5 (page 237)
1. ¼   **3.** ½   **5.** ¼   **7.** 10 in.
9. 9

Unit 9 Review Exercises (pages 238–240)
1. 175 sq ft   **3.** 19.215 sq in.   **5.** 11.78 sq in.   **7.** 4.084 in.
9. 0.0614 cu in.

Exercises 10-1 (page 245)
1. 44°   **3.** 87°   **5.** cot   **7.** tan
9. 0.600   **11.** 0.750   **13.** 0.800   **15.** 1.334
17. 7°29′   **19.** 0.61894

Exercises 10-2 (page 248)
1. 0.22722   **3.** 3.84377   **5.** 22°44′   **7.** 0.88117
9. 3.07768   **11.** 0.14263   **13.** 0.85660   **15.** 0.19652
17. 4.38970   **19.** 0.88835   **21.** 0.91099   **23.** 0.99892
25. 2.50440   **27.** 0.30292   **29.** 26°   **31.** 60°20′
33. 41°40′   **35.** 23°   **37.** 29°30′   **39.** 84°8′
41. 39°20′   **43.** 75°5′   **45.** 13°30′

Exercises 10-3 (pages 251–254)
1. 4.265 in.   **3.** $x = 1.28$, $y = 0.75$ **5.** 2.53 mm
7. 1.174 in.   **9.** 2.49 mm
11. $A = 33°17′$, $c = 7.52$ in.
13. 45°   **15.** 1°43′

Exercises 10-4 (page 257)
1. 35°30'   **3.** 1.635 in.   **5.** 16°30'

Exercises 10-5 (pages 259–262)
1. 8.4853   **3.** 4.6353   **5.** 1.2928
7. 4.14 in.   **9.** $+x = 8.112$ in.; $-y = 4.313$ in.
11. 5.113 in., or 5⁷⁄₆₄ in.   **13.** 1.172 in.
15. 11.7143 in., or 11²³⁄₃₂ in.
17. 5.232 in.   **19.** 18°26'
21. $x = 1.217$ in., $y = 1\frac{1}{8}$ in

Exercises 10-6 (pages 265–266)
1. $C = 75°$; $c = 34.28$ in.; $d = 34.95$ in.
3. $D = 42°15'$; $b = 5.23$ mm; $c = 1.97$ mm
5. 110°   **7.** $B = 52°33'$; $D = 67°27'$
9. 92°4'

Exercises 10-7 (pages 269–270)
1. $b = 8.074$ in.; $C = 38°16'$; $D = 111°44'$
3. $c = 5.058$ cm; $B = 99°8'$; $D = 53°52'$
5. $d = 9.64$ in.; $B = 57°31'$; $C = 79°49'$
7. $B = 31°47'$; $C = 81°1'$; $D = 67°12'$
9. $B = 58°32'$; $C = 84°18'$; $D = 37°10'$

Unit 10 Review Exercises (pages 270–272)
1. $\sin X = x/r$; $\cos X = y/r$; $\tan X = x/y$; $\cot X = y/x$
3. 0.48608   **5.** $b = 11.05$; $c = 12.69$
7. 1.294   **9.** 0.293

Exercises 11-1 (pages 280–282)
1. 0.0208 in. per inch   **3.** 0.038; 0.02097 in. per inch
5. 0.437 in.   **7.** No. 16; 1.6 in.
9. 0.03125 in., or ¹⁄₃₂ in.

Exercises 11-2 (pages 285–286)
1. 0.563 in., or ⁹⁄₁₆ in.   **3.** 0.3 in. per ft
5. 0.109 in., or ⁷⁄₆₄ in.   **7.** 0.042 in.

Exercises 11-3 (pages 289–290)
1. 4°40'   **3.** 9°32'   **5.** 4°10'   **7.** 54'

Exercises 11-4 (pages 296–297)
1. 0.0018 in. per inch   **3.** 0.0031 in. per inch   **5.** 0.0025 in. per inch
7. 0.007 in. per inch   **9.** 0.044 in. per inch   **11.** 0.042 in.
13. 41°

Unit 11 Review Exercises (page 298)

1. 1½ in.   **3.** 0.260   **5.** 0.5938 in.
7. 7°26′   **9.** 30°; 7°30′; 12°47.5′

Exercises 12-1 (pages 302–304)

1. 100 rpm   **3.** 32 teeth   **5.** 8:3   **7.** 112 teeth
9. 108 rpm   **11.** 400 rpm   **13.** ¾ turn   **15.** 36 rpm

Exercises 12-2 (pages 306–309)

1. 800 rpm   **3.** 6 turns   **5.** 1⅔ turns   **7.** 4⅕ turns
9. 33 teeth   **11.** 200 rpm

Unit 12 Review Exercises (pages 309–310)

1. 80 rpm   **3.** idler: counterclockwise; driven gear: clockwise
5. 60 rpm   **7.** 48 teeth   **9.** 75 rpm

Exercises 13-1 (page 315)

1. 560   **3.** 201   **5.** 91   **7.** 1200
9. 100   **11.** 750   **13.** 1029   **15.** 300
17. 225

Exercises 13-2 (pages 320–321)

1. 2.67   **3.** 6.06   **5.** 0.47
7. 3.14   **9.** 3.18   **11.** 100 rpm; 2.4 in.
13. 50 rpm; 4.9 in.   **15.** 111 rpm; 11 in.   **17.** 1200 rpm; 7.2 in.
19. 201 rpm; 4.02   **21.** 961 rpm; 199.89   **23.** 111 rpm; 27.75
25. 0.06 min, or 3.4 s   **27.** 40$^{in}/_{min}$

Unit 13 Review Exercises (page 322)

1. 4000 rpm   **3.** 68.58$^{m}/_{min}$ **5.** 34$^{in}/_{min}$
7. 400 rpm   **9.** 312.5 rpm

Exercises 14-1 (page 326)

1. 1 in.; ½ in., ¼ in.   **3.** ¹⁄₁₆, ¹⁄₁₀, ⅛, ¼   **5.** 0.0833; 0.0833

Exercises 14-2 (pages 334–335)

h	P	P.D.	T.D.S.
1. 0.023	0.036	0.227	0.214
3. 0.027	0.042	0.348	0.333
5. 0.059	0.091	0.566	0.534
7. 0.032	0.050	0.218	0.200
9. 0.0080	**11.** 0.0103; 0.0481	**13.** 0.0131; 0.0577	**15.** 0.0160; 0.0722
17. 0.0206; 0.0962	**19.** 0.0289; 0.1283	**21.** 0.0357	

Exercises 14-3 (pages 336–337)
1. 0.060; ³⁄₆₄ **3.** 0.073; #53 **5.** 0.086; #50
7. 0.099; #45 **9.** 0.112; #42 **11.** 0.125; #37
13. 0.138; #33 **15.** 0.164; #29 or #28 **17.** 0.190; #21
19. 0.216; #14

Exercises 14-4 (page 340)
1. h = 0.063; T.D.S. = 0.520; G = 0.0609
3. h = 0.100; T.D.S. = 0.820; G = 0.0974
5. h = 0.036; T.D.S. = 0.386; G = 0.0348
7. h = 0.125; T.D.S. = 1.270; G = 0.1218
9. h = 0.042; T.D.S. = 0.312; G = 0.0406

Exercises 14-5 (page 343)
1. w = 0.0238; h = 0.0528; w = 0.0388; h = 0.0858
3. w = 0.0517, h = 0.1144; w = 0.0310; h = 0.0687
5. w = 0.0221, h = 0.0490; w = 0.0194, h = 0.0429
7. 0.1655 in., or #19

Unit 14 Review Exercises (pages 343–344)
1. ⁵⁄₆₄, or .0781 **3.** 0.0271; 0.0962

	P	h	P.D.	T.D.S.
5.	0.056	0.0364	0.2761	0.2565
7.	0.063	0.0409	0.3341	0.3120
9.	0.111	0.056	0.819	0.784

Exercises 15-1 (pages 350–351)
1. 5 **3.** 2.556 **5.** 1.583 **7.** 74
9. 11.750 **11.** 6¼ in. **13.** 4⁷⁄₁₂ in. **15.** 6 in.
17. 5⅝ in. **19.** 2.150 in.

Exercises 15-2 (pages 353–354)
1. 0.500 **3.** 0.375, or ⅜ **5.** N_p = 42; N_g = 75
7. N_p = 40; N_g = 115 **9.** 0.589 **11.** 0.628
13. 1.257

Exercises 15-3 (page 359)
1. 16; 44; 2¾:1 **3.** 20; 42; 2.1:1 **5.** 25; 27; 1.08:1

Exercises 15-4 (page 361)
1. 25.4; 0.124; 22 mm; 0.866 in. **3.** 3.175; 0.989; 192 mm; 7.559 in.
5. 8.467; 0.371; 66 mm; 2.598 in.

Exercises 15-5 (page 365)

1. 16-hole circle; 1 turn; 4 spaces
5. 1 turn
9. 15-hole circle; 1 turn; 9 spaces
13. 17-hole circle; 1 turn; 3 spaces

3. 15-hole circle; 10 spaces
7. 33-hole circle; 3 turns, 21 spaces
11. 51-hole circle; 20 spaces
15. 15-hole circle; 8 spaces

Exercises 15-6 (pages 371–372)

3. $^{11}/_{18}$
5. $^{12}/_{18}$ or $^{18}/_{27}$
7. $1^{3}/_{18}$
9. $^{14}/_{18}$ or $^{21}/_{27}$
11. $^{15}/_{18}$
13. $^{16}/_{18}$ or $^{24}/_{27}$
15. $^{17}/_{18}$
17. $^{18}/_{18}$, or 1 turn
19. $1^{1}/_{18}$
21. $1^{3}/_{27}$
23. $1^{3}/_{18}$
25. 16'22"; 11'29"; 8'43"
27. 53 holes in a 58-hole circle

Unit 15 Review Exercises (page 372)

1. 8.375 in.
5. 200 mm
9. 28-hole circle; 1 turn; 12 spaces

3. 10.7 in
7. 20 and 9 teeth

Index

Acme thread, 326, 338
Acute angle, 180
Addend, defined, 5
Addition
  carrying (renaming) numbers, 6
  of decimals, 67–68
  of fractions, 40–42
  of measurements, 8–9, 102
  of whole numbers, 5–9
Allen wrench, 199
American National Standard thread,
    327–334
Angles, 179–185
  acute, 180
  alternate exterior, 184–185
  alternate interior, 184–185
  complementary, 180
  central, 223–224
  corresponding, 184–185
    to a taper, 287–289
  decimal equivalent of minutes and
    seconds, 181–182
  dividing, 181–182
  function of, 242–244
  identification of, 179–182
  included, 180. *See also* Tapers;
    Threads, screw
  measuring, 171–173, 179–180
  obtuse, 180, 264
  right, 180
  straight, 180
  subtracting, 181

supplementary, 180, 185
vertical, 184–185
See also Triangles
Arc of circle, 202, 223–224
Area
  of circle, 115, 222
  of hexagon (ex.), 232
  of parallelogram, 217–218
  of rectangle, 217–218
  of sphere's surface, 231–232
  of square, 218
  of trapezoid, 219
  of triangle, 215–217
Axioms (properties of equations),
    107–109

Bilateral tolerance, 175–176
Bolt circle, 241, 259 (ex.), 271 (ex.)
Brown & Sharpe taper, 275

Calculator
  metric-English conversion with, 99
  subtraction with, 8
  square root with, 120–121
Calipers, 159, 166–168
Capacity, 86–87, 99
Carrying (renaming) in addition, 6
Cast iron
  composition, 157 (ex.)
  weight, 233 (ex.)
Centi-, 90–104
Central angle, 223, 224